Literature and Medicine

Ronald Schleifer • Jerry B. Vannatta
Authors

Literature and Medicine
A Practical and Pedagogical Guide

Authors
Ronald Schleifer
University of Oklahoma
Norman, OK, USA

Jerry B. Vannatta
Oklahoma City University
Oklahoma City, OK, USA

ISBN 978-3-030-19127-6   ISBN 978-3-030-19128-3   (eBook)
https://doi.org/10.1007/978-3-030-19128-3

© The Editor(s) (if applicable) and The Author(s), under exclusive licence to Springer Nature Switzerland AG 2019
This work is subject to copyright. All rights are solely and exclusively licensed by the Publisher, whether the whole or part of the material is concerned, specifically the rights of translation, reprinting, reuse of illustrations, recitation, broadcasting, reproduction on microfilms or in any other physical way, and transmission or information storage and retrieval, electronic adaptation, computer software, or by similar or dissimilar methodology now known or hereafter developed.
The use of general descriptive names, registered names, trademarks, service marks, etc. in this publication does not imply, even in the absence of a specific statement, that such names are exempt from the relevant protective laws and regulations and therefore free for general use.
The publisher, the authors and the editors are safe to assume that the advice and information in this book are believed to be true and accurate at the date of publication. Neither the publisher nor the authors or the editors give a warranty, express or implied, with respect to the material contained herein or for any errors or omissions that may have been made. The publisher remains neutral with regard to jurisdictional claims in published maps and institutional affiliations.

This Palgrave Macmillan imprint is published by the registered company Springer Nature Switzerland AG.
The registered company address is: Gewerbestrasse 11, 6330 Cham, Switzerland

Preface

Goals of the Book, Audience and Background, How to Use This Book

Literature and Medicine: A Practical and Pedagogical Guide is a book that grows out of more than 18 years of teaching and collaboration between the authors, Professor Ronald Schleifer, a George Lynn Cross Distinguished Research Professor of twentieth-century literature and culture, literary aesthetics, and semiotics at the University of Oklahoma, and Dr. Jerry Vannatta, a David Ross Boyd Professor of Internal Medicine (retired), former Executive Dean of the University of Oklahoma College of Medicine with a long career as a practicing physician, a researcher, and an award-winning classroom professor. The goals for this book—as they have been for authors' classes for pre-med and medical students, and for their workshops for practicing physicians and healthcare professionals—are very specific. They are:

- To help develop in physicians and healthcare professionals through the study of literature and narrative *habits of attentive listening* with the patients and others with whom they work. Among other things, these forms of attention will contribute to more precise and more efficient understandings of the medical conditions and personal concerns that brought the patient to the healthcare provider, which, in turn, will lead to more accurate diagnoses on the part of physicians and healthcare providers.
- To help develop in physicians and healthcare professionals through the study of literature and narrative *habits of responsive engagement* with their patients. Among other things, these forms of interaction will lead to a greater sense of empathy on the part of healthcare providers, a greater commitment to treatment plans on the part of patients, and a greater sense of satisfaction on the parts of *both* patients and healthcare providers.
- To help develop in physicians and healthcare professionals through the study of literature and narrative *habits of critical thinking*. Among other things, these forms of reflection will lead to everyday behaviors that will create a greater sense of professionalism and a more habitual practice of basic ethical responses such as simple decency and good will.

The goals of this text-anthology itself grow out of a generation of scholarly work by physicians, psychologists, anthropologists, and literary critics in the United States aimed at developing the ways that engagement with the humanities in general, and literature in particular, can create better and more fulfilled physicians and caretakers. As we note in our book, *The Chief Concern of Medicine: The Integration of the Medical Humanities and Narrative Knowledge into Medical Practices* (a book, like this text-anthology, greatly indebted to this generation of work),

> by "better physicians," we mean better diagnosticians in listening to and understanding the patient's story; better and more fulfilled professionals in developing powerful relationships with patients; more sensitively responsible doctors in the actions of everyday practice; and, perhaps encompassing all of these, people who will bring greater care to those who come to them ailing or in fear or faced with terrible suffering. (2013: 2)

Literature and Medicine, like our earlier work, is based upon the assumption that storytelling and narrative are centrally important in the patient-physician—and, more generally, in the patient-caretaker—relationship. In her groundbreaking book, *Narrative Medicine*, Dr. Rita Charon eloquently describes the role of narrative understanding in healthcare as "medicine practiced with the narrative competence to recognize, absorb, interpret, and be moved by the stories of illness" that patients bring to physicians. She adds that

> a medicine practiced with narrative competence will more ably recognize patients and diseases, convey knowledge and regard, join humbly with colleagues, and accompany patients and their families through the ordeals of illness. These capacities will lead to more humane, more ethical, and perhaps more effective care. (2006: vii)

In a similar fashion, in *The Chief Concern of Medicine*, we assume that "through the practice, analysis, and discussion of narrative (and particularly of literary or 'art' narratives) physicians—and, indeed, all of us—can become better at recognizing stories, comprehending their parts, rearranging them in new contexts, responding to them, and acting on the knowledge we have gained" (2013: 1). And this, of course, is what healthcare providers do every day—listen to patients' stories. It is the purpose of *Literature and Medicine* to contribute to the creation of practical pedagogical programs for this practice, analysis, and discussion in our colleges and medical schools.

The Goals of the Book

Although the work of *Literature and Medicine* falls under the wider umbrella of "medical humanities," this text is designed for a *practical* rather than a "conceptual" or "intellectual" engagement with the humanities. That is, this book, growing out of the successful team-teaching by the authors for many years—including a systematic follow-up study tracing the subsequent medical careers of our students (see Shakir et al. 2017)—aims at demonstrating the practical usefulness in medicine for engagements with literary and everyday narratives by students who, in large numbers, have not experienced the formal study of narrative.

Our overall purpose, then, is not simply to create a collection of stories and other materials related to practicing medicine. Rather, we aim to encourage appropriate pedagogical strategies that will allow both instructors and students to develop methods of engagement that are emphasized in the humanities in general and in literary studies more specifically—what we are calling here humanistic "resources"— which are often excluded from necessarily intensely focused study in biomedical education. Such strategies emphasize interpretation, inference, and grasping the "wholeness" of meaning rather than the apprehension of demonstrable—and often isolated—"facts" of biomedical understanding. For this reason, the goals of this book are to instill, in people aspiring or pursuing careers as healthcare providers, the understanding that formal patient-caretaker encounters, such as the medical interview, call for a set of skills different from those necessary for biomedical understanding. Over the years, we have found that engagement with and classroom discussion of literary narrative allows students to think of these skills as professional resources. The basis for this assumption, as we had hoped the short survey of cognitive psychology in Chap. 1 and in Appendix 1 might suggest, is that these skills of interpersonal engagement and understanding, including the ability to grasp "narrative knowledge," are available to all as part of our human inheritances. Thus, while it might well be that, as some advocates of including humanities courses in a medical education suggest, close reading and creative writing require very intensive and personal training over years, our aim and experience has been that a basic set of "elementary" questions and focuses in relation to narrative makes students and instructors aware of the practical usefulness of existing human resources for professional use in a relatively shorter time. Many recent studies in cognitive psychology have corroborated this experience.

It has been our experience that the vast majority of pre-med and medical and healthcare students have had scant experience with the humanities in general and engagements with literary narrative in particular. One very bright second-year medical student announced in class one day that the four novels we read in our three-week class double the number of books he had read in his lifetime. (This student went on to become part of a student-team that established a literary/arts journal for the University of Oklahoma Medical College, *Blood and Thunder*.) Moreover, it has been our experience that this majority of pre-med and healthcare students consistently reads on the level of content without consideration of form. In the section of the Introduction focused on Medicine, we argue that attention to the form of patient narratives as well as their content plays an important role in clinical medicine. Moreover, we discuss, at some length, the relation of form and content in Chap. 1, and in our pedagogical appendices, a majority of questions for both discussion and daily writing asks readers to focus on *how* narrative texts create their effects. Such questions help people understand their own responses to narrative, which include, studies in cognitive psychology have found, ethical discernment, attention to metaphorical language (which necessarily focuses on motive as well as fact), and awareness of ambiguity, among other things.

Thus, the goal of the book is to help instructors to create a course in which the realization of the importance and usefulness of these resources is made clear. For almost 20 years, each fall, starting in late August, the authors have taught this class to junior and senior undergraduates, mostly pre-med students; and both individually and together, we have also taught this class to second-year and fourth-year medical students and students in other healthcare programs. Every year, usually after about six weeks (in early October), students begin to internalize and habituate certain strategies of judgment and interaction, which they hadn't thought of as part of the professional work they had chosen for themselves. Needless to say, there were always some students who didn't "get it"; but significant majorities of our students came to think that interacting with patient stories *as narrative*, that pursuing rather than dismissing empathetic responses, that relating their patient encounters to stories they had "vicariously" experienced in reading literary texts, all could make their work in healthcare more successful for patients and more fulfilling for themselves as practitioners. To put it succinctly,

> The overall purpose of *Literature and Medicine* is to help structure classroom engagements which will encourage those studying to become physicians and other healthcare professionals to come to see that humanistic understanding is, in many instances in their professional lives, as important as biomedical understanding in caring for those in distress. This is achieved by encouraging and training people to be "careful" readers and listeners to narrative just as the rigorous study of biomedicine trains people to widely knowledgeable about illness and its causes.

In relation to this goal, the audience for this book includes instructors as well as students and private readers. That is, as a "text-anthology," *Literature and Medicine* aims at giving instructors in healthcare education some elementary tools with which they can more fully describe skills in attentive listening, responsive engagement, and critical thinking while at the same time giving instructors in humanities education a set of the elementary issues and questions that arise in the everyday practice of medicine. We are calling this book a "text-anthology" because it combines an anthology of literary works and medical vignettes and a textbook designed to present the basic methods and concepts that govern the study of narrative and the teaching of literature to students, instructors, and readers who might not be trained in literary studies or in the medical humanities. Our years of team-teaching and team-writing have made us acutely aware of these skills and how they can benefit both medical and humanities educators. That is, insofar as the humanistic "resources" are social and personal behaviors that most of us participate in our everyday lives, and insofar as these experiences respond to recurrent situations and problems in healthcare, *Literature and Medicine* hopes to facilitate in all its audiences useful interpersonal strategies by making these strategies explicit and explicitly experienced in relation to literary readings. For this reason, the aim of the book—and the courses it will enable—is not conceptual engagement with the humanities and literature, but *practical engagement* of storytelling on all levels for people hoping and trying to do the best for the patients they encounter.

We want to reiterate: by a "careful" reader we mean someone who might notice things that otherwise would not be seen. We believe that classroom engagements and discussions occasioned by this text-anthology can and will result in many participants discovering, and even habituating in their action, resources for care and understanding that they hadn't realized were available. In fact, it is our great hope that practicing teaching physicians might read Tolstoy with an eye toward making explicit in classroom discussion the ways in which Tolstoy's narrative can help them and their students more fully understand their successes and short-comings in their professional practice; and, equally so, that talented humanities teachers will find practical uses in the context of pre-med and medical education for the careful engagements with texts they have pursued throughout their careers. Additionally, because of the kinds of *focuses of attention* that *Literature and Medicine* promotes in (a) its aim of explicitly articulating narrative "features" and in (b) its aim of explicitly isolating particularly recurring "medical topics" and "medical issues" (including errors and mistakes, cultural awareness, strategies for listening), it is our reasonable hope that committed, if inexperienced, readers can discover important professional resources in their engagements with the strategies and content of this book. This has been our experience in co-teaching together with our complementary commitments to strengthening engagements with literary texts and developing successfully efficient clinical practices.

Finally, we want to make explicitly clear that it is not our purpose to create a text-formula for the reproduction of our courses, which in fact have greatly changed over the years. Rather the hope and goal of the book's content and organization is to make clearer to all participants an array of professional strategies and understandings that engagements with narrative repeatedly instill in students and readers over the course of a single semester or even over the shorter time span of medical-school courses. We believe—and have found in many years of undergraduate and graduate courses and professional workshops—that focusing on narrative *skills* strengthens listening, engagement, and critical thinking simply by making explicit in focused discussions attitudes and engagement-strategies of interaction that are often easily overlooked.

To these ends, this text-anthology focuses on five ways that "narrative knowledge" contributes to medical education and successful medical practices:

1. It contributes to an awareness of medical professionalism.
2. It contributes to the establishment of a strong patient-caretaker relationship.
3. It helps build a conscious awareness and a habitual sense of everyday ethics of medical practice (including, specifically, an awareness of mistakes in medicine).
4. It helps readers acquire an understanding of the ways that literary narrative provokes empathy and vicarious experience that can help negotiate differences in basic life experiences that healthcare providers and patients bring to the patient-caretaker relationship.
5. It demonstrates a systematic understanding of the logic of making a diagnosis.

Audience and Background

This text-anthology is meant for students and instructors of health sciences but also practitioners who are searching for strategies of connecting more meaningfully with their patients. It is meant to appeal to medical professionals who are interested in further developing their skills and understanding their interactions with patients by reading, reflecting on, and occasionally writing about literary narratives—stories of excellent quality that make important points in relation to the practice of medicine. To this end, the text-anthology makes studying literary art narratives more effective by providing, in Chap. 1, outlines of features of literary narrative which create an efficient (and easily habituated) understanding of how stories are structured and function. In addition, in each chapter of this text-anthology, we offer literary narratives and poems related to the chapter's focus and theme, and we offer and analyze "everyday" narratives ("vignettes") taken from actual clinical situations to help define and delimit each chapter's theme. As we note in Part II of Chap. 1, an important difference between everyday narratives and literary narratives (both of which are ubiquitous in human cultures) is that the former most often function to promote action in the world (see Boyd 2009) while the latter, as the narratologist James Phelan notes, promotes strong "focus on teller, technique, story, situation, audience, and purpose: all the elements that help determine the shape and effect of the story" (1996: 4).

For this reason, as we argue in Part I of Chap. 1, engagements with literary narrative make us more sensitive to narratives in general, and throughout *Literature and Medicine*, we emphasize this by juxtaposing literary and everyday narratives in every chapter. This is also apparent in the two-part "Introduction" below, which creates frameworks in both literature and medicine for people engaged with this book. In addition, the appendices in Part II of this book provide supplementary materials for the topics of this text-anthology: particular guides for writing assignments (if desired); and a particular guide for discussing diagnosis and diagnosis errors for instructors; discussion questions for instructors in working with these narratives in medical school; a detailed program of the use of literature in inculcating professionalism in medical students and physicians, which includes, in its bibliography, a number of empirical studies in cognitive psychology validating the effectiveness of literary reading; and finally, a guide to the "unsaid" that is encountered both in literature and in the clinic.

How to Use This Book

Literature and Medicine is designed to provide clear examples of strategies of reading and analysis of texts that are focuses of literature courses in ways that will be *practically useful* for students and instructors of medicine and other healthcare fields. A notable feature of the text-anthology is the fact that each chapter begins by analyzing a vignette from medical practice which helps define a field of concern that engagement with a literary text can help illuminate. That is, built into the structure of the book itself is the practical relationship between the situations and practices of healthcare and strategies of understanding and behavior that literature helps foster. We have organized this text-anthology so that it can be the basis of both long and short courses on literature and medicine in undergraduate and postgraduate courses. Its double Introduction plus 14 chapters can organize a full semester of study—especially if instructors incorporate discussions of poetry as well as fiction and if *Literature and Medicine* is supplemented by the full-length novels, Roddy Doyle's *The Woman Who Walked into Doors* and Toni Morrison's *Beloved*, discussed in two chapters, and perhaps also supplemented by a full-length "vignette" such as Anne Fadiman's *The Spirit Catches You and You Fall Down*. (In addition, we discuss five texts, which are readily available (see Bibliography), that might supplement the work of this volume: Ernest Hemingway's short story, "Indian Camp," discussed in Chap. 5 and Appendix 6; Anatole Broyard's narrative of being a patient in "Doctor Talk to Me," discussed in Chap. 6; Dr. David Cassel's "The Nature of Suffering and the Goals of Medicine," discussed in Chap. 11; Dr. David Hilfiker's "Facing Our Mistakes," discussed in Chap. 13; and Dr. William Carlos Williams' "The Use of Force," also discussed in Chap. 13.) At the same time, however, this text-anthology also lends itself to shorter, four- or eight-session courses following its section topics represented by the roman numerals in the Contents. Thus, the texts and topics of the book—after the short introductions and opening chapter focused on the ways literary narrative works in relation to clinical medicine—present a wide range of global issues facing physicians that we have already listed: systematic diagnosis, professionalism, the patient-physician relationship, ethics in medicine, the diversity of patients for most healthcare providers, mistakes in medicine, and death and dying. And we end with a short Postscript, which returns readers to the fulfillments of a career in healthcare, something that can be lost in the welter of "problems" that arise in the professional caring for others.

The book is structured to allow great flexibility for instructors and readers. Thus, while the two-part Introduction and Chap. 1 create a framework for the examination of the relationship between literature and medicine, instructors and readers can focus on subsequent chapters in any order that serves their purposes. Moreover, the choices of sub-topics focused on the patient-caretaker relationship and the vicarious experience of literature also allow for great flexibility. And the addition of poems in each chapter—along with the appendices—again offers instructors and readers the ability to organize the use of *Literature and Medicine* with their own goals in mind.

Norman, OK, USA Ronald Schleifer
Oklahoma City, OK, USA Jerry B. Vannatta

Bibliography

Boyd, Brian. 2009. *On the Origin of Stories: Evolution, Cognition, and Fiction*. Cambridge: Harvard University Press.
Charon, Rita. 2006. *Narrative Medicine: Honoring the Stories of Illness*. New York: Oxford University Press.

Phelan, James. 1996. *Narrative as Rhetoric: Techniques, Audiences, Ethics, Ideology*. Columbus: Ohio State University Press.

Schleifer, Ronald, and Jerry Vannatta. 2013. *The Chief Concern of Medicine: The Integration of the Medical Humanities and Narrative Knowledge into Medical Practices*. Ann Arbor: University of Michigan Press.

Shakir, Mubeen, Jerry Vannatta, and Ronald Schleifer. 2017. Effect of College *Literature and Medicine* on the Practice of Medicine. *Journal of the Oklahoma State Medical Association* 110 (November 2017): 593–600.

Acknowledgments

With years of working together with one another, the authors have a great number of people to acknowledge. Our team-teaching began when Dr. Vannatta asked University of Oklahoma Provost Nancy Mergler in 1997 how best to organize a class designed to bring together literature and medicine. Provost Mergler, who had been Director of the Honors Program at our university, suggested that he find an English Professor and team-teach a course for the Honors College. When asked, she suggested Professor Schleifer, whose work she knew because he was her husband. That more-or-less chance meeting began years of collaboration and friendship. Our work and friendship was also aided by the work and encouragement of Marianne Vannatta, Jerry's wife, who died in 2009. Marianne even participated as a simulated patient in the DVD-ROM that Schleifer and Vannatta wrote and produced with the aid of another friend, Sheila Crow, *Medicine and Humanistic Understanding: The Significance of Narrative in Medical Practices*, published by the University of Pennsylvania Press in 2005. In addition, Melinda Lyon has greatly supported Dr. Vannatta throughout the writing of this book.

Over the years, students in our classes contributed to our growing understanding of our work as well. We are particularly grateful to Dr. Jason Sanders, currently Provost of the University of Oklahoma Health Sciences Center; Sarah Swenson, PhD and MD; and Dr. Mubeen Shakir, all three of whom went to Oxford as Rhodes Scholars to pursue the study of medicine and the humanities after participating in our classes. Sarah's husband, Dr. Spencer Kinzie, worked closely with us in developing the pedagogical appendices to this book. In addition, our work with Dr. Casey Hester, with whom we wrote Appendix 5 to this volume, has been a constant aid in our thinking about the human sides of medicine since the time we commenced work on this book.

Others, with whom we worked on *Medicine and Humanistic Understanding*—many of whom appear in the pages of this book—Dr. Rafael Campo, Dr. Rita Charon, Anne Hunsaker Hawkins, Katherine Montgomery, Dr. Oliver Sacks, Dr. Richard Selzer, Dr. John Stone, Dr. Abraham Verghese, all contributed to our thinking about this book with their insight and thoughtfulness. Others in the humanities—James Bono, Lennie Davis, Daniela Garofalo, Kyle Harper, Susan Kates, Robert Markley, David Morris, Rebecah Pulsifer, Russ Reising, Robert Schleifer, Seth Vannatta, Nancy West—have helped us to more fully realize our vision of interdisciplinary work. And foreign friends—Jen Crawford and Jen Webb (Canberra University),

Ruoping Li (Shanxi Medical University), Begga Kristjánsdóttir (the University of Iceland), Kedung Liu and Tiao Wang (Harbin Institute of Technology), Ulrika Maude (University of Bristol), Neil Murphy (Nanyang Technological Institute [Singapore]), Boris Vejvodsky (Lausanne University), Tania Venediktova (Moscow State University), Yong Cai (Lanzhou University), Zhu Liya (Shanghai Maritime University)—have given us the opportunity of discussing strategies of medical education with scholars and healthcare providers at various places around the world. Allie Troyanos and Rachel Jacobe at Palgrave provided us with unstinting support from start to finish. And Courtney Jacobs, a graduate assistant at the University of Oklahoma, supplemented our goals and sharpened our presentation. In addition to her diligent and innovative work on the nuts and bolts of bringing this book to publication, she engaged with the editors in every chapter with responses, suggestions, and understanding that made this book better. Finally, we gratefully acknowledge support of the Vice President for Research of the University of Oklahoma that allowed the timely completion of this book.

At a late stage of composition, Annie Cloud and Jared Conaster, two students (now graduates) of the Physician Assistant Program at Oklahoma City University, which Dr. Vannatta recently helped to develop, read through the complete late draft of this book with great care and acumen. They have offered us wonderful insights on almost every page into ways to make our book more friendly and useful for people committed to careers in healthcare.

As this suggests, in many ways, our interdisciplinary project has been the work of friendship, and working together with friends and colleagues—as well as many other, students, workshop participants, patients, and colleagues we have encountered in our years of thinking about strategies of human caring—has made pursuing this project among the most exhilarating and fulfilling experiences of our careers. We are deeply grateful for the opportunity of pursuing this work with friends and strangers who understand the importance of caring and caretaking in our lives.

The editors would like to acknowledge the following persons and publishers for the kind permission to reproduce literary works and medical vignettes in *Literature and Medicine*:

- **Dr. Dannie Abse**. "In the Theatre" from *New and Collected Poems* by Dannie Abse. Copyright © 2003 by Dannie Abse. Used by permission of United Artists for the estate of the author.
- **Dr. Rafael Campo**. "The Couple" from *Landscape with Human Figure* by Rafael Campo. Copyright © 2002 by Rafael Campo. Reprinted by permission Georges Borchardt, Inc. for the author.
- **Alicia Gaspar de Alba**. "Making Tortillas" from *Three Times a Woman* by Alicia Gaspar de Alba. Copyright © 1989 by Alicia Gaspar de Alba. Reprinted by permission of the author.
- **Lous Heshusius**. Material reprinted from "That Which Has No Words, That Which Cannot Be Seen," from *Inside Chronic Pain: An Intimate and Critical Account*, by Lous Heshusius. Copyright © 2009 by Lous Heshusius. Used by permission of the publisher, Cornell University Press.

- **Dr. Casey Hester, Dr. Jerry Vannatta, and Ronald Schleifer, Ph.D.** "Medical Professionalism: Using Literary Narrative to Explore and Evaluate Medical Professionalism" from *New Directions in Literature and Medicine Studies*, ed. Stephanie Hilger (New York: Palgrave, 2017). Used by permission of Palgrave and the authors.
- **Audre Lorde**. From: *The Cancer Journals* by Audre Lorde. Copyright © 1980 by Audre Lorde. Reprinted by permission of the Charlotte Sheedy Literary Agency.
- **Derek Mahon**. "Everything is Going to be All Right" from *New Collected Poems* by Derek Mahon copyright © 2011. Reprinted by kind permission by the author and the Gallery Press, Loughcrew, Oldcastle, County Meath, Ireland.
- **Demetria Martinez**. "The Annunciation: Lupe" from *The Block Captain's Daughter*. Copyright © 2012 by Demetria Martinez. Used by permission of the publisher, the University of Oklahoma Press.
- **Grace Paley**. "A Conversation with My Father" from *Enormous Changes at the Last Minute* by Grace Paley. Copyright © 1971, 1974 by Grace Paley. Reprinted by permission of Farrar, Straus and Giroux; in addition, reprinted by permission of Sally-Wofford-Girand, Union Literary Agents.
- **Dr. Richard Selzer**. "Imelda" from by Richard Selzer *Letters to a Young Doctor*. Copyright © 1982 David Goldman and Janet Selzer, trustees. Reprinted by permission Georges Borchardt, Inc. for the author.
- **Dr. Audrey Shafer**. "Monday Morning" from *Sleep Talker: Poems by a Doctor/Mother* by Audrey Shafer. Copyright © 1992. The poem originally appeared in *Annals of Internal Medicine* (117 (1992): 167).
- **Dr. John Stone**. "He Makes a House Call" from *In All This Rain*. Copyright © 1980 by John Stone. Reprinted by permission Louisiana State University Press.
- **Dr. Damon Tweedy.** Excerpt from "When Doctors Discriminate" from *Black Man in a White Coat: A Doctor's Reflection on Race and Medicine* by Damon Tweedy. Copyright © 2015 by Damon Tweedy. Reprinted by permission of Picador.

Contents

Section I   Narrative and Medicine   1

1   **Narrative and Cognitive Science; Literature and Medicine**   3
    Encountering an HIV Patient: A Vignette (A Passage from Dr. Abraham
    Verghese, *My Own Country* [1994])   8
    Literary Narrative: "A Conversation with My Father" (1972) by Grace
    Paley   10

Section II   The Logic of Making a Diagnosis   31

2   **The Narrative Structure of Diagnosis**   33
    The Woman with Hyponatremia: A Vignette (Excerpt from *The Chief
    Concern of Medicine*)   34
    Literary Narrative: "The Resident Patient" (1893) by Dr. Arthur
    Conan Doyle   37
    Poem: Sonnet 73, "That Time of Year Thou Mayst in Me Behold" (1609)
    by William Shakespeare   49

Section III   Professionalism   51

3   **Literature and Professionalism in Medicine**   53
    Playing God: A Vignette (A Passage from Dr. Michael LaCombe,
    "Playing God," in *Bedside: The Art of Medicine* [2010])   54
    Literary Narrative: "Imelda" (1982) by Dr. Richard Selzer   56
    Poem: "Monday Morning" (1992) by Dr. Audrey Shafer.   64

Section IV   Building the Patient-Provider Relationship   67

4   **Rapport and Empathy in Medicine**   69
    An Elderly African American Patient: A Vignette (Excerpt from
    The Chief Concern of Medicine)   70
    Literary Narrative: "A Doctor's Visit" (1898) by Dr. Anton Chekhov . . .   71
    Poem: "He Makes a House Call" (1980) by Dr. John Stone   78

5	**Listening to Patients**	81
	Young Mother with Abdominal Pain: A Vignette (Excerpt from *The Chief Concern of Medicine*)	82
	Literary Narrative: "Araby" (1914) by James Joyce	84
	Poem: "The Red Wheelbarrow" (1923) by Dr. William Carlos Williams	88
6	**The Patient**	91
	Introduction to *The Cancer Journals*: A Vignette (Excerpt from Audre Lorde, *The Cancer Journals* [1980])	92
	Literary Narrative: "The Yellow Wallpaper" (1892) by Charlotte Perkins Gilman	97
	Poem: "The Couple" (2002) by Dr. Rafael Campo	107
7	**The Doctor**	109
	When Doctors Discriminate: A Vignette (Excerpt from Dr. Damon Tweedy *Black, Man in a White Coat: A Doctor's Reflections on Race and Medicine* [2015])	110
	Offhand Paragraph: A Vignette-Discussion (Excerpt from Dr. Michael LaCombe "Diagnosis," in *Bedside: The Art of Medicine* [2010])	113
	Literary Narrative: "The Lynching of Jube Benson" (1904) by Paul Laurence Dunbar	114
	Poem: "Sometimes I Feel Like A Motherless Child" (Probably Nineteenth Century), Traditional Spiritual	119

Section V	**Everyday Ethics of Medical Practices**	**121**
8	**Everyday Ethics of Medical Practices**	123
	The Patient with Diabetic Ketoacidosis: A Vignette (Excerpt *The Chief Concern of Medicine*)	124
	Literary Narrative: "Enemies" (1887) by Dr. Anton Chekhov	126
	Poem: "A Poison Tree" (1794) by William Blake	134

Section VI	**Vicarious Experiences**	**137**
9	**Culture**	139
	The Patient's Chief Concern: A Vignette (Excerpt from *The Chief Concern of Medicine*)	140
	Literary Narrative: "The Annunciation: Lupe" (2012) by Demetria Martinez	142
	Poem: "Making Tortillas" (1989) by Alicia Gaspar de Alba	145
10	**Sexual and Domestic Abuse**	147
	You Don't Deserve This: A Vignette by Dr. Jerry Vannatta	148
	Literary Narrative: "Berenice – A Tale" (1835) by Edgar Allan Poe	150
	"Leda and the Swan" (1924) by W. B. Yeats	156
11	**Pain**	159
	"That Which Has No Words, That Which Cannot Be Seen": A Vignette (Excerpt from Lous Heshusius, *Inside Chronic Pain: An Intimate and Critical Account* [2009])	160

Literary Narrative: "The Operation" from *White Jacket* (1850) by
Herman Melville.. 162
Poem: "Pain has an Element of Blank" (1890; poem #650) by Emily
Dickinson 168

12 Ageing.. 171
Treating a Very Old Woman: A Vignette by Dr. Jerry Vannatta 171
Literary Narrative: "The Little Shop Window," from *The House of Seven
Gables* (1851) by Nathaniel Hawthorne............................... 173
Poem: "I Look into My Glass" (1898) by Thomas Hardy 179

Section VII   Mistakes in Medicine   181

13 Mistakes in Medicine 183
Mistakes: Enough to Spread Around: A Vignette by Dr. Jerry Vannatta . 184
Literary Narrative: Chapter Eleven from *Madame Bovary* (1857)
by Gustav Flaubert 187
Poem: "In the Theatre" (1983) by Dr. Dannie Abse.................. 194

Section VIII   Death and Dying   197

14 Death and Dying 199
The Good Death: A Vignette by Dr. Jerry Vannatta 200
Frenzy Facing Death: A Vignette by Dr. Jerry Vannatta.............. 201
Literary Narrative: "*The Death of Ivan Ilych*" (1886) by Leo Tolstoy ... 204
Poem: "Death Be Not Proud" (1609) by John Donne 235

Section IX   Postscript: The Fulfillments of Healthcare   237

15 Afterword... 239
Poem: "Everything Is Going to Be All Right" (1979) by Derek Mahon . 240

**Appendix 1: Experimental Results: The Cognitive Science of Literary
Reading** 243

Appendix 2: Discussion Questions for the Chapters 247

Appendix 3: Daily Writing Assignment 255

Appendix 4: Guide for Discussing Diagnosis and Diagnosis Errors 259

**Appendix 5: Medical Professionalism: Using Literary Narrative to
Explore and Evaluate Medical Professionalism** 263

**Appendix 6: Teaching Literature to Medical Students: Ernest Hemingway,
Nick Adams, and the "Unsaid" in Narrative** 279

Bibliography 287

Index... 291

Literary Texts

Dr. Dannie Abse, "In the Theatre" (poem)
William Blake, "A Poison Tree" (poem)
Dr. Rafael Campo, "The Couple" (poem)
Dr. Anton Chekhov, "A Doctor's Visit"
Dr. Anton Chekhov, "Enemies"
Dr. Arthur Conan Doyle, "The Resident Patient"
Emily Dickinson, "Pain has an element of blank" (poem)
John Donne, "Death Be Not Proud" (poem)
Gustav Flaubert, Chapter 11 from *Madame Bovary*
Alicia Gaspar de Alba, "Making Tortillas" (poem)
Charlotte Perkins Gilman, "The Yellow Wallpaper"
Thomas Hardy, "I Look into My Glass" (poem)
Nathaniel Hawthorne, Chapter 2 from *The House of Seven Gables*
James Joyce, "Araby"
Demetria Martinez, "The Anunciation: Lupe"
Derek Mahon, "Everything is Going to be All Right" (poem)
Herman Melville, "The Operation" from *White Jacket*
Grace Paley, "A Conversation with My Father"
Dr. Richard Selzer, "Imelda"
Dr. Audrey Shafer, "Monday Morning" (poem)
William Shakespeare, "Sonnet 73, That Time of Year" (poem)
Dr. John Stone, "He Makes a House Call" (poem)
Leo Tolstoy, *The Death of Ivan Ilych*
Traditional Spiritual, "Sometimes I Feel like a Motherless Child" (poem)
Dr. William Carlos Williams, "The Red Wheelbarrow" (poem)
W. B. Yeats, "Leda and the Swan" (poem)

Vignettes and Vignette-Discussions

Lous Heshusius, *Inside Chronic Pain: An Intimate ad Critical Account* [excerpt]
Dr. Michael LaCombe, *Bedside: The Art of Medicine* [2 analysis discussions]
Audre Lorde, *The Cancer Journals* [excerpt]
Dr. Damon Tweedy, *Black Man in a White Coat* [excerpt]
Dr. Jerry Vannatta, ten various vignettes (5 vignettes excerpted from *The Chief Concern of Medicine* and 5 new vignettes)
Dr. Abraham Verghese, *My Own Country* [analysis discussion]

Introduction to *Literature and Medicine*

Part I

Literature Introduction: Structure and Focus of the Chapters

In the Medicine Introduction, Part II of the book's Introduction, we set forth the complexity of clinical medicine—namely the combination of cognitive and affective skills and responses necessary for facilitating the most efficient, informative, and medically effective narrative from a patient—and also set forth the more general goals of clinical medicine. By training and, most usually, by temperament, healthcare students—especially medical students—and practicing healthcare professionals focus most fully on the valorization of one order of understanding in the patient-physician encounter, that of apprehending the patient's story primarily in biomedical terms. We argue in this book that soliciting and understanding the patient's concern about her condition or ailment and acknowledging the emotions that accompany that concern lead to a fuller understanding of the patient's plight, clearer and more efficient insight into the nature of his condition, and development of strategies to respond to that situation which more fully involves the patient. Among other things, this last phenomenon also reduces the pressure and stress physicians sometimes feel. The structure of this book pursues these goals by focusing on a small number of "topics" or themes outlined in the Preface and spelled out in the Table of Contents:

 I. Narrative and Medicine (the overarching framework for the book, so not technically a focused topic)
 II. The Logic of Making a Diagnosis
 III. Professionalism
 IV. Building the Patient-Provider Relationship
 V. Everyday Ethics of Medical Practices
 VI. Vicarious Experiences
 VII. Mistakes in Medicine
 VIII. Death and Dying
 IX. Postscript: The Fulfillments of Healthcare

While this is not an exhaustive list of topics with which to explore the ways that narrative competence in healthcare practices can contribute to an education in healthcare and greater success in practicing, we have found that students are particularly responsive to the items on this list and that these areas of concern engender particularly acute discussions in class and in writing assignments, and—we have found in a long-term study (see Shakir et al. 2017)—particularly richer engagements in the clinic. Before introducing these topic-sections, the Medicine Introduction offers a short argument for the parallels between skills useful in clinical practice and skills useful in critical reading; in the general topic of "Narrative and Medicine" in Chap. 1, we examine a number of "features" of literary narrative that help habituate the related skills of clinical practice and critical reading. The topic of Chap. 1, then, offers a broad overview of the "textbook" aspect of *Literature and Medicine*, which the individual chapters explore more specifically.

The Chapters: Vignettes, Stories, Poems

Here, though, in the Literature Introduction, let us set forth a chart of the organization of topics and narrative texts in this book.

Topic I. Narrative and Medicine	**1. Cognitive Psychology and Reading Literature** **2. 13 Features of Narrative** Literary text: Grace Paley, "Conversation with My Father"
Topic II. The Logic of Making a Diagnosis	**Systematic Hypothesis Formation** Literary Text: Dr. Arthur Conan Doyle, "The Resident Patient"
Topic III. Professionalism	**The Nature of Professionalism** Literary Text: Dr. Richard Selzer, "Imelda"
Topic IV. Building the Patient-Provider Relationship 1. Rapport and Empathy 2. Listening to Patients 3. The Patient 4. The Doctor	Literary texts: 1. Dr. Anton Chekhov, "The Doctor's Visit" 2. James Joyce, "Araby" 3. Charlotte Gilman, "The Yellow Wallpaper" 4. Paul Laurence Dunbar, "The Lynching of Jube Benson"
Topic V. Everyday Ethics of Medical Practices	**Virtue Ethics and Narrative** Literary text: Dr. Anton Chekhov, "Enemies"
Topic VI. Vicarious Experiences 1. Culture 2. Sexual Abuse 3. Pain 4. Ageing	Literary texts: 1. Demetria Martinez, "The Annunciation: Lupe" 2. Edgar Allan Poe, "Berenice" 3. Herman Melville, "The Operation" 4. Nathaniel Hawthorne, excerpt from *House of Seven Gables*
Topic VII. Mistakes in Medicine	**Systematically Understanding Mistakes** Literary text: Gustav Flaubert, excerpt from *Madame Bovary*
Topic VIII. Death and Dying	**Facing Mortality in Healthcare** Literary text: Leo Tolstoy, *The Death of Ivan Ilych*
Topic IX. Postscript: The Fulfillments of Healthcare	**Good news in Healthcare** Literary text: Derek Mahon, "Everything is Going to be All Right"

Topic I After the consideration of "The Complexity of Clinical Medicine" later in this Introduction, which examines parallels between clinical practice and critical reading, in Chap. 1 we offer an exploration of the working relationship between narrative and medicine by first presenting and analyzing recent work in cognitive psychology; and second, we offer a thorough analysis of a medical vignette describing the first encounter of patient and physician and Grace Paley's short story, "A Conversation with My Father" (Paley's father was a physician). In analyzing the story, we set forth 13 primary "features" involved in literary narrative that are useful in developing narrative competence; these features create the basic, "elementary" focus for all the readings—short stories, vignettes, poems—of *Literature and Medicine*. Chap. 1 also offers strong suggestions that narrative organization of experience is an evolutionary adaptation of human beings, and to this end, it reviews extensive studies in cognitive psychology that demonstrate the measurable effectiveness of engagements with literary narrative in fostering ordinary human interactions. Specifically, cognitive psychology has demonstrated that engagement with literary texts (1) creates empathy, (2) refines Theory of Mind (psychologists' description of the powerful skill of human beings—and to a lesser extent, other primates—that enable us to apprehend that other creatures have cognitive and affective responses to the world that may be different from our own), and (3) "transports" readers into the world of a narrative that, among other things, produces vicarious experience. Each chapter presents an actual "vignette" from medical practice, and in each chapter, the vignette and its analysis come first to create a real-world context for the engagement with the literary texts.

Topic II After the first, longest chapter, the next two chapters offer a general sense of narrative reasoning. The first of these topics examines diagnosis in relation to the popular literary form of the detective story. Chap. 2 presents the engaged fun of reading in the form of a Sherlock Holmes story. In so doing, it also offers an exploration of the "logic" used in making a medical diagnosis that is rooted in a strong sense of narrative. This chapter—along with Chaps. 3 and 8—presents a "strategy" of comprehension and action that supplements the features of literary narrative presented in Chap. 1 with a focus on narrative concerns that are generally important to medicine: diagnosis (Chap. 2), professionalism (Chap. 3), and interpersonal ethics (Chap. 8). The logic of diagnosis was first described by the nineteenth-century American polymath, Charles Sanders Pierce. He described it as "abduction," a word he coined that is sometimes defined as "hypothesis formation" or "inference to the best explanation." Students are commonly surprised to know that this logic has been described not only philosophically but narratively in detective stories, such as the Sherlock Holmes stories created by the physician-writer Arthur Conan Doyle at the same time Peirce was writing.

This chapter presents one of Dr. Arthur Conan Doyle's Holmes stories, "The Resident Patient," and develops in a short analysis the analogy of medical detective work in making a diagnosis and criminal detective work in solving a crime to demonstrate, even to beginning students, that the cognitive and affective information obtained in a patient interview can, and should, lead to the systematic examination of possible diagnoses and actions in relation to the patient's condition. The vignette associated with this chapter, "The Woman with Hyponatremia," examines the ways that an attending physician displays many of the qualities of a classical detective. The related poem in this chapter, Shakespeare's sonnet, "That Time of Year," allows, early in the course—if this chapter is taken up by instructors early in the semester—for an instructor to turn students into literary "detectives" by asking them to recompose the sonnet, printed in the anthology with its lines out of order. In order to do this, students have to develop (hypothetical) order out of disordered phenomenal evidence, a problem that often faces physicians in the clinic. The strategies of assuming the systematic order of the poem and its various levels of meaning and structure offer a fine parallel to the systematic confrontation with symptoms to the end of figuring out what's going on. We include in Appendix 4 pedagogical strategies we have found useful in creating the link between literary narrative and practical diagnostic strategies.

Topic III Chapter 3 focuses on a specific general topic for medical education and medical practice, "professionalization" in healthcare practices. More specifically, it focuses on the ACGME (Accreditation Council for Graduate Medical Education) program of developing specialty criteria for professionalization in medicine. It describes the ways that an analysis of Dr. Richard Selzer's "Imelda"—particularly focusing upon the features of "patterned repetition," "the unsaid," "narrative agents and concern," and "narrative as moral education" outlined in Chap. 1—can materially help students comprehend what is at stake in medical professionalization. The vignette analyzed in this chapter, a short excerpt from "Playing God" by Dr. Michael LaCombe, allows students to examine a particular patient-physician encounter in light of the understanding of both professionalization and narrative that can result from careful reading of Selzer's short story. In this chapter, we present and discuss Dr. Audrey Shafer's poem, "Monday Morning," to examine the relation of family life to professional life in relation to this topic. We include in Appendix 5 a study of the programmatic use of this story in a Professionalization Workshop run by our colleague, Dr. Casey Hester.

Topic IV The next of the book's topic-sections, "Building the Patient-Provider Relationship," suggests how literature can help students develop or understand four different aspects of that relationship, namely the ability to cultivate a powerful sense of empathy and rapport by experiencing what Dr. Vannatta describes as a "flood" of emotion in relation to a patient in the clinic (Chap. 4), the ability to carefully listen to what we describe in Chap. 1 as the "unsaid" in patient narratives (Chap. 5), the ability to engage with the experience of being a patient in a world of cultural stereotypes (Chap. 6), and finally, the ability to engage with the experience of being a doctor in a world of cultural stereotypes (Chap. 7).

Chapter 4, "Rapport and Empathy in Medicine," examines how reading Toni Morrison's novel *Beloved* affected Dr. Vannatta's encounter with a patient and, indeed, convinced him of the value of a literary education in the practice of medicine. The chapter goes on to present a powerful story by Dr. Anton Chekhov, which narrates and provokes empathy simultaneously. The poem associated with the chapter, "He Makes a House Call" by Dr. John Stone, reinforces the way that two time-frames of narrative (features discussed in Chap. 1) provoke the kind of rapport and empathy Chekhov enacts in his story and Dr. Vannatta describes in his practice.

The next chapter under this heading (Chap. 5), "Listening to Patients," presents James Joyce's short story of a young boy experiencing "puppy love" for the first time, "Araby." The discussion of this story asks readers to "fill in" what is unsaid by the young boy in the story. (Appendix 6, related to this chapter, offers an analysis of Ernest Hemingway's well-known short story, "Indian Camp." Appendix 6 offers a systematic analysis of the narrative unsaid in relation to clinical medicine that could supplement discussion of Joyce's story with that of Hemingway.) The vignette analyzed in this chapter, "Young Mother with Abdominal Pain," allows students and readers to see how attending to the "unsaid" works in clinical medicine. The poem associated with this chapter is Dr. William Carlos Williams' famous poem, "The Red Wheelbarrow," which strongly connects the narrative form of literature to acts of discernment in a poem that requires its readers to "fill in" its implicit narrative and thematic purport.

The third chapter of this section, "The Patient," presents Charlotte Perkins Gilman's harrowing representation of a patient suffering from postpartum psychosis descending into irrationality. The vignette associated with this story, an excerpt from Audre Lorde's powerful memoir, *The Cancer Journals*, offers a highly detailed first-person account of a patient's encounter with life-transforming and life threatening illness. In the course of this discussion, we also touch upon Anatole Broyard's detailed examination of the patient-physician encounter from the vantage of the patient entitled "Doctor Talk to Me" (a text that is widely available: see Bibliography). The poem related to this chapter, Dr. Rafael Campo's "The Couple," expands the role of "patient" to include the family of the ill person.

The final chapter in this section, "The Doctor," examines stereotyping—in this case, racist stereotyping in both the vignette and the short story and sexist stereotyping analyzed in a second vignette—in the practice of medicine. Its literary selection, Paul Laurence Dunbar's short story "The Lynching of Jube Benson" is a classic and powerful examination of deadly racist stereotyping narrated by a physician. This chapter takes the unusual strategy of including two vignettes (Chap. 14 does so as well): one from Dr. Damon Tweedy's memoir *Black Man in a White Coat*, which presents the unrelenting ordinariness and banality of prejudice, and a second a single paragraph from Dr. Michael LaCombe's first-person narrative of a woman physician's work that is hardly part of the overall trajectory of his narrative description of diagnosis, but an off-hand description of ordinary everyday sexist encounters of a female physician with colleagues. The offhand nature of this paragraph describes ordinary "everyday" prejudice, what Dr. Tweedy calls "unconscious (implicit) bias" (2015: 270). Such "implicit" bias does not seem to be intentional prejudice, but habitual, unreflected-upon behavior, and the fact that LaCombe mentions it seemingly outside the significant trajectory of his narrative underlines—as a "formal" rather than "content" feature of his narrative—how it is unconscious and implicit.

Topic V Chapter 8, "Everyday Ethics in Medicine," could also fall under the category of "Building the Patient-Provider Relationship," but we feel it is necessary to offer it as a topic section on its own because the examination of the way that narrative knowledge can inform ethical behavior is an important part of what the study of literature in a medical education has to offer. Rather than principle-based (or "normative") ethics of proper behavior that is usually part of medical education or utilitarian (or "cost-benefit") ethics that governs much discussion in epidemiology, a discussion of the ethics of everyday practice of medicine, organized around "virtue ethics" as introduced by Aristotle, underlines the relationship of narrative events to habitual ethical behavior. That is, Aristotle's virtues are organized around the *actions* of agents—people acting in actual situations, described in Chap. 1 as the feature of "narrative agents and concerns"—rather than more or less abstract propositions about right and wrong found in the principle-based ethics often associated with medicine or cost-benefit analyses associated with public health concerns. This chapter presents Dr. Anton Chekhov's short story "Enemies," which examines the behavior of a physician weighing the plight of a patient against his own personal tragedy. This chapter examines the story in relation to a schema of six virtues particularly appropriate to the patient-physician relationship that physicians and caregivers can easily remember in order to reflect upon and judge their own actions as they go about their everyday practice. (Chekhov's story is also a powerful example of the feature of "twice told stories" discussed in Chap. 1.) The vignette analyzed in this chapter, "The Patient with Diabetic Ketoacidosis," demonstrates what occurs in the absence of these everyday virtues. The related poem in this chapter is William Blake's "A Poison Tree," a poem that examines ethical behavior in relation to emotion and agency. (As mentioned earlier, this chapter, like Chap. 2, "The Logic of Making a Diagnosis," and Chap. 3, "Professionalism," focuses upon a "strategy" of comprehension and action that supplements the features presented in Chap. 1. All three of these chapters are a little longer than other chapters: Chap. 8 presents an acronym related to virtue ethics, and the earlier two stories have related appendices in Part II.)

Topic VI The next topic-section of the text-anthology, consisting of four chapters, examines particular examples of "vicarious experience" that can contribute to an education in medicine, namely experiences of cultural differences (Chap. 9), sexual abuse (Chap. 10), pain (Chap. 11), and ageing (Chap. 12). This section encompasses many of the qualities of engaging with literature discussed in Chap. 1, which analyze the ability of literature to provoke vicarious experiences in readers/listeners by means of the various features of literary texts discussed in the earlier chapter.

Chapter 9 presents the experience of living within the Mexican American subculture in the United States in the presentation of Demetria Martinez's story "The Annunciation: Lupe." Given this story, one could particularly focus upon the features of literature—"patterned repetition," "the unsaid," "narrative genres"—that condition the vicarious experience of Mexican American culture the story provokes. But we have chosen this story for other reasons as well: the manner in which it presents biomedicine as a belief-system (the belief in cholesterol as parallel to religious belief) and the way it brings up birthing in relation to medicine. The vignette analyzed in this chapter, "The Patient's Chief Concern," portrays two doctors engaging with a patient with a very different cultural sense of illness than the physicians possess. The related poem is Alicia Gaspar de Alba's poem "Making Tortillas." Besides describing everyday actions that constitute what we describe as the "hum" of culture, the story undramatically presents the naturalness of nonheterosexual relationships.

The analysis of the vignette of Chap. 10—Dr. Vannatta's narrative of his encounter with a long-term patient who presents symptoms of spousal abuse—analyzes this sadly not unusual aspect of healthcare in relation to Roddy Doyle's *The Woman Who Walked into Doors*, a novel about long-term spousal abuse written by a man in the first-person voice of a woman. The novel uses many of the features of literature described in Chap. 1 to provoke *vicarious experience* of a woman's terrible ordeal, not only at home but in the clinic, particularly "the dynamic of form and content," "defamiliarization and style," and "patterned repetition," which create the voice of the main character, Paula Spencer. The vignette analyzed in this chapter, *You Don't Deserve This*, presents a physician who—having read Doyle's novel—is confronted with a patient whose bruises do not fit with the narrative she presents. The fictional text of this chapter is one of Edgar Allan Poe's horror stories, "Berenice," that presents domestic abuse from the vantage of an obsessed narrator who, as in much abuse, reduces a person to body parts. The related poem in this chapter is Yeats's great poem, "Leda and the Swan," which confronts, philosophically and poetically, violence in the world with the example of the terrible crime of rape.

Chapter 11 presents Herman Melville's "The Operation" from *White Jacket*, which describes the wide difference between a patient's terrible apprehension of an operation and its pompous description by the surgeon. Here, again, in reading Melville's narrative—and especially the satirical use of proper names—one could pursue a particular focus on the "patterned repetition in language" of literary texts. The vignette analyzed in this chapter is an excerpt from Lous Heshusius's memoir *Inside Chronic Pain: An Intimate and Critical Account*. The poem associated with this chapter is Emily Dickinson's "Pain has an Element of Blank."

Finally, the last chapter of this section, "Ageing," presents an early chapter of Nathaniel Hawthorne's *House of Seven Gables* that examines a day in the life of an ageing woman. In reading this chapter, one could focus on how implicit "patterned repetition of events" allows the main character, Hepzibah Pyncheon, to internalize governing prejudices about ageing in a manner that allows young readers to reflect on their own assumptions and prejudices. The vignette analyzed in this chapter examines a physician's encounter with a very old man. The related poem in this chapter is Thomas Hardy's "I Look into My Glass," a poem that presents the lack of any sense of future in its "endless rest" that is foreign to most young students, and indeed, to mid-career professionals.

Topic VII Chapter 13 presents another topic that could fit under the category of "Building the Patient-Provider Relationship" or the category "Everyday Ethics of Medicine," but it does so here by describing the "destruction," rather than the building, of that relationship by means of "Mistakes in Medicine." The vignette of this chapter, presenting a systematic and horrifying mistake encountered in Dr. Vannatta's medical practice, allows him to reflect the important chapter of Dr. David Hilfiker's *Healing the Wounds*, entitled "Facing Our Mistakes," that many years ago (1984) offered a powerful catalogue of kinds of mistakes in medicine, which were then—and often still remain—unreflected upon in the profession. The literary text of this chapter, an excerpt from Gustav Flaubert's *Madame Bovary*, describes professional and social pressures that lead to mistakes, another area (besides the "systematic" mistakes Dr. Vannatta describes) supplementing Dr. Hilfiker's catalog. The related poem in this chapter is Dr. Dannie Abse's poem about the terrible consequences of badly executed brain surgery.

Topic VIII The final topic in Part I, "Death and Dying," presents another matter that could fit under the category of "Vicarious Experience." This chapter presents Leo Tolstoy's novella, *The Death of Ivan Ilych*, which, in fact, gathers up together all the topics of Part I and offers as well a fine sense of the working of literary narrative by offering a strong example of the "defamiliarization" that literature creates, discussed in Chap. 1. The vignettes analyzed in this chapter describe physicians encountering both a "good" dying—where a patient and her family has come to terms with end of life—and the panic and frenzy associated with a patient and his family who haven't been able to do so. The poem related to this chapter is John Donne's "Death Be Not Proud," that engages death and dying from the vantage of religious contemplation.

Postscript Still, the topics of Part I do not, adequately enough in our opinion, emphasize the great motive of caretaking for those who pursue careers as healthcare providers, so we end Part I with a short Postscript that focuses on the good news medical care often brings to those suffering and in distress. Here, the literary text is a short poem by Derek Mahon, "Everthing is Going to be All Right." Although there is not a vignette as such to the postscript, we do quote Dr. John Stone's wonderful description of the usefulness of this literary text in his work in healthcare. In the Postscript we are trying not to lose sight of the goodness in caretaking, a powerful motivating energy for those pursuing healthcare and, indeed, for this book.

Each chapter of Part I ends with a short section entitled "Lessons for Providers." Needless to say, these are not the only lessons our readings might provoke, but they touch upon each chapter's readings with a gesture toward the clinical and medical benefits that might arise in engaging with them. In fact, readers might want to look at them before going through the chapter as a whole. And in group discussion, readers might want to supplement the "lessons" with their own. In many cases, these lessons at the end of each chapter offer questions and observations specific to healthcare providers.

The appendices of *Literature and Medicine* offer readers'/teachers' guides (Appendix 2) for discussion questions, (Appendix 3) for daily writing questions, and (Appendix 4) for a specific class handout associated with the chapter/class focused on diagnosis. Appendix 5 reproduces the systematic presentation of a professionalization workshop written by the authors and their colleague Dr. Casey Hester, which describes the practical usefulness of literature in physician training and also offers an outline for the "elements" or "structure" of narrative and narrative genres. Finally, Appendix 6 offers a strictly clinical analysis of Ernest Hemingway's story, "Indian Camp," that complements the literary analysis of Joyce in Chap. 5 in a manner, we hope, that offers a good sense of how literary and clinical education work together. Appendix 1 supplements Part I of Chap. 1 with further considerations of "The Cognitive Science of Literary Reading."

Part II

Medicine Introduction: The Complexity of Clinical Medicine

In Part I, the Literature Introduction, we have offered a narrative of the trajectory of *Literature and Medicine*, its local tactics and its overall strategy in pursuing the goals and purposes of this book we described in the Preface. In Part II, the second "Medicine" Introduction, we articulate those tactics and strategies in relation to the complex nature of clinical medicine. Thus, before we turn to the specific examination of the "features" of literary narrative that can contribute to training for a career in medicine and healthcare, it is important to situate this program in the context of clinical medicine more generally. This is certainly the aim of such recent works as Dr. Rita Charon's *Narrative Medicine*, Dr. John Biro's *Listening to Pain: Finding Words, Compassion, and Relief*, or our own *The Chief Concern of Medicine*.

The Complexity of Clinical Medicine

One of the notable features of clinical medicine that the study of narrative and literature helps make clear is its complexity. The patient-physician interaction—both the initial interview with new patients that results in the History and Physical Exam and ongoing patient-physician encounters in general – involves three different orders of cognition and interpersonal relationship at the same time: that of *biomedical understanding*, that of *the patient's understanding*, and finally that of *the affective engagement*—an emotional rather than cognitive order—of both patient and physician. The first order of cognitive understanding—that which is most readily perceivable by people whose training has prepared them for work as physicians and other healthcare providers—is that of the simple biomedical "facts" that patients present, the patient's *chief complaint*. The work of clinical medicine, and especially the History of Present Illness (HPI), is to solicit the *narrative* of facts that led the patient to consult the physician on this particular occasion with these particular problems or symptoms. It is then the job of the physician to "translate" the narrative patients bring into another narrative, that of biomedical knowledge and understanding. Such a translation needs also to be retranslated back to the patient in a vocabulary that she can comprehend and understand.

A second order of cognitive understanding in clinical medicine that is often given less emphasis than the recognition of biomedical facts discerned in the patient's narrative is the *patient's agenda*. That is, patients bring particular concerns and desires to the patient-physician encounter, which are not always fully congruent with the *physician's biomedical agenda* (even though quite often the two agendas are fully congruent). Thus, along with biomedical facts, the patient-physician interview entails a different order of cognitive understanding by the physician. In this text-anthology—and also more fully in *The Chief Concern of Medicine*—we describe this as the patient's *chief concern*. The chief concern is the patient's understanding of the consequences of his chief complaint: death or suffering, the loss of friends or job, the destruction of personal relations or goals, and so on. If the chief complaint exists within the physician's sphere of expertise—after all, she is trained in a vast amount of biomedical facts, histories, and procedures—then the chief concern exists within the patient's sphere of expertise: after all, the patient knows what the physician can only guess, what his particular condition or ailment *means* to him. It is the task of the initial medical interview not only to define or delineate the patient's chief complaint, but also to solicit and engage the patient's chief concern. Such solicitation demonstrates that the physician knows the difference between the patient's symptoms and the patient's understanding and concern *as a patient*, and it also can serve as the basis of the deliberation and negotiation between patient and physician concerning what is to be done in relation to the ailment or condition that brings the patient to her doctor.

It is our contention in this book that the systematic study of literature can aid the physician in accomplishing these goals. As we more completely explain in Chap. 1, reading literature helps the reader improve empathy. An empathic caregiver is more likely to habituate wondering about the patient's concern, and more specifically, about their affective response to that concern. Second, the study of literature focuses on the characters' actions and motives in order to define literary genres, which conditions the apprehension of the *whole* meaning of a narrative. Thus, training and practice in grasping the local and global meanings in a literary narrative is excellent practice for listening to and comprehending patients' histories of present illness. It is also excellent practice at listening for the unsaid—often important in uncovering subconscious motivations in a patient's telling of the story or motivations in the patient's life journey. There are other ways—revealed throughout the chapters of the book—that careful study and reflection on literary narrative can help a physician in her care of patients. Third, the study of literary narratives helps students of medicine develop critical thinking skills. The curriculum of a medical education—at least in the early years—requires rote memorization of huge amounts of necessary biomedical facts. This process does not call for critical thinking skills. However, the most important diagnostic information the physician will have access to is *narrative* in nature, namely the patient's story, which calls for critical thinking skills along with empirical knowledge. It is, therefore, our claim that understanding how literary narratives do their work—their structure, their genre, the motives and actions of their characters—will aid the physician in facilitating and understanding this knowledge, which can be called *narrative knowledge*, in the service of making a correct diagnosis.

For the past several years, at the University of Oklahoma College of Medicine, students have been trained to formally elicit the patient's "chief concern" as well as her "chief complaint," and our students and instructors have found that this addition to the protocols of the history and physical has *habituated* both the attentive listening and the responsive engagement we mentioned at the beginning of the Preface. An emphasis on these two orders of cognitive engagement—the physician's agenda and the patient's agenda—grows out of the engagements with literature and narrative we pursue in this text-anthology, and in large part, the purpose of fostering a relationship between literature and medicine is to encourage attention to these two orders of understanding by demonstrating—in practice as well as theory—how engagements with literature can create a broader sense of the relationship between a patient's ailment and his wider life. Specifically, this is clear in the first of the features of literature we examine in Chap. 1, the dynamic relationship between content and form that is emphasized in literary narrative. "Content" corresponds to the biomedical facts discernable in the patient's narrative, and "form" corresponds to the "style" of a patient's story and more generally to the present encounter of storyteller and listener, patient and caretaker.

The third order of the patient-caretaker relationship in the complexity of clinical medicine entails the task of creating a foundation of trust, honesty, and goodwill. The acknowledgment of the complementary nature of the cognitive orders we have described—that is, the active solicitation of the patient's agenda in connection with the physician's agenda—goes a long way toward establishing this foundation. But this order of interpersonal relationship is less a cognitive aspect of interactions between patients and physicians and more an *affective* aspect of what goes on between a patient and his caretaker. Part of the solicitation of the chief concern involves the physician's *conscious awareness* of the patient's emotional state that always accompanies the patient's cognitive understanding of her ailment, and much recent work in cognitive psychology (examined in Chap. 1 and Appendix 1 of this text) demonstrates that engagements with literary texts make people more sensitive to the affective states of others. As we discuss in more detail in Chap. 1, research teams associated with Raymond Mar, David Miall, and Melanie Green, among others, have demonstrated, using empirical research methods, that reading literary narrative both enhances the ability to acknowledge that other people can understand the world differently from the way one does and enhances a greater awareness of the emotional states of others, both necessary to increase empathy. This enhanced empathy as well as the expanded understanding of lives different from one's own (accomplished in literature through the provocation of vicarious experience) enhances the education of the student and also of the practicing healthcare provider.

Thus, it is our contention in this text-anthology that explicit training in narrative understanding and narrative knowledge, with particular attention to the manner in which authors and patients articulate their histories, can help healthcare students and physicians develop the combination of skills and comprehension that the complexity of clinical medicine requires. As we have suggested, traditionally courses in clinical medicine have focused upon the first cognitive order of understanding, while not attending as fully as possible to the second cognitive order and the third affective order that are also involved in narrative. In fact, studies continue to demonstrate that patients consistently complain that their physicians do not seem to care enough—they do not *listen* to their patients—and that they almost never complain that the physicians do not know enough. The overall purpose of serious engagements with literature, as we note in Chap. 1, is the apprehension of a text's overall meaning or purport, what we describe as the author's or the text's "overall meaning"—which is not necessarily conscious or fully intentional, but which is, nevertheless, apprehensible. Moreover, it is our contention—based upon years of teaching students committed to healthcare—that the systematic engagement with literary texts can develop habits of attention to this *overall* sense of concern, meaning, and purpose in the patient-caretaker encounter.

It is our further contention that familiarity with even a small number of literary narratives such as presented in this book (as well as "everyday" narratives that patients bring to their physicians, which are also presented in the various analyses of "vignettes" in this book) can allow students and instructors to integrate the complex levels of cognition and affect into habitual practices in their engagements with patients. Grace Paley's story "A Conversation with My Father" examined in Chap. 1, demonstrates the complexity of narrative corresponding to the complexity of the medical interview. Specifically, the story presents explicitly the dynamic of form and content we have already mentioned. In *Narrative Medicine*, Dr. Charon describes this as focusing on both the "what" of a story and "how" it is told (2006: 109). As we note in Chap. 1, one of the features of literary narrative—that gives rise to the proven effects of Theory of Mind, empathy, and the "transportation" of readers into the world of narrative (i.e., a version of vicarious experience) that recent experiments in cognitive psychology have demonstrated—is that literary texts are almost always "twice-told tales," and learning to attend to how stories are told and retold allow physicians and other caretakers to more fully engage with their patients. After all, the stories patients bring to their physicians are usually "re-told" by their caretakers, and awareness of the *features* of literary narrative we describe in Chap. 1 lends to this process greater attention, precision, and efficiency. As we mention in Chap. 1, these two tellings often reflect the general fact of *two time-frames* in narrative, the time of the events that are recounted (usually in the past) and the time of the telling in the engagement of storyteller and listener (in the present). The first order of cognition of the patient's HPI emphasizes and focuses upon the first of these time-frames; but the patient's present concern and present emotional state—the second and third orders of cognition and interaction—emphasize the second (the time that the patient is in the office telling the tale). We believe the achievement of the *simultaneous* cognitive and affective comprehension of all three orders in the patient narratives that constitute in good part the patient-caretaker relationship will be aided by the systematic understanding of literary and everyday narrative. Such achievement will allow physicians and healthcare providers more effectively and efficiently to

Introduction to *Literature and Medicine*

1. obtain the necessary biomedical information from the patient that will allow a more accurate diagnosis;
2. obtain the necessary personal information ("concern") from the patient that will allow the physician and patient together to determine the best "goals" to be achieved in this particular situation; and
3. create a better and more fulfilling interpersonal relationship between patient and physician that will forge trust, honesty, and goodwill that will, again, allow a more efficient, more effective, and satisfying response to the situation at hand *by both the patient and the caretaker*.

The Goals of Clinical Medicine

As we have suggested, the aim of this text-anthology is to create practical engagements with literary and "everyday" narratives, which help *habituate* practical forms of attention and action that more readily instill in medical students the goals of medicine of defining and promoting good health. There are four major goals of narrative medicine implicit in the patient-provider relationship that further specify the goals of *Literature and Medicine* we described in the Preface. In Chap. 1, the features of literary narrative we describe further specify objects of attention that can practically promote these goals in reading literary texts:

1. The facilitation of detailed information concerning the patient's **chief complaint**. The physician facilitates the narration of events and circumstances surrounding the particular ailment or condition that occasioned the patient's seeking out medical help.
2. The facilitation of detailed information concerning the patient's **chief concern**. The physician solicits the patient's worry or concern that accompanies the chief complaint. Implicit here is the goal of developing through joint deliberation and negotiation between patient and physician what would constitute "health" or some other practical goal or end toward which the medical consultation strives.
3. The development of a relationship of **trust and honesty** between patient and physician. Such a relationship is based upon mutual respect, the physician's recognition of the dignity of the patient, and the explicit acknowledgment on the part of the physician or healthcare provider that the patient's story—both the complaint and the concern—is "honored" (see Charon 2006). Active engagement with literature and the vicarious experience it provokes substantially promotes all three of these goals.
4. The development, on the basis of the information that grows out of the patient-physician interview, of a **differential diagnosis**. Clinical medicine, as it is conceived here and, in fact, as it is conceived, at least implicitly, in courses focused on clinical medicine, strives to develop *usable information* toward the end of addressing, and, when possible, alleviating the patient's condition or ailment that has concerned the patient sufficiently to seek out medical aid. Needless to say,

clinical medicine courses in the first years of medical education cannot call upon a large database of biomedical information that students have yet to master, but even so the overall purpose of clinical medicine is precisely to demonstrate even to novices that diagnosis lends itself to *systematic practices* of hypothesis formation that can be efficiently habituated.

These goals are always implicit and sometimes explicit aspects of clinical medicine, and one overriding purpose of *Literature and Medicine* is to clearly define these goals in the education and practice of physicians and other healthcare providers.

Bibliography

Charon, Rita. 2006. *Narrative Medicine: Honoring the Stories of Illness*. New York: Oxford University Press.

Hilfiker, David. 1984. Facing Our Mistakes. *New England Journal of Medicine*. http://www.davidhilfiker.com/index.php?option=com_content&view=article&id=51:facing-our-mistakes&Itemid=41

Shakir, Mubeen, Jerry Vannatta, and Ronald Schleifer. 2017. Effect of College *Literature and Medicine* on the Practice of Medicine. *Journal of the Oklahoma State Medical Association* 110 (November 2017): 593–600.

Tweedy, Damon. 2015. *Black Man in a White Coat: A Doctor's Reflections on Race and Medicine*. New York: Picador.

Section I

Narrative and Medicine

Narrative and Cognitive Science; Literature and Medicine

Part I: Narrative and Cognitive Science

Cognitive/Affective Responses and Literature

There is good evidence that episodic memory systems (the ability to remember specific details of situations and life events) and procedural memory systems (the ability to remember the generality of action across events such as how to perform tasks like birds singing birdsongs or people riding bicycles) evolved separately and involve different neurological processes (see Schleifer and Vannatta 2013: 412–13; see also Donald 1991; Sherry and Schacter 1987; Milner 1966, 1975). That is, there is good evidence that organizing experience in terms of narrative is a "natural" function of human beings and other animals. More to our point, in the last 25 years, there has been a considerable number of studies in cognitive psychology and other related fields demonstrating, by means of quantitative and qualitative experiments, the effectiveness of the engagement with literary narratives in enhancing a number of cognitive/affective responses that are particularly useful in building the strong patient-caretaker relationship we described in Part II of the Introduction. These responses include *empathy*, enhanced practicing of *Theory of Mind*, and the *transportation* that literature gives rise to.

Empathy The psychoanalyst Roy Schafer has defined "generative empathy ... as the inner experience of sharing in and comprehending the momentary psychological state of another person" (cited in Charon 2006: 133). In an interview, Rita Charon elaborates on this:

> Empathy is the method, or the tool, that gets you toward engagement. Empathy is that ability to recognize the plight of another person and to be moved by it. Empathy does not require that I have experienced what the patient is experiencing. It doesn't require that I imagine it happening to myself necessarily. I mean, I can't really say, "I'm a 98-year-old demented woman." That doesn't work, but it does require that I can imagine the whole situation, and if she's 98 and demented, I have to say as I use my imagination, "Well, what

© The Author(s) 2019
R. Schleifer, J. B. Vannatta, *Literature and Medicine*,
https://doi.org/10.1007/978-3-030-19128-3_1

does that mean?" Probably she can't do very much cooking in the house if she can't remember where she put the rice, and perhaps that means she can't use the telephone any more, and those very practical things. So, this ability to imagine the predicament or the plight of the patient puts us in a position to treat them all the more effectively.

I think it's worth pointing out, especially to inexperienced doctors, that having empathy does not mean that you are weeping all the time, or it doesn't turn you into a sort of passive, sympathetic observer, do you know? It's rather a very, sort of, lean and muscular thing. Empathy is very muscular; it takes a lot of work, you know? What do you see in a day, twenty patients in your office? And, the conceptual effort, almost the physical effort of doing this twenty times in a day is exhausting. And to always be saying, "Well, if that, then what?" And to enter the world as told by the patient, always sort of looking in the corners, as we said before, trying to hear the unsaid, trying to see the unseen. Man, is it ever exhausting! (Vannatta et al. 2005: Chap. 1, screen 35 [video]).

Moreover, empathy, as Charon describes it, belongs to the domain of emotional and narrative *understanding*. It does not spring forth from the logico-scientific study of medicine: rather, empathy is an affective as well as cognitive understanding of another's feelings, pain, or concern. As such, empathy is a skill of great use in medical care. Furthermore, as the scientific work we survey in this chapter suggests, empathy is a skill that can be learned and enhanced through engagements with narrative literature.

In recent years, neuroscientists have discovered a physiological basis for empathy, what they have called "mirror neurons." In the 1990s, neurologists discovered that such "mirror neurons" fire not only when a person (and primates as well) perform a certain action, such as eating or stubbing a toe. In addition, "when we see someone else suffering or in pain," Marco Iacoboni observes, "mirror neurons help us to read her or his facial expression and actually make us feel the suffering or pain of the other person" (Iacoboni 2009: 4). Furthermore, Iacoboni—in whose laboratory mirror neurons were first discovered—notes another neurological study that demonstrated that "areas in the brain known to control the movements of particular body parts (i.e., the hand or the mouth) were activated not only when subjects watched the movement on video but also when subjects read sentences about the movement" (2009: 94). This scientific work seems to account, by means of neurophysiological events, for the phenomena of empathy and of vicarious experience. (Nevertheless, there have been some striking critiques of the larger claims for mirror neurons: for a broad discussion, see Leys 2012 and Hickok 2014.)

Theory of Mind Several of the experiments we examine in this chapter discuss "Theory of Mind" (sometimes abbreviated ToM). Theory of Mind is a technical term in cognitive psychology that describes the ability, in human and other primates, to imagine what another person/cohort thinks or feels. When Charon imagines the experience of "a 98-year-old demented woman," she is exercising ToM. (As this suggests, "Theory of Other People's Minds" might better describe this phenomenon.) Technically defined, ToM is the ability to attribute mental states—beliefs, intentions, desires, pretending, knowledge, and so on—to others and to understand that others have beliefs, desires, and intentions that are different from one's own. (As a corollary of this, it is also the ability to attribute mental states to oneself as well.) There is good experimental evidence that ToM appears in human beings around the age of 4, before which children usually assume any knowledge they have is possessed by others as well. The anthropologist Robin Dunbar asserts that "no living species will ever aspire to producing literature as we have it. This is not simply because no other species has a language capacity that would enable it to do this, but because no other species has a sufficiently well-developed theory of mind to be able to explore the mental worlds of others" (1996: 102). Several of the scientific studies we cite demonstrate that engaging with literary narrative enhances and refines this ability to attribute mental states to others.

Narrative Transportation Theory Using a metaphor of traveling, Transportation Theory posits that experiences of narrative—usually reading and storytelling experiences, but also experiences of drama and cinema—create a mental state in which the experiencing subjects reading, hearing, or watching narrative "lose themselves" in the story they encounter. In a recent review of both experimental and interpretational studies of Transportation Theory, Tom Van Laer et al. succinctly define it from the point of view of marketing. In this description, we have replaced the term "consumers" by "readers":

> Narrative transportation theory proposes that when [readers] lose themselves in a story, their attitudes and intentions change to reflect that story (Green 2004). The mental state of narrative transportation can explain the persuasive effect of stories on [readers] (Gerrig 1993), who may experience narrative transportation when certain contextual and personal preconditions are met, as Green and Brock (2000) postulate for the transportation-imagery model. As we elaborate further subsequently, narrative transportation occurs whenever the [reader] experiences a feeling of entering a world evoked by the narrative because of empathy for the story characters and imagination of the story plot. (2014: 798; the citations mentioned in this paragraph appear in Appendix 1.)

The reader is in a state of detachment from the world of origin, and Transportation Theory suggests, her attitudes and intentions change in relation to the narrative. In this way, Transportation Theory is dependent upon empathy, Theory of Mind, and features of narrative to achieve its effects. Such features are defined as mental imagery by Green and others, but as we suggest in Part II of this chapter, aural qualities, patterns of repetition, and large and small narrative structures also contribute to "transportation." As this might suggest, Transportation Theory describes what has been traditionally designated as "vicarious experience," but it does so in order to develop the scientific testing—through computer modeling as well as qualitative and quantitative experiments—of this mental state. Thus, Van Laer et al. note that "narrative transportation is a form of experiential response to narratives and thus similar to other constructs [of cognitive psychology], such as absorption, optimal experience or flow, and immersion" (2014: 800), and they even explicitly note that one study of narrative persuasion "propose[s] that identifiable characters affect narrative transportation because story receivers vicariously experience characters' beliefs and emotions, empathize with them, and become engrossed in the story" (2014: 802).

In Appendix 1, interested readers can find a sampling of a number of rigorous scientific studies demonstrating that engagement with fiction produces effects in readers that contribute to skills and attitudes that create more efficient, precise, and fulfilling engagements between physicians and patients. When this research is considered as a "whole," it seems reasonable to infer that there exists a causal relationship that follows a path that might be drawn in the following ways:

Reading literary fiction that exhibits the *features* of literary narrative we describe in Part II of this chapter (e.g., styles of foregrounding, defamiliarization, narrative structures, etc.) →
  emotional triggers (including empathic emotions) →
   transportation into the story →
    enhanced ToM →
     increased empathy.

Moreover, as we note in Part II of this chapter, the understanding and "vicarious" experience of literary narrative contribute to the *moral education* of its readers.

Part II: Literature and Medicine

The evidence developed by cognitive psychology demonstrates the effectiveness of literary narrative in developing both empathetic responses to experience and models of ethical behavior. However, most of the experimental work of cognitive psychology we note in Part I and Appendix 1 records the results of engagements with literature by means of standardized tests without fully analyzing *how* literary narrative achieves these ends. Even those studies that do pursue the linguistic, artistic, and narrative aspects of literature that help people develop empathy, Theory of Mind, and "transportation" (e.g., Miall and Kuiken), do not systematically discuss the features of narrative that can help instructors and students to develop *habits of attention* that can benefit clinical medicine. That is, a systematic understanding of a small number of features of literary narrative can organize within an undergraduate or medical-school course (or a private course of readings at home) the engagement with literature that cognitive psychology has demonstrated produces stronger empathetic responses, a great utilization of Theory of Mind, and the phenomenon of "transportation" experience in reading. In this way, it allows instructors and students to focus on particular aspects of narrative presentations to the end of habituating particular *strategies of attention and engagement* useful in the clinic. Finally, such attention and engagement, as we note in the Preface, leads to a third goal of *Literature and Medicine*, the *critical thinking* that the attention and engagement necessary for the systematic study of literature fosters. Charon makes the relationship between close reading in medicine and in literature particularly clear, and she notes that in both cases we are focused upon *teachable skills*. (The empirical studies in cognitive science also demonstrate that these skills are "teachable.") "Training for close reading of literary texts," she writes, is not unlike training for more clinical kinds of reading that health professionals assimilate. If I were to put a normal chest X-ray up on a view box …, any doctor would say something like the following: "This is a well-penetrated, nonrotated film. The inspiration is adequate. The bony structures are unremarkable. The mediastinum is normal. The cardiac silhouette is normal. The lung parenchyma is without consolidation. There are no effusions." The reader has learned to pay attention to various features of the visual text, moving sequentially through a drill of specific aspects so as to capture all the news that the chest X-ray has to offer. (2006: 113)

In a similar fashion, we can teach students what features of a narrative are most likely to capture "all the news [it] has to offer," what aspects of storytelling most fully repay attention in terms of understanding, engagement, and developing the best possible actions that the situation of storytelling calls for. In both medicine and literary studies, students can and should acquire habits of attention by means of "a drill" that habituates what "specific aspects" to attend to in order to most fully comprehend the significance of the information encountered.

We begin our discussion of the elements of narrative that are particularly pronounced in literary narrative giving rise to empathetic engagement with an analysis of a short vignette of Dr. Abraham Verghese's excitement that arose when he interviewed a new HIV patient. Here, as in all the chapters of this book, a real-life vignette creates a practical context for the engagement with various narrative features that can benefit understanding and judgment. In this chapter, the vignette presents a practical example of anticipated empathy, which cognitive psychology has demonstrated is increased through engagements with literary narrative. After analyzing this practical beginning—a simple, everyday example of empathetic engagement at the very beginning of a medical interview—we then turn to a literary narrative, Grace Paley's short story "A Conversation with My Father," to offer a rather schematic reading of a literary text that allows us to delineate 13 particular features of narrative that can be brought to bear in relation to other literary and everyday narratives throughout *Literature and Medicine*. Although there are a good number of additional "features" or aspects of narrative discourse, we have found through years of teaching that focusing on these particular narrative features is consistently useful in developing "careful" habitual responses to narrative in students who have not had extensive engagements with literature.

Empathetic Engagement: A Vignette

As we suggested in Part II of the Introduction, the patient's "chief concern" is her emotional and cognitive responses to her symptoms. The symptoms themselves constitute her chief complaint. Asking the patient her chief concern soon after the chief complaint allows the patient to express her fears, anxieties, sadness, and sometimes anger. When the doctor takes the opportunity to verbally empathize with this emotion—while attending to the patient's "concern"—a rapport begins to develop much more quickly than if the physician stays with her own agenda instead of the patient's.

Encountering an HIV Patient: A Vignette (A Passage from Dr. Abraham Verghese, *My Own Country* [1994])

Author Note: Dr. Abraham Verghese (b. 1955) is an Indian American physician-author, Professor for the Theory and Practice of Medicine at Stanford University Medical School, and Senior Associate Chair of the Department of Internal Medicine. He is the author of three best-selling books, two memoirs (*My Own Country*, *The Tennis Partner*), and a novel (*Cutting for Stone*). In 2015, he received the National Humanities Medal from President Barack Obama.

Here, in an example from clinical practice—and its remarkably fine memoir, *My Own Country*—Dr. Abraham Verghese narrates his first encounter with an HIV patient, Gordon Vines, early in the AIDS epidemic in the United States:

> When Essie [Gordon's sister] left, I began to take Gordon's history. As I interviewed him, I instinctively sized him up, trying to pick out as many clues as possible to who he was and to his condition. The patient encounter is traditionally divided into the history and the physical. But in actual fact, the examination begins the moment patients enter the room. One is alert to whether their hands are cold or warm and sweaty (which could indicate hypothyroidism or hyperthyroidism). One notes whether they are dressed shabbily or with glaring mistakes such as mismatched sock or clothing inappropriate for the season, a sign of dementia or delirium. Do they have the normal inflections in their speech or is it a dry monotone, as in Parkinson's disease? Is their facial expression or "affect" appropriate to their emotional state? A discrepancy between the sadness the patient expresses in speech and the hearty smile on their face is a clue to schizophrenia.
>
> To me the history and the physical are the epitome of the internist's skill, our equivalent of the surgeon's operating room. Like Sherlock Holmes—a character based on a superb clinician, Dr. Bell—the good internist should miss no clue, and should make the correct inference from the clues provided. (1994: 80–81)

As Verghese enters the room, he "sizes up" the patient and introduces himself. As he notes, the division between the (verbal) "history" of the patient and the ("hands-on") physical examination of the patient is superseded by a holistic "impression" of the patient. Most important in this description is Dr. Verghese's attention to the "emotional state" of his patient. As we note in the Medicine Introduction and in Appendix 6, it is important to focus on the emotional state of patients for the sake of both engagement and obtaining as much information as possible: even Sherlock Holmes (whom Verghese mentions) makes the emotional state of the resident patient in Chap. 2 a central object of his attention.

After the initial encounter with Gordon, Dr. Verghese begins questioning him:

> "Do you have fever every day?"
>
> "Not too bad. I stay cold all the time. I can barely shower. Or wet my head, 'cause I freeze to death. And then I'll suddenly break a sweat."
>
> "And what else?"
>
> "Otherwise I'm doing pretty well."
>
> "Any shortness of breath?" He was speaking in clipped sentences because he was so short of breath. And yet he did not volunteer this symptom.
>
> "I guess."
>
> Gordon was being a reluctant patient. Passive, as if he recognized the fever and other symptoms, but was only marginally conscious that it was happening to him. As if he had already separated from his body in some way. (1994: 82).

Here, then, he listens carefully as Gordon tells his story regarding his fever. He recognizes that the primary emotion here is ambiguous: Gordon seems "hearty" and sad at the same time, a "reluctant" patient. Toward the end of this interview Verghese diagnoses Kaposi sarcoma lesions in Gordon's mouth and pneumonia more generally, and he recommends a hospital stay. He also notes that Gordon had "a stroke of luck" (p. 83) with the nurse assigned to him, who brought him special food and music and even visited him after his discharge.

In this account in his memoir of dealing with AIDS in rural Tennessee early in the epidemic, Verghese offers a masterful psychosocial portrait to understand the patient's plight. His particular plight is that he has lost his life-world—he tells Verghese that two of his partners died of AIDS. The doctor recognizes and acknowledges his "reluctance"—that his answers leave much "unsaid"—and recognizes that his job, like that of the nurse Gordon luckily encounters, is caretaking in the face of disaster. (Among other things, Verghese suggests that Gordon suffers from "early HIV dementia" [p. 82]). Verghese responds to all of the information his patient presents, both verbal and nonverbal, before pursuing and facilitating his biomedical story. He understands that this demonstration of empathy—that is, verbal and nonverbal responses to his narrative that demonstrate the *cognitive* apprehension of the patient's feelings and *concern*—is important in creating rapport with his patient and in further elucidating his story.

These responses to his patient in this vignette-discussion in many ways follow the pattern of responses that accompany serious engagements with literature (and, as we have seen, Verghese even mentions Sherlock Holmes, whom we encounter in Chap. 2). That is, Verghese's focus on his patient's psychosocial situation—on the framework of his "concern"—is very much like the way that engaging with literary narrative calls attention to the overall meaning of a text. Such an overall literary meaning is parallel to the patient's concern; moreover, it is the final "feature" of literary narrative described in this chapter. A poem that explicitly pursues such overall meaning is Dr. William Carlos Williams' "The Red Wheelbarrow," which is presented in Chap. 5. The poem's first line, "so much depends" nicely offers an abstract meditation on the importance of the details to be found in narrative, just as Verghese un-self-consciously expresses how much his engagement with patients "depends" on the level of energy and empathetic interest he presents in his narrative without explicitly focusing on it. (Appendix 2 offers suggested topics for short one-page responses that, as we have found, helps instill this kind of responsive engagement in students.) In this way, the kinds of responsive engagements to literature that *Literature and Medicine* encourages, both in discussion and writing, help foster important understanding and behavior in the clinic.

The goal of the discussion of Paley's story in this chapter—and, indeed, of teaching narrative to medical students and healthcare professionals more generally—is to set forth *teachable* strategies of reading and listening to narrative that can help instill the "habits of attention" that we have described and that Verghese exhibits as an important part of the patient-physician relationship. By "habit" we do not mean routine mechanical responses, but rather a set of explicit expectations for markers or features of narrative to be regularly ("habitually") noticed, just as the reading of an

X-ray benefits from the particular attention to specific details. Most importantly, this is the "habit" of following the lead of the patient in the clinic, just as experienced readers follow the lead of the authors and of the language of the texts they read. Many believe that empathy is a character trait that one either possesses or does not possess, and a large part of the burden of the cognitive psychology we discussed at the beginning of this chapter is to make clear that while empathy is a "natural" predisposition in people, this does not mean that it is not a learned (and learnable) skill. That is, empathy is an event or practice which takes place within a relationship and, like learning to play a musical instrument—or learning attentive engagements with literary and everyday narratives—it becomes better with practice and thrives on feedback and interchange. Such feedback and interchange, cognitive psychology has demonstrated, is the basis of serious engagements with literary narrative; and such engagements, many strictly conducted experiments have also demonstrated, strengthen and increase empathy, the cognitive and affective apprehensions of Theory of Mind, and the vicarious experience of the "transportation" that literature provokes. Strengthening skills in feedback and interchange should and can be an important part of an education in healthcare.

The Features of Narrative

We can help make encounters with patients and their stories more cognitively effective, more efficient, and, yes, more humane, by studying the array of features that condition the effects of narrative that cognitive psychology has measured. To do so, let us read a short story by Grace Paley about a woman's repeated conversations with her dying father. Her father, as the story tells us, had pursued a career in medicine and later as a painter.

Literary Narrative: "A Conversation with My Father" (1972) by Grace Paley

> *Author Note*: Grace Paley (1922–2007) was an American short story writer, poet, teacher, and political activist. She taught at Sarah Lawrence College. In 1980 she was elected to the National Academy of Arts and Letters, and in 1989 the Governor of New York made her the first official New York State Writer. Her volumes of short stories include *The Little Disturbances of Man* (1959), *Enormous Changes at the Last Minute* (1974), and *Later the Same Day* (1985). *The Collected Stories* appeared in 1994.

A Conversation with My Father

> My father is eighty-six years old and in bed. His heart, that bloody motor, is equally old and will not do certain jobs any more. It still floods his head with brainy light. But it won't let his legs carry the weight of his body around the house. Despite my metaphors, this muscle failure is not due to his old heart, he says, but to a potassium shortage. Sitting on one pillow, leaning on three, he offers last-minute advice and makes a request.

"I would like you to write a simple story just once more," he says, "the kind de Maupassant wrote, or Chekhov, the kind you used to write. Just recognizable people and then write down what happened to them next."

I say, "Yes, why not? That's possible." I want to please him, though I don't remember writing that way. I *would* like to try to tell such a story, if he means the kind that begins: "There was a woman ..." followed by plot, the absolute line between two points which I've always despised. Not for literary reasons, but because it takes all hope away. Everyone, real or invented, deserves the open destiny of life.

Finally I thought of a story that had been happening for a couple of years right across the street. I wrote it down, then read it aloud. "Pa," I said, "how about this? Do you mean something like this?"

> Once in my time there was a woman and she had a son. They lived nicely, in a small apartment in Manhattan. This boy at about fifteen became a junkie, which is not unusual in our neighborhood. In order to maintain her close friendship with him, she became a junkie too. She said it was part of the youth culture, with which she felt very much at home. After a while, for a number or reasons, the boy gave it all up and left the city and his mother in disgust. Hopeless and alone, she grieved. We all visit her.

"O.K., Pa, that's it," I said, "an unadorned and miserable tale."

"But that's not what I mean," my father said. "You misunderstood me on purpose. You know there's a lot more to it. You know that. You left everything out. Turgenev wouldn't do that. Chekhov wouldn't do that. There are in fact Russian writers you never heard of, you don't have an inkling of, as good as anyone, who can write a plain ordinary story, who would not leave out what you have left out. I object not to facts but to people sitting in trees talking senselessly, voices from who knows where ...."

"Forget that one, Pa, what have I left out now? In this one?"

"Her looks, for instance."

"Oh. Quite handsome, I think. Yes."

"Her hair?"

"Dark, with heavy braids, as though she were a girl or a foreigner."

"What were her parents like, her stock? That she became such a person. It's interesting, you know."

"From out of town. Professional people. The first to be divorced in their county. How's that? Enough?" I asked.

"With you, it's all a joke," he said. "What about the boy's father? Why didn't you mention him? Who was he? Or was the boy born out of wedlock?"

"Yes," I said. "He was born out of wedlock."

"For Godsakes, doesn't anyone in your stories get married? Doesn't anyone have the time to run down to City Hall before they jump into bed?"

"No," I said. "In real life, yes. But in my stories, no."

"Why do you answer me like that?"

"Oh, Pa, this is a simple story about a smart woman who came to N.Y.C. full of interest love trust excitement very up to date, and about her son, what a hard time she had in this world. Married or not, it's of small consequence."

"It is of great consequence," he said.

"O.K.," I said.

"O.K. O.K. yourself," he said, "but listen. I believe you that she's good-looking, but I don't think she was so smart."

"That's true," I said. "Actually that's the trouble with stories. People start out fantastic, you think they're extraordinary, but it turns out as the work goes along, they're just average with a good education. Sometimes the other way around, the person's a kind of dumb innocent, but he outwits you and you can't even think of an ending good enough."

"What do you do then?" he asked. He had been a doctor for a couple of decades and then an artist for a couple of decades and he's still interested in details, craft, technique.

"Well, you just have to let the story lie around till some agreement can be reached between you and the stubborn hero."

"Aren't you talking silly, now?" he asked. "Start again," he said. "It so happens I'm not going out this evening. Tell the story again. See what you can do this time."

"O.K.," I said. "But it's not a five-minute job." Second attempt:

> Once, across the street from us, there was a fine handsome woman, our neighbor. She had a son whom she loved because she'd known him since birth (in helpless chubby infancy, and in the wrestling, hugging ages, seven to ten, as well as earlier and later). This boy, when he fell into the fist of adolescence, became a junkie. He was not a hopeless one. He was in fact hopeful, an ideologue and successful converter. With his busy brilliance, he wrote persuasive articles for his high-school newspaper. Seeking a wider audience, using important connections, he drummed into Lower Manhattan newsstand distribution a periodical called *Oh! Golden Horse!*
>
> In order to keep him from feeling guilty (because guilt is the stony heart of nine tenths of all clinically diagnosed cancers in America today, she said), and because she had always believed in giving bad habits room at home where one could keep an eye on them, she too became a junkie. Her kitchen was famous for a while – a center for intellectual addicts who knew what they were doing. A few felt artistic like Coleridge and others were scientific and revolutionary like Leary. Although she was often high herself, certain good mothering reflexes remained, and she saw to it that there was lots of orange juice around and honey and milk and vitamin pills. However, she never cooked anything but chili, and that no more than once a week. She explained, when we talked to her, seriously, with neighborly concern, that it was her part in the youth culture and she would rather be with the young, it was an honor, than with her own generation.
>
> One week, while nodding through an Antonioni film, this boy was severely jabbed by the elbow of a stern and proselytizing girl, sitting beside him. She offered immediate apricots and nuts for his sugar level, spoke to him sharply, and took him home.
>
> She had heard of him and his work and she herself published, edited, and wrote a competitive journal called *Man Does Live By Bread Alone*. In the organic heat of her continuous presence he could not help but become interested once more in his muscles, his arteries, and nerve connections. In fact he began to love them, treasure them, praise them with funny little songs in *Man Does Live* ...
>
>> the fingers of my flesh transcend
>> my transcendental soul
>> the tightness in my shoulders end
>> my teeth have made me whole
>
> To the mouth of his head (that glory of will and determination) he brought hard apples, nuts, wheat germ, and soy-bean oil. He said to his old friends, From now on, I guess I'll keep my wits about me. I'm going on the natch. He said he was about to begin a spiritual deep-breathing journey. How about you too, Mom? he asked kindly.
>
> His conversion was so radiant, splendid, that neighborhood kids his age began to say that he had never been a real addict at all, only a journalist along for the smell of the story. The mother tried several times to give up what had become without her son and his friends a lonely habit. This effort only brought it to supportable levels. The boy and his girl took their electronic mimeograph and moved to the bushy edge of another borough. They were very strict. They said they would not see her again until she had been off drugs for sixty days.
>
> At home alone in the evening, weeping, the mother read and reread the seven issues of *Oh! Golden Horse!* They seemed to her as truthful as ever. We often crossed the street to visit and console. But if we mentioned any of our children who were at college or in the hospital or dropouts at home, she would cry out, My baby! My baby! and burst into terrible, face-scarring, time-consuming tears. The End.

First my father was silent, then he said, "Number One: You have a nice sense of humor. Number Two: I see you can't tell a plain story. So don't waste time." Then he said sadly, "Number Three: I suppose that means she was alone, she was left like that, his mother. Alone. Probably sick?"

I said, "Yes."

"Poor woman. Poor girl, to be born in a time of fools, to live among fools. The end. The end. You were right to put that down. The end."

I didn't want to argue, but I had to say, "Well, it is not necessarily the end, Pa."

"Yes," he said, "what a tragedy. The end of a person."

"No, Pa," I begged him. "It doesn't have to be. She's only about forty. She could be a hundred different things in this world as time goes on. A teacher or a social worker. An ex-junkie! Sometimes it's better than having a master's in education."

"Jokes," he said. "As a writer that's your main trouble. You don't want to recognize it, Tragedy! Plain tragedy! Historical tragedy! No hope. The end."

"Oh, Pa," I said, "She could change."

"In your own life, too, you have to look it in the face." He took a couple of nitroglycerin, "Turn to five," he said, pointing to the dial on the oxygen tank. He inserted the tubes into his nostrils and breathed deep. He closed his eyes and said, "No."

I had promised the family to always let him have the last word when arguing, but in this case I had a different responsibility. That woman lives across the street. She's my knowledge and my invention. I'm sorry for her. I'm not going to leave her there in that house crying, (Actually neither would Life, which unlike me has no pity.)

Therefore: She did change. Of course her son never came home again, But right now, she's the receptionist in a storefront community clinic in the East Village. Most of the customers are young people, some old friends. The head doctor has said to her, "If we only had three people in this clinic with your experiences …."

"The doctor said that?" My father took the oxygen tubes out of his nostrils and said, "Jokes, Jokes again." "No, Pa, it could really happen that way, it's a funny world nowadays."

"No," he said, "Truth first, She will slide back. A person must have character, She does not."

"No, Pa," I said. "That's it. She's got a job, Forget it. She's in that storefront working."

"How long will it be?" he asked. "Tragedy! You too. When will you look it in the face?"

Feature Analysis

Some research in cognitive psychology suggests that literary narrative is more effective than "popular" narrative and that fiction is more effective than non-fiction in promoting the "skills" of empathy and Theory of Mind (e.g., Kidd and Castanol 2013, Mar 2009, Djkic et al. 2009, Bal and Vektkamp 2013). It is our argument throughout this book that the salient features of literary narrative—some of which have been experimentally examined (e.g., Miall and Kuiken 1994, Green 2004)—are important contributors to these effects because they are more readily discernible in literary ("art") narrative rather than in popular narrative insofar as art narrative more self-consciously takes up narrative features to create aesthetic patterns and provoke aesthetic responses. Scholars of general narrative have isolated some of the basic features of any narrative. Among other features, narratives are stories that have a teller, a listener, recognizable agents, a sequence of events (plot), an overall meaning, and a witness who learns. While Charon argues that, "unlike scientific knowledge or epidemiological knowledge, which tries to discover things about the natural world that are universally true or at least appear to be true to any observer, narrative knowledge enables one individual to understand particular events befalling another individual not as an instance of something that is universally true but as a singular and meaningful situation" (2006: 6), we have disputed the implication here that narrative knowledge is as "singular" as Charon suggests it is (see *The Chief Concern*: 95–100)—after all, the overall sense of a narrative often appears to be true to many independent observers. Still, we agree with Charon—and there is a pretty good consensus among scholars of narrative—that there are a small number of features of narrative that allows even very young children to distinguish between well-formed and ill-formed stories (see Polkinghorne 1988: 20).[1] In the article reproduced in Appendix 5, we list six distinctive features of narrative that are useful for the study of literature mentioned above (see Section A, "Narrative Structure," p. 266). Several of these features appear in this chapter as well, in our discussion of Paley's story, and it is our contention, set forth in the "features" of this chapter, that literary narrative presents more starkly recognizable features than everyday general narrative. That is, Paley's art narrative offers more precise narrative details that create a more pronounced utilization of the features of narrative than the everyday encounter of Gordon Vines and Dr. Verghese (even if *My Own Country* verges on the writerly presentation that Verghese has developed as a novelist). Such detail and emphasis allow for a more specific understanding of how a story articulates and conveys narrative meaning and engagement. Such understanding is particularly useful in clinical encounters between patients and healthcare providers.

1. The Dynamic of Form and Content

This first feature, like the related category of the *patterned repetition* of literature discussed below, is a general feature of the *art* of literary narrative, attention to which enhances the experience of reading for all and can be particularly useful for healthcare providers encountering patient stories. Before all else, "A Conversation with My Father" makes this dynamic feature, implicit in all storytelling, explicit: the relationship between what is being discussed—the details of situations and life events of episodic memory—and the manner in which these details are presented. Moreover, the focus on the *features* of literary narrative in *Literature and Medicine* asks instructors and students to attend to the form of presentation as fully as they do to the content in the same way that literary narratives in general depend upon and emphasize—or "foreground," as cognitive psychology suggests (e.g., Miall and Kuiken 1994)—discursive structures and features to attain their meaning and power. (The "power" of narrative is its ability to provoke the kinds of effects in its readers described in Part I of this chapter. Such power is conveyed as much by the form [or

---

[1] While many of these features have not been focus of experimental studies in cognitive psychology, they have been developed through careful analyses of discourse and narrative by narratologists and literary scholars throughout the twentieth century. A good number of the experimental studies reviewed in Part I of this chapter (e.g., Miall and Kuiken 1994, Green 2004) base their analyses of these features on this work of literary scholarship.

"features"] of narrative as it is by its content.) Thus, many literary scholars point out the importance of these two aspects of narrative by noting the distinction between the "story," which consists of content-events or a narrative, and the "discourse," which is the manner in which it is told. In the short citation from *My Own Country*, Verghese relates both the events of his meeting with Gordon and his observations and generalizations, which are addressed to a different audience than his patient. Thus, the dynamic relationship between the "story" and the "discourse" usually describes a relationship between (past-time) events and the (always present-time) presentation of those events. Moreover, the cognizance of the simple complex presentation of two times—especially *earlier* time of the onset of symptoms and the patient's *present* mode of presenting those facts to her medical provider—allows healthcare providers to attend to the different orders of cognition and affect in relation to their patients examined in the Part II of the Introduction. In Dr. Verghese's interview with Gordon, he begins by focusing on general anomalies of patient presentations that call for interpretation (or diagnosis): the dementia, Parkinson's, and schizophrenia he mentions.

Inexperienced readers of literature often focus primarily on the content rather than the form, thinking that what a text "is about" exhausts its meaning and power, and in so doing they neglect to focus upon how it is structured and organized, and what meanings and feelings its structure suggests and provokes. Thus, inexperienced readers often assume that "comprehension" is no more than a simple paraphrase of a narrative, and such limited comprehension neglects the social skills—sources of empathy, Theory of Mind, "transportation"—that cognitive psychology suggests literary narrative promotes in its readers. Like inexperienced readers, in the clinic inexperienced healthcare providers often almost exclusively focus on the content of the story rather than the form of its telling: after all, they assume that the *facts* of the case took place in the past, and that the comprehension of those past events might well form the basis of a reasonable diagnosis of the patient's condition. In the story, Paley's father—as a trained physician and later as an artist—is particularly concerned with *factual details* his daughter's story leaves out, both *descriptive* facts (her character's "looks" and the quality of her hair) and also *causal* facts ("what were her parents like, her stock" he asks; and he ends by relating action to a person's "character"). These two different kinds of "fact" that can be discovered in the content of story-events, descriptive and causal, are nicely gathered together in the story when Paley's father remarks that he believes her heroine was "good-looking, but I don't think she was so smart." That is, the content of the story events can be simply matters-of-fact (e.g., "she's good-looking") or implicit facts (e.g., the kind of intelligence we can infer from matters of fact).

The systematic study of literary narrative—particularly in conjunction with the short writing assignments set forth in Appendix 2, but even in group discussions of works that articulate different points of view—asks readers to attend to features of literary narrative beyond its simple content. Thus, Verghese attends to and responds to Gordon's affect, the "reluctance" he presents as he answers the doctor's questions; and, in a similar fashion, Paley's relationship with her father, 86 years old and frail from heart trouble, is *oddly* depicted by juxtaposing two stories, that of the mother-neighbor narrated in "A Conversation with My Father," and that of the relationship of Paley and her father that "frames" the stories she tells. Moreover, the juxtaposition of two stories is fraught with interpersonal information. Thus, her father accuses her of misunderstanding him "on purpose," and he speaks to her in a kind of private language. (In fact, one of Paley's earlier stories has people sitting in trees in Central Park talking to one another, though someone unacquainted with her work would have little idea what her father meant when he said "I object not to facts but to people sitting in trees talking senselessly, voices from who knows where....") Similarly, Dr. Verghese's short interview with Gordon is primarily about creating a sense and "event" of shared understanding—not only with his patient, but with the audience of his memoir—so that, in the process of narrating the interview, recounting its past events, he also comments more generally on the "skill" of the internist to his readers encountering the interview just now. One of the first things engagements with literary narrative teach is to be attentive to aspects (or features) of narrative beyond the mere chronological facts, which is how Verghese begins his account of Gordon. Content (past-tense "facts") seem to be the most valuable parts of a patient's narrative for a healthcare provider: it allows for—and in fact calls for—attention to descriptive facts and contemplation of implicit (causal) facts. But the (present-tense) encounter between story-teller and story-listener is always an integral part of storytelling, and to ignore this aspect of narrative ignores as well a second function of narrative telling, to build social community, which in this case is the patient-caretaker relationship, as well as convey factual information. (A good body of arguments and evidence for the biological adaptation of storytelling as a ubiquitous human institution confirms this social-building function of language. For a good overview, see Dunbar [1996].) "A Conversation with My Father" makes this distinction between what a story is about and how it is conveyed a *functional part of how the story works*: how attention to the deployment of language, which is what we mean by the narrative's "structure," is crucial to a full engagement with the story and its meanings. (In listing of the elements of narrative in Appendix 5, p. 266, we note how different features emphasize the past-tense content of narrative and while others emphasize its present-tense form.)

This is clear in the fact that in a complex way "A Conversation with My Father" presents *parallel acts of telling* between a character in the story and the story itself. Moreover, in the story itself, Paley's father repeatedly focuses on the ways the form of storytelling—how characters are described, for instance—are important. Such attention creates a kind of detachment from the events being recounted—like the detached attention that an art museum promotes in its visitors—even while it increases attention on the "effects" (or the *emotional power*) of a story. Thus, while everyday narrative, as Brian Boyd argues in a study of narrative in the context of evolutionary biology, functions to get the listener or listeners to behave in a certain way by creating a framework for action as well as understanding, the attention that literary narrative requires encourages the discernment of the "cognitive, emotive, and ethical responses" that narrative provokes and the discernment of "the complexity of the relationship between facts, hypotheses, and theories" (Phelan 1996: 14, 15).

2. Twice-Told Stories

Literary narrative very often emphasizes the dynamic between form and content in a notably obvious way by telling the same story twice. While this is not a feature that is always present in literary narrative in the way that the more general feature of "patterned repetition" is a defining feature of discursive art, it is remarkably true for "A Conversation with My Father." In fact, Nathaniel Hawthorne entitled one of his collections of stories *Twice Told Tales*, and the twentieth-century philosopher, Walter Benjamin, makes this explicit when he *defines* storytelling as narratives that create the necessity of being told over again (1969). By repeating the same story in two different manners, literary narrative makes clear the relationship between the dynamic of form and content and the second general quality of literature discussed below, the various features of *patterned repetition*. This "global" feature of repeated telling is perhaps most clear in the final literary narrative of *Literature and Medicine*, Leo Tolstoy's *The Death of Ivan Ilych*, whose first chapter narrates the death and funeral of Ivan, while the rest of the novella re-tells the same story as a short biography of Ivan's life. But other texts in this book are "twice told" as well: the doctor's double visit to his rural patient in Chekhov's "A Doctor's Visit"; the imagined and then the actual operation in Gustav Flaubert's *Madame Bovary*; and, of course, Arthur Conan Doyle's Sherlock Holmes story, which, like *all* mystery stories, tells its story twice in the narrative of events recounted by the crime victim—often in the same kind of fragmented narrative that physicians hears from patients—and a second narrative-explanation recounted by Holmes himself.

Most elaborately, in Paley's story, the narrator tells the narrative of the woman across the street twice, and her story makes it clear that she does so—as Benjamin notes all good stories do—*because* stories always imply an interlocutor who will question what is going on. (This emphasizes the *two time-frames* of narrative in general: the time of the events of the story and the time of the telling.) But in addition, Paley's story narrates the interchange between father and daughter twice, once after each of the stories. It is this insistent doubling—twice-telling both the past-tense story and the present-tense encounter—that allows readers to more fully experience and grasp the dynamic of form and content in narrative.

The feature of twice-told stories is a specific example of the *patterned repetition* that, like the dynamic of form and content, is a general feature of literary narrative. We describe three specific features of patterned repetition in describing the next three features of narrative. Patterned repetition might well be, in fact, the distinctive feature of *all* art forms: the patterned repetition of painting, music, sculpture, architecture, all of which emphasize qualities of the medium ("form")—language, color and line, sound, objects in space—to provoke aesthetic as well as cognitive responses (the "foregrounding" that experimental studies discern [Miall and Kuiken 1994]). That is, patterned repetition, implicit in literary acts of telling stories twice, is the larger source of the discursive *art* of literature that creates all kinds of "parallelisms" of *language*, *event*, and *meaning*. (Thus, the negative parallelism between the matter-of-fact opening chapter of *The Death of Ivan Ilych* and the following chapters exploring Ivan's understanding and feelings teaches attentive readers to look twice at narrative information and attend to different orders of engagement with the story.) One of the great linguists of the twentieth century, the Russian/Czech/American Roman Jakobson describes the patterned repetition of literary art ("poetics") under the category of linguistic "parallelism" in his essay, "Linguistics and Poetics." There, he argues that "rhyme is only a particular, condensed case of a much more general, we may even say the fundamental problem of poetry, namely parallelism" (1987: 82). By "poetry" he means the wider sense of creating artistic patterns from language (in the same way, painterly artists—like Paley's father—create artistic patterns from color and line), and we can note three *kinds* of patterned repetition (or "parallelism") that are pronounced in literature (although they may also be present in everyday narratives as well) in the following three feature of literary narrative.

3. Patterned Repetition in the Sounds of Language (Phonics)

One important feature of literature—perhaps most pronounced in poetry, but also present in prose narrative—is the patterns of the sounds of words. This is perhaps most clear in poetic rhymes, and Jakobson quotes the British poet Gerard Manley Hopkins describing poetic rhyme as a form of "marked parallelism [which] is concerned with structure in verse—in rhythm, the recurrence of certain syllables, in metre the recurrence of certain rhythm, in alliteration, in assonance, and in rhyme." In *alliteration* (the repetition of consonant beginnings), for instance, we can note how much more memorable and pleasing is Hawthorne's phrase "twice told tales" than our phrase "twice told stories." *Assonance* is repetition of vowel sounds, such as the "o" sound in "hopeless and alone," a phrase in the story Paley tells her father; and *rhyme*, of course, is the repetition of both of these aural features of language: "I like Ike" is one of Jakobson's examples. Literature, in its aim at creating aesthetic effects, emphasizes and calls attention to these features of language-sound while they fade from prominence in everyday discourse, including narratives people tell one another, which aims at more practical ends. (These three features of patterned repetition in literary narrative have been tested under the category of "foregrounding" in work discussed in Part I of this chapter and in Appendix 1.)

4. Patterned Repetition of Narrative Events (Syntax)

As well as patterned repetition on the level of language-sounds, literature creates patterns of the structures of language and the larger structures of narrative events. When Edgar Allan Poe's narrator repeatedly describes his "undue, intense, and morbid attention" excited by "frivolous" objects—the typography of a book, a summer shadow, the smell of a flower—that attention transforms itself into the obsessive fascination with his wife's teeth so that "in the multiplied objects of the external world I had no thoughts but for the teeth." In this, Poe is presenting the patterned repetition of syntax. Similarly, in the novel Dr. Vannatta mentions in that chapter's vignette, Roddy Doyle's *The Woman Who Walked into Doors*, his main character, Paula Spencer, repeatedly silently says "Ask me. Ask me. Ask me" when her physicians seem to willfully ignore how she could have gotten her terrible bruises, she is also presenting the patterned repetition of syntax. And when Doyle repeats this scene several times in his novel, he is participating in the "twice telling" of story events we already discussed. The spousal violence, which Paula suffers from her husband and tells over and again, is found in Poe's obsessive language as well—obsessive repetitions—as both narratives present terrible events. The patterned repetition of parallel features in a text can be seen in details of narrative: the way a story like "The Yellow Wallpaper" (Chap. 6) parallels the form and content of the story in the narrator's use of language, which makes her progressive shortening of paragraphs "parallel" with her increasing psychosis; or the way that Arthur Conan Doyle creates a "parallel" between the surprised reactions of his narrator, Dr. Watson, and the reactions of his readers so that we, as readers, experience the same magical surprise at Holmes's detective work that his friend and narrator experiences. In this, we can see how patterned repetition—like Jakobson's parallelism—is a chief strategy for the ways that literature creates vicarious experience, the "transportation" of experience that cognitive psychologists describe. Paley makes this explicit: she self-consciously creates two parallel stories (1) of daughter and dying father and (2) of a mother and drug-addicted son. Another example of parallelism in narrative event is literary "allusion," when an author purposefully calls up a parallelism between the present text and an earlier one: we see later that Sherlock Holmes explicitly alludes to Edgar Allan Poe in "The Resident Patient"; we saw in the present chapter how Paley's father implicitly alludes to one of her earlier stories where people are talking in trees.

5. Patterned Repetition in Narrative Themes (Semantics)

Besides patterned repetition that calls upon the (aural) qualities of language and the (represented) events of story, literature also creates parallelism of themes (larger ideas, assumptions about the world, the text's overall meaning, which we discuss later as another feature of narrative). This is clearly seen in *narrative genres*, such as the "tragedy" that Paley's father seeks—as opposed to the implicit "comedy" that the narrator seeks in the world—that we discuss in more detail below. But it is also clear in the "themes" of scientific versus humanistic forms of attention suggested by the metaphors we describe below in Paley's story or the different ways father and daughter seem to need to understand the world in the story as a whole. Finally, linguistic metaphors enact narrative themes on the level of semantics (meaning) rather than phonics (sound) or syntax (narrative action). That is, Paley's parallel descriptions, metaphorical and literal, of her father's condition are also an implicit form of the patterned repetition of narrative events, but now on the level of word-meanings rather than events or sounds. Thus, she describes her father's heart's action as "flood[ing] his head with brainy light." Here is a metaphoric description that makes his heart the "agent" of action—it makes the heart the active subject of the sentence, another feature of narrative discussed below. Moreover, metaphor enacts patterns of meaning: that the circulation of blood "flood[s] his head with brainy light" uses a parallel meanings of "flood" (which can be attributed to the literal liquidity of blood) and "brainy light" (where the adjective "brainy" is figuratively coupled with light) as an implicit pattern of repetition. Moreover, the repetitions continue when she notes that "despite my metaphors," as she says of her description of her father's heart as a "bloody motor," "this muscle failure is not due to his old heart, he says, but to a potassium shortage." Here, as throughout the story, the parallel is between two ways of describing the world, her father's more or less literal scientific explanation, and her own more or less metaphorical, figurative explanation.

6. The Unsaid

Artistic parallelism suggests another feature of literary narrative that is regularly found in the narratives that patients bring to healthcare providers, namely the "unsaid" in storytelling. We saw this explicitly noted in Verghese's narration of his engagement with Gordon. That is, parallelism—like the two narrative frames of storytelling in general, or the ways that authors sometimes seem to tell the same story twice—is not explicitly mentioned by authors but *enacted* in the manner of telling, and one of the strategies that literary narrative depends upon (and teaches its readers to pursue) is to "notice" what is not said in a story, the "elephant in the room." In Paley's story, what is unsaid is the fact that both narrator and her father *know* he is dying and yet they carry on an argument about literature that does not acknowledge this fact—and, in fact, takes place precisely because neither one *wants* to acknowledge the fact. As we shall see in Chap. 5, the great master of the "unsaid" is James Joyce, who never tells his readers what his characters do not have to explicitly think about. But this *skill* of attending to the unsaid is one that all literature teaches. As with parallelism and the other features of literature we are describing, when a reader can habitually take up the unspoken assumptions of authors and characters, she achieves the kind of "transportation" cognitive psychology describes, so this *skill*, we are arguing, is of great importance in a healthcare provider's interactions with his patients. Verghese explicitly attends to the unsaid with his patient and Paley organizes her story so that readers do that with her characters.

In the sub-story of "A Conversation with My Father," in the face of her son's addiction, the mother in the narrator's story begins using drugs herself to be close to her son—something that the narrator cannot do with her dying father—and in the end, the son overcomes his addiction and won't see his addicted mother. Hearing the story, her father repeatedly asks for realistic details, such as we find in the stories of Chekhov and de Maupassant. The narrator attributes this to the fact that her father "had been a doctor for a couple of decades and then an artist for a couple of decades and he's still interested in details, craft, technique." But the story itself suggests, as the literary scholar D. S. Neff has noted, that the narrator's reluctance to offer "realistic detail" can also be attributed to the fact that "the daughter retreats to the comforting realm of metaphor while the father strives to demystify her evasions in an attempt to help her accept his imminent death" (1983: 119). In any case, "A Conversation with My

Father" is a narrative that explores the nature of storytelling itself. Thus, the daughter notes that she often misjudges her characters, thinking them more extraordinary than they are, and "you just have to let the story lie around till some agreement can be reached between you and the stubborn hero." But this act of re-judgment is "unsaid," enacted but not pointed out by character or author.

7. Relational "Facts"

The "unsaid" may be discerned by attending to *relationships* among facts rather than simply to facts themselves. Such relationships are, in fact, another instance of the "parallelism" that constitutes the "art" of literature, and they point out a significant difference between logical-scientific training of biomedicine and the aesthetic-holistic training of a literary education. A great example of such relationships of facts can be seen in the first detective story in English, Edgar Allan Poe's "The Murders in the Rue Morgue" that Holmes explicitly alludes to (as "one of Poe's sketches") in "The Resident Patient" (Chap. 2). In Poe's story, the police report a number of witnesses who heard the purported killers talking in the course of the crime, but each witness assumed one of the killers spoke a language that the witness didn't know. Other witnesses were native speakers of this assumed language, and each of them suggested the killer spoke a different language, again one that the witness didn't know. The detective, Auguste Dupin, put these separate reports together and concludes that the killer was not speaking any language at all. That is, he discovers what is "unsaid" in focusing on parallels between witness reports. To be more specific, he is dealing with different *kinds* of facts: in this case, the difference between *sounds* and *language*. Another *difference in kind* is the difference between "murder"—used in Poe's title—and "killing," which is in fact what occurs, since the killings in the story were perpetrated by an orangutan and by definition animals cannot "murder." We are putting "facts" in quotation marks here, because the kind of fact we are describing is not simply a matter of fact, but a factual understanding that is grasped by arraying or "configuring" a set of facts together (for such "configuration," see Schleifer 2018). In this way, Poe's detective is assuming—as literary art does—that *the whole is greater than the sum of the parts*, and that one can apprehend a theme, or what we later call "the overall meaning," by attending to the whole in focusing on parallel relationships among the parts. (Biomedical science generally assumes *the whole equals the sum of the parts*, and it is precisely this difference, apprehensible in training in literary reading, that nicely complements the training many medical students and healthcare professionals have.)

Paley's story also calls upon its readers to create relationships among the "facts" it presents, even if it does so in a manner that is less explicit than in Poe or Arthur Conan Doyle. That is, the parallel stories she presents leave us with a need to figure out what is going on, *why* Paley tells this somewhat inconsequential story to her father altogether. As Neff notes, while the father and daughter in this story "never fully understand each other, … it is the initiative shown by both parties that matters": both father and daughter are trying to find consolation and love in the face of death by employing narrative. "The expressive wars of Paley's characters," he concludes, "make us realize that love exists beyond the confines of tragedy and comedy, and that the most mature art, like the most ethical physician, must revel in a capacity for self-transcendence in an endless quest to encompass life's inexplicable mixture of endings and renewals" (1983: 123–24). This story, and the miscommunication between daughter and father that it presents, can also be used as a demonstration of how the patient and the healthcare provider can miscommunicate—try as they may to understand one another. It gives the instructor an opportunity to explore how and why the characters in the story misunderstood each other, what different "agendas"—to use a term from our discussion of clinical medicine in the Introduction—they bring to their encounter.

8. Narrative Genres

The father, Paley's stubborn hero, and his daughter mention two of the genres of narrative that are noted in the article reproduced in Appendix 5, Section C "Genres of Narrative" (p. 267): *tragedy* and *comedy*. In the discussion of genre in the Appendix we note that the distinctive element of tragedy and comedy is designated by who receives the cultural value at the story's end. In tragedy—and the father explicitly exclaims, "Tragedy! Plain tragedy! Historical tragedy! No hope. The end"—it is the *helper*-actor, which in Paley's story is the daughter, who is left to carry on at the story's putative end (the death of her father) without the *hero*. In comedy, it is the *heroine* (or the ungendered object of desire) who receives the cultural value. In the story, the daughter tells her dying father, it is the boy—the object of his mother's desire—recovering his true ("healthy") self in the presence of his new girlfriend who receives the wished-for good. In her second more elaborate narrative that the daughter tells her father, the narrative depicts the son's addiction as "not hopeless," and in fact he meets a young woman and "in the organic heat of her continuous presence he could not help become interested once more in his muscles, his arteries, and nerve connections. In fact, love them, treasure them." In this opposition of two narrative genres, as Neff has written in a fine account of this story from the vantage of end-of-life medicine,

> the father's subjective experience as a dying physician is validated by seemingly objective patterns of tragic art. The daughter nurtures a comic perspective because her age and health enable her to comfort herself with half-evasions of ultimate truth that help humankind to live with death and survive with hope. (1983: 123)

The story as a whole, however, in its *aesthetic* enactment of this opposition, "concludes," as Neff says, with "ironic stasis"—the very kind of the modern genre of *irony* that the discussion of narrative genres in Appendix 5 suggests—where the *opponent*, death, seems irresistible.

9. Narrative Agents and Concern

As well as genres, narrative is organized around "agents," which is to say around characters who act in the (fictional) world. In narrative, character is defined by action: how a character *acts* defines the kind of person she is. As we note in Appendix 5 (B "Roles of Narrative," pp. 266–267), there are a small number of agent-characters in narrative that are defined by their action in relation to other agents, and an awareness of this *feature* of narrative allows listeners to hear the "unsaid" (see the vignette in Chap. 5 for a good example of this in the clinic). In "Conversation with My Father" there are two sets of agents: those of the main story and those of the sub-story Paley writes for her father. In the main story, her father can be understood as the "stubborn hero," and the daughter takes on the role of the hero's helper. In the sub-story, the mother can be seen as the hero. But in both of these cases, the characters—Paley's father, the neighbor mother—are *tragic*, their lives ending in death and failure. But if father and neighbor mother are understood as "objects of desire"—which is to say, as a person whom the active hero strives to join—then the story can be understood as *comedy*: in the main story Paley "recovers" her father; and in Paley's last re-telling of the sub-story when she imagines the neighbor mother as "the receptionist in a storefront community clinic in the East Village," she sees both the son and the mother herself as recovering health. By attributing a particular *role* to a character or agent—which is sometimes the work of the author, sometimes a character (one can note the different roles Ivan attributes to himself in Tolstoy's novella), and sometimes the work of the reader/listener (as when a physician imagines himself to be the patient's "helper" or, alternatively, the "hero" of the patient-physician encounter)—can materially affect the value and outcome of the story as a whole.

That is, in storytelling, the "point" or "end" of a story is negotiated between listener and teller so that the *genre* of the story itself (which is related to its "overall meaning") is something that is not given once and for all, but the object of balancing different ways of attending to the story. (Similarly, in clinical medicine the "end" or "goal" of medicine—namely, a definition of "health" for the particularly *situation* of the patient—is sometimes negotiated between patient and physician.) The balancing of different ways of attending to the story is clear in the ways Dupin and Holmes bring different forms of attention to the "evidence" they encounter: Holmes is primarily visual in his relation to evidence—this is clear in *The Resident Patient* and his standard representation holding a magnifying glass—while Dupin, as we note in Chap. 2 and have already seen in his attention to the spoken language of the putative killers in "The Murders in the Rue Morgue," reacts to aural evidence. In Paley's story, both father and daughter each clarifies and complicates the story he or she hears and questions, just as a healthcare provider hears and questions the stories of patients. And what physicians and healthcare students, taking on the role of *agents of significance* we see in Dupin and Holmes, learn is the very *concern* that is at the heart and at the "end" of a story: how a dying person tells and hears a different story from the living; how the absoluteness of endings can be inflected within the community of narrative.

That is, patients generally come to physicians with a basic demand,
- "Make me well."

And along with this demand, patients bring three basic questions:
- "What is my condition?" (i.e., "what do I have?")
- "What should I do?" and, often much less explicitly, particularly when there is no easy answer to the demand
- "What, specifically, do I want?"

Paley's story does not address the patient demand ("make me well"), but it does address the patient's question ("what is my condition?"), by noting, but not dwelling upon the fact that her father has a terminal illness. Moreover, this condition gives rise to a response to the second question, "what should I do [in the face of an imminent death]?": what this patient and his *helper* (the daughter in the story, but the role of *helper* is often that of the physician in the clinic) should do is to talk about the imminent death directly or indirectly. Paley's story does so indirectly, in "parallel" dialogues and deliberations concerning both hope and hopelessness, love and loss, ending life and ongoing life. The answer to the last question—"what do I want?"—hovers around the two narratives Paley presents. The main story ends when her father says: "truth first … Tragedy! You too. When will you look it in the face?" and he means, among other things, the truth of his dying condition. But when Paley published this story in *Enormous Changes at the Last Minute*—it had appeared earlier in a literary magazine—she added a note at the beginning of the collection of stories that underlines the agency of author, including the text's "overall meaning," which we later discuss. Thus, she wrote: "Everyone in this book is imagined into life except the father. No matter what story he has to live in, he's my father, I. Goodside, M.D., artist, and storyteller.—G. P." Here, she breaks up the "illusion" of fiction by finding purpose *outside fiction* that helps create the "transportation" cognitive psychology examines. (We examine this in greater length in the later feature "Narrative and Moral Education.")

10. The Witness Who Learns

Another feature of narrative, related to narrative agency, is the witness who learns: the learner can be the protagonist, the antagonist, or the reader. While the agent is part of the *story* time-frame (the agent in the already-completed narrative), the witness who learns can be part of the *discourse* time-frame (a witness who learns from the story at hand). In Sherlock Holmes stories Dr. Watson (and the readers) learn the truth in the end; in Joyce, the young boy (the protagonist of the story) learns something for which he doesn't quite have a vocabulary to articulate; in "Conversation with My Father" in fact there are two witnesses who learn, both the daughter and father (protagonist and antagonist). In this narrative, the daughter rewrites her story in response to the dialogues with her father; and the father also learns from experience: "He inserted the tubes into his nostrils and breathed deep. He closed his eyes and said, 'No.'" But the larger story's readers—lay readers and healthcare providers—are also witnesses who learn from the "experience" of this narrative. Of any narrative—including that of the patient in the clinic—it is proper and often illuminating to ask the simple question: "who learns from this experience? what does she learn? what difference (in behavior) does it make?" In engagements with literature, the witness who learns often encounters straight-on the author's or text's overall meaning, the final feature described in this chapter.

11. Defamiliarization and Style

When Paley names her father in the Preface to her book, which is *outside* the story as it is presented in this text-anthology, she makes an authorial gesture that leads to three final features of narrative. Awareness of these features can help readers attend more fully to texts and help healthcare providers attend more fully in their engagements with patients. Again, as in "patterned repetition," they can be understood on the level of language, narrative events, and narrative themes. On the level of language, authors present stories in particular "styles" of writing that identify particular writers and—most importantly—that can be imitated and parodied. (This is important because the possibility of parody indicates that *particular linguistic elements of style* can be isolated and imitated; it indicates that style is not simply "singular" and idiosyncratic but can be discerned through attentive analysis.) Perhaps the most pronounced styles in the fiction writers in this collection are those of James Joyce and Edgar Allan Poe. In their different ways, these authors leave expected discursive strategies out: Joyce does not *explain* what is going on, and Poe spends a great deal of time describing things "around" his narrative focus that creates implications of meanings that are rarely explicitly presented, the "frivolous" objects mentioned earlier. In Paley's story, this feature of style is part of the narrative insofar as the author herself is a character in her story. In the story itself, the narrator's father repeatedly focuses on his daughter's writing style.

    A strictly *literary* technique associated with style—and with the sense of the "art" of literature more generally—is *defamiliarization*, a term coined by literary scholars in Russia in the early twentieth century who were seeking to describe a particular feature of literary discourse that distinguishes it from non-literary discourse. The Russian Formalists, as they were called, wanted to isolate the quality of "literariness" that can be found in literature. (Roman Jakobson was part of this movement, and he later redefined "literariness" as the quality of "poeticity" that can be found in all discourse but is emphasized in literature. For a discussion of Jakobson's model of speech communication, see *The Chief Concern*: 215–20.) They claimed that one function of literature is to renew readers' sense of the newness of experience by disrupting habitual ways of reacting to or perceiving experience. Such disruption works to undo habitual *familiar* responses to the world: it *de-familiarizes* experience. Thus, Viktor Shklovsky, who explicitly argued for this idea, notes that perception "becomes habitual, it becomes automatic"; the habits of ordinary speech "devours works, clothes, furniture, one's wife, and the fear of war." For this reason, he goes on, "art exists that one may recover the sensation of life; it exists to make one feel things, to make the stone *stoney*." To be made new and poetically useful, language must be "defamiliarized" and "made strange," as Shklovsky says, through linguistic displacement, which means deploying language in an unusual context or effecting its presentation in a novel way. (Such deployment, as Miall and Kuiken [1994] argue, "foreground" language use rather than what language represents and thus "arrests" attention and habitual responses.) Defamiliarization is, therefore, the manner in which poetry and literary narrative function to rejuvenate and to revivify language. Thus, Shklovsky notes that "Tolstoy makes the familiar seem strange by not naming the familiar object. He describes an object as if he were seeing it for the first time, an event as if it were happening for the first time. In describing something he avoids the accepted names of its parts and instead names corresponding parts of other objects … [so that] the familiar … is made unfamiliar both by the description and by the proposal to change its form without changing its nature" (1989: 59). In Chap. 14, Tolstoy's novella *The Death of Ivan Ilych* repeatedly offers examples of this phenomenon. His narrative is a good place to see and feel the workings of literary narrative. In the present chapter, a good example of defamiliarization is Paley's description of her father's heart as a "bloody motor" in the first sentence of the story. There, she changes the "form" of description, by no longer using the vocabulary of anatomy, without changing its "nature," since a mechanical description of the heart muscle corresponds to its physiological function. More generally, though, attention to the *quality* of linguistic description gives rise to insight and energy. In the vignette in Chap. 4, Dr. Vannatta describes one moment of such attention to the language of a patient's story in the clinic and the manner in which it transforms a routine ("habitual") consultation into powerful and productive interchange. Attention to defamiliarization helps train healthcare professionals to attend to the *anomalies* in the stories they hear. Attending to anomalies—rather than dismissing them—as we note in Chap. 2, is a significant trait in detective stories and in the "abductive" reasoning of diagnosis.

12. Narrative as Moral Education

A second feature of narrative that focuses on the border between everyday life and narrative knowledge—after the feature of the provocation of feeling by means of defamiliarlization—is the way in which narratives lend themselves to moral education. This is a crucial aspect of the "transportation" literature gives rise to insofar as it allows readers to test out their judgments and responses to experiences they haven't encountered in real life. Francis Steen (a scholar of narrative) suggests that such "testing out"—he describes it as the ability to "construe" possible outcomes of action in the world—is the evolutionary-adaptive function of narrative, and he argues that one can discern the structure of narrative (very much like the structure described in Appendix 5) in the playfights of rhesus macaque monkeys (Steen 2005). The function of such playfights, he argues, is to teach younger monkeys what to expect from action in the world by means of playfighting with older cohorts.

Playfighting exhibits and rehearses predictable structures of action. In a related fashion, Scott Stroud (a scholar of rhetoric) has suggested that the power of literary narrative is to create vicarious experience—closely related to the "transportation" we discussed early in this chapter—that provides the "subjects" of that experience (i.e., readers and listeners) with a "type of knowledge, ... gained by virtue of the literary narrative's aesthetic qualities, which result in a certain type of activity in the reader" including the reader's "*identification* with the values, beliefs, and/or behaviors of the simulated agent" (2008: 20). The "aesthetic qualities" he describes are precisely the result of the features of literary narrative and of literature more generally that we have been outlining throughout this second part of this chapter in relation to Paley's short story, and in fact his analysis explicitly examines several of these features. The features of literature and the "aesthetic qualities" they give rise to, as Stroud (and also the cognitive psychologists we surveyed earlier) argue, create a "simulation" of experience from which a reader can "construe" possible endings and concerns for fictional stories. Like Steen, Stroud sees literary works serving life beyond the particular knowledge, experiences, and emotions they provoke by means of defamiliarization. That is, the reader uses "the imagination to test the viability of certain values and goals in terms of what results they would have for one's life and its flourishing." "Powerful fiction," Stroud concludes, "is useful in getting one to possibly revise, strengthen, or change one's values. Literary narrative, therefore, holds important cognitive value in enabling readers to grow and develop morally" (2008: 26). The vicarious experience that Stroud describes is a result of all the techniques of literary style—the features of literary narrative—we have been outlining here, many of which have been demonstrated to create these effects in empirical studies.

In any case, the moral education Stroud describes can be found throughout all the stories in *Literature and Medicine*, from the questions concerning professionalism that are engaged by Dr. Richard Selzer's "Imelda" (and its analysis in terms of the formal measurements of professionalism described in Appendix 5), to racism in Paul Laurence Dunbar and Demetria Martinez, and to sexism in Gilman; and the larger questions throughout our text-anthology of encountering people with different experiences and values, of encountering pain and suffering, and the explicit problems, as we see in Chekhov's Enemies" and Flaubert's *Madame Bovary*, of ethics and mistakes in medicine and everyday life. In Paley's story, the issue of a moral education arises with the questions we suggest are always implicitly present in a patient-physician encounter, questions concerning matters of fact ("what condition do I have?"), matters of practical behavior ("what should I do?"), and matters of overall meaning or desire ("what do I want?"). Paley's story, in this overall *concern*, as we noted above, focuses on her relationship with her father *outside* the "aesthetics" of the story: by naming her father in the Preface to the collection of stories which included "A Conversation with My Father," she calls into question the strict division between fact and fiction. A grasping of this story as both aesthetic and extra-aesthetic, like the teaching of literature in the context of medicine, leads to a special kind of reading and attention, which the features and protocols of reading literary narrative help attain. These protocols are outlined in this chapter and throughout *Literature and Medicine*, and they can teach engagement, understanding, and discernment to healthcare providers striving—as we see in the various physicians represented in literary works and vignettes throughout this book and in the profession by and large—to do the best by their patients.

13. The Overall Meaning

A final feature of literature we want to mention, closer to "theme" than to narrative events related to moral education or the functioning of language related to style, is the overall meaning of a literary text, which is something more than an author's "intended" meaning. Of course, this is closely related to the "point" or "concern" of Paley's story. Such overall meaning is not (simply) personal but something that arises by means of many of the *linguistic and discursive* features we are describing, and for this reason, it might better be described, in rather abstract terms, as a literary text's "claims" on its readers. Those claims manifest themselves by focusing on what readers take away from a text and, even more generally, why they begin and continue to read in the first place. Charon describes this feature by noting that "the casual reader reads for relaxation or distraction, fulfilling only a desire for entertainment or rest. The close reader deploys full powers of intellect, concentration, imagination, metaphorical thinking, and moral confrontation, fulfilling desires for identity, self-examination, facing up to challenge, and attaining new clarity about the world and self and other" (2006: 124–25). Charon's list of the deployment of powers provoked by close reading faithfully align themselves with the features of literature we have described in relation to Paley's story, features that recur throughout our text-anthology.

The purpose of explicitly enumerating these features is to create an outline of forms of attention that students and instructors can bring to their encounters with literary works and with patients in the clinic. The overall meaning of a text is parallel to the patient's chief concern: it is the overriding framework of value—aesthetic/experiential, practical, and ethical—that governs the profession of healthcare. In Dr. Verghese's clinic, Gordon presents, without explicitly articulating, his chief concern as a sense of loss of the community of friends rather than, say, his fear of decrepitude; and such a concern may call for different medical *treatment* than a fear of decrepitude. Paley's story suggests that its chief concern—its *overall meaning*—manifests itself in the struggle to comprehend some balance between love and death. In other words, the chief concern is a framework of value, of what is important. Encountering this framework of value in the ("vicarious") experience of literature, in the practices of attentive reading, and in the values that both establish these experiences and practices and also grow out of them, can properly enlarge an education in medicine. This is the aim and goal of this chapter—with its list of enumerated features to guide reading—and of this book as a whole: that the kinds of habits of attention that leads to "close reading" lead as well to more efficient, effective, and fulfilling encounters of patients and healthcare providers. To promote such attention, we believe, is the *overall meaning* animating the literary authors collected here, the author-editors presenting these texts and strategies for engaging them, and the students and teachers who will share this book with one another.

Conclusion

The 13 features of literary narrative can each be the source of a discussion question in reading the literary narratives—and in engaging the non-literary vignettes—in this text-anthology. One can ask of any text: what is its genre? who is the witness that learns? can students discover instances of patterned repetition or linguistic parallelism in the writing? what is the text's "overall meaning"? In other words, just as Charon notes that reading an X-ray is enhanced by "drills" of attention that habituate the task, so drills of attention in relation to the narratives of *Literature and Medicine* can habituate certain kinds of attention, engagement, and critical thinking for people who are faced with narratives on a daily basis. In Chap. 10—a chapter in this book that presents a terrible social problem that is also a medical problem—we return to these features to describe them in detail in relation to real-life encounters with violence and literary representations of violence in Poe's "horror" story and W. B. Yeats's great poem, "Leda and the Swan."

Lessons for Providers
1. Close reading of literary narrative improves skills and behaviors we seek to habituate: empathy, analysis, and search for meaning, among others.
2. There is growing evidence that suggests how apprehending the patient's story leads one toward engagement with the patient, as opposed to "detached concern" (i.e., concern without emotional involvement).

Bibliography

Bal, P. Matthijs, and Martijn Vektkamp. 2013. How Does Fiction Reading Influence Empathy? An Experimental Investigation of the Role of Emotional Transportation. *PLoS One* 8 (1): E55341.
Benjamin, Walter. 1969. *Illuminations*. Trans. Harry Zohn. New York: Schoeken.
Charon, Rita. 2006. *Narrative Medicine: Honoring the Stories of Illness*. New York: Oxford University Press.
Djkic, Maja, Keith Oatley, Sara Zoeterman, and Jordan Peterson. 2009. On Being Moved by Art: How Reading Fiction Transforms the Self. *Creativity Research Journal* 21 (1): 24–29.
Donald, Merlin. 1991. *The Origin of the Modern Mind*. Cambridge: Harvard University Press.
Dunbar, Robin. 1996. *Grooming, Gossip, and the Evolution of Language*. Cambridge: Harvard University Press.
Gerrig, R.J. 1993. *Experiencing Narrative Worlds: On the Psychological Activities of Reading*. New Haven: Yale University Press.
Green, Melanie. 2004. Transportation into Narrative Worlds: The Role of Prior Knowledge and Perceived Realism. *Discourse Processes* 38 (2004): 247–266.

Green, Melanie and Timothy Brock. 2000. The Role of Transportation in the Persuasiveness of Public Narratives. *Journal of Personality and Social Psychology* 79: 701–721.

Hickok, Gregory. 2014. *The Myth of Mirror Neurons*. New York: Norton.

Iacoboni, Marco. 2009. *Mirroring People: The Science of Empathy and How We Connect with Others*. New York: Picador.

Jakobson, Roman. 1987. Linguistics and Poetics. In *Language in Literature*, ed. Krystyna Promorska and Stephen Rudy, 62–94. Cambridge: Harvard University Press.

Kidd, David and Emanuele Castano. 2013. Reading Literary Fiction Improves Theory of Mind. *Sciencexpress*. http://www.sciencemag.org/content/early/recent/3october2013/Page1/10.1126/science.1239918. Accessed 8 Mar 2015.

Leys, Ruth. 2012. 'Both of Us Disgusted in My Insula': Mirror Neuron Theory and Emotional Empathy. *Nonsite* 5, March 18, 2012. At http://nonsite/org/article/"both-of-us-disgusted-in-my-insula"-mirror-neuron-theory-and-emotional-empathy. Accessed 16 May 2015.

Miall, D., and D. Kuiken. 1994. Foregrounding Defamiliarization, and Affect: Response to Literary Stories. *Poetics* 22: 389–407.

Milner, B. 1966. Amnesia Following Operations on the Temporal Lobes. In *Amnesia*, ed. C. Whitty and O. Zangwill. New York: Butterworth.

———. 1975. Psychological Aspects of Focal Epilepsy and Its Neurosurgical Treatment. In *Advances in Eurology*, ed. D.O. Purpura, J.K. Penry, and R.D. Walter, vol. 8. New York: Raven Press.

Neff, D.S. 1983. 'Extraordinary means': Healers and Healing in 'A Conversation with My Father'. *Literature and Medicine* 2: 118–124.

Phelan, James. 1996. *Narrative as Rhetoric: Techniques, Audiences, Ethics, Ideology*. Columbus: Ohio State University Press.

Polkinghorne, Donald. 1988. *Narrative Knowing and the Human Sciences*. Albany: SUNY Press.

Schleifer, Ronald. 2018. *A Political Economy of Modernism: Literature, Post-Classical Economics, and the Lower Middle-Class*. Cambridge: Cambridge University Press.

Schleifer, Ronald, and Jerry Vannatta. 2013. *The Chief Concern of Medicine: The Integration of the Medical Humanities and Narrative Knowledge into Medical Practices*. Ann Arbor: University of Michigan Press.

Sherry, D.F., and D.L. Schacter. 1987. The Evolution of Multiple Memory Systems. *Psychology Review* 94: 439–454.

Shklovsky, Viktor. 1989. Art as Technique, Trans. Lee T. Lemon and Marion J. Reis. In *Contemporary Literary Criticism*, 2nd ed., ed. Robert Con Davis and Ronald Schleifer. New York: Longman.

Steen, Francis. 2005. The Paradox of Narrative Thinking. *Journal of Cultural and Evolutionary Psychology* 3: 87–105.

Stroud, Scott R. 2008. Simulation, Subjective Knowledge, and the Cognitive Value of Literary Narrative. *Journal of Aesthetic Education* 42: 19–41.

Van Laer, Tom, Ko de Ruyter, Luca Visconti, and Martin Wetzels. 2014. The Extended Transportation-Imagery Model: A Meta-Analysis of the Antecedents and Consequences of Consumers' Narrative Transportation. *Journal of Consumer Research* 40: 797–817.

Vannatta, Jerry, Ronald Schleifer, and Sheila Crow. 2005. *Medicine and Humanistic Understanding: The Significance of Narrative in Medical Practices*. Philadelphia: University of Pennsylvania Press. A DVD-Rom publication.

Verghese, Abraham. 1994. *My Own Country*. New York: Viking.

Section II

The Logic of Making a Diagnosis

The Narrative Structure of Diagnosis

The specific logic of making a diagnosis, or diagnostic reasoning, is seldom taught in a medical school education. This is odd. Instead of a systematic examination of procedures of diagnosis—of practical hypothesis formation—it is supposed that most medical students will learn diagnosis simply by observation (role modeling by residents and attending physicians). However, there is a reasoning process of diagnosis, and it is identical to the reasoning in detective stories and novels, and so it is relatively easy to teach diagnosis *as a procedure* using literary narratives, specifically the detective stories. The feature of literary narrative closely associated with diagnostic procedures is the literary feature of "relational facts" examined in Chap. 1.

The American philosopher Charles Sanders Peirce (1839–1914) offered a systematic account of what we described as "relational facts" under the term "abduction," which outlined a logical process of forming an explanatory hypothesis. Abduction seeks an explanation of a particular fact that allows it to be explained by some more general causal principle. One of Pierce's most quoted descriptions of abduction is:

> The surprising fact C is observed.
> But if A were true,
> C would be a matter of course. (1931–58, 5.189)

As cognitive psychology has demonstrated, literary narrative often presents "surprising facts" (under the categories of "foregrounding" and "defamiliarization"), and Peirce's account here offers a small narrative sequence of facts or events. Let us spell out this narrative: the surprising fact (C) that red blisters appear all over the body of a young boy. But if the boy (A) has contracted chicken pox, the red blisters would be a matter of course. (See *The Chief Concern*: Chap. 4 for a full examination and analysis of Peirce's conception of abduction in relation to the practice of medicine.) When Peirce suggests we attend to a "surprising fact" to begin with—which we might also describe as some *anomaly* in a narrative or text—he is suggesting that we *begin* with an element or part of a narrative or description that is not self-evidently important, but that is precisely one that seems to *disrupt* coherent wholeness of a story or an idea. Reading literature teaches us to attend to such facts, and it offers a method that can be brought to the patient-physician encounter. Here we present a short clinical narrative vignette in order to examine the formal strategies of hypothesis formation that Peirce sets forth.

The Woman with Hyponatremia: A Vignette (Excerpt from *The Chief Concern of Medicine*)

A resident admitted a middle-aged woman from Wewoka, Oklahoma, during an extraordinarily busy night. As he entered her room, the woman was buried in covers, her face without expression, skin sallow in appearance as if she were chronically ill or depressed. There were no family members – they complained her thinking was "messed up" and had left at midnight – and the room was barren except for a small pile of mostly worn-out clothes and a pair of rayon stretch pants pulled through the rounded handles of a vinyl purse sitting on a chair. Her responses were short, usually not to the point, and irritated. Feeling angry at her failure to help in the diagnosis, the resident hurried through her narrative of recent events, past history, and systems review. When asked, she specifically denied taking any medication.

Still, it was clear to the resident that the patient's problem was that of hyponatremia, the dilution of the sodium concentration in the blood. Since blood is basically a salt water solution, reasons for the decreased sodium concentration should not be an unsolvable riddle: medications, and their effects on brain hormones or kidney, an under-active thyroid gland, psychogenic water drinking, adrenal insufficiency, congestive heart failure, chronic renal failure, low albumin levels in the blood, ascites (free water in the abdominal cavity), stress, pain, vomiting, diarrhea, the syndrome of inappropriate antidiuretic hormone, each was a possible "cause" of her condition. Yet, neither the resident's questions nor his physical examination provided an answer to the cause – and subsequent treatment – of her sodium dilution.

When he met with the attending physician, Dr. C. G. Gunn, the next morning, all the resident could report was that the patient was a "bad historian." "Was she a bad historian," Dr. Gunn asked, "or were you an inadequate interviewer?" The attending physician questioned the patient again, and again asked if she were taking medication and whether she had something to add to her history. "Why do you keep asking me all these questions?" she asked as she glanced at the nightstand, then down toward the end of the bed. Afterwards, Dr. Gunn mysteriously announced to the resident and interns that in her purse she had chlorthalidone, a diuretic that in this age group commonly causes hyponatremia. When they returned to the patient, she was curled up under the bedclothes with the blanket over her head, and the purse that was evident the night before was nowhere in sight. Dr. Gunn asked the patient to get her purse "so I can look at the pills in it." She rolled over toward the bedside table, pointed to it and told the intern she could get it out for her. The purse contained, among other things, four pill bottles, a thyroid medication, a tranquillizer, a pain medication, and the diuretic, chlorthalidone. "Do you take these pills every day?" "Most days. I didn't take any yesterday because I felt bad." She lay back down, pulled the covers over her head, and said she wanted to be alone.

In this vignette the resident had completed a history and physical examination and had access to the laboratory report that indicated his patient had hyponatremia. Yet even in the face of a complete knowledge of the biomedical causes of hyponatremia he was unable to make a full diagnosis. Dr. Gunn was more experienced than the resident and knew that the most common cause—by far in this particular practice environment—was diuretics, and the most common diuretic prescribed here was chlorthalidone. The patient had denied taking any medications to the resident—a piece of evidence the resident took at face value. Acting on the above knowledge about hyponatremia, the attending physician asked the patient twice if she took any medications. Twice she looked at the night stand instead of at him and answered no. The surprising fact C here is that the patient is hyponatremic without any known cause. If A were true (and A happened to be that her meds were in her purse which was in the night stand, and the meds included chlorthalidone), then C (her hyponatremia) would be a matter of course. This indeed was what turned out to be the case.

When most students read this case, they are shocked to learn that she had chlorthalidone in her purse but are most distressed by the fact that the attending physician's pronouncement seems more like magic. It isn't magic, but on the other hand the attending physician could have been wrong. Abduction is an explanatory hypothesis—which means it could always be wrong, in which event one has to return to the process and abduce all over again. Still, there is a method to this procedure, and it approximates the method of grasping the *meaningful whole* of a narrative, putting together a fragmented narrative (such as the fragmented narrative described in Chap. 5) into something that feels like a coherent series of events. Moreover, the lesson for physicians here is important. One should be sensitive to anomalies in the way the attending physician is in testing out the importance of his patient's glance at the nightstand, the anomaly of her verbal language being contradicted by her body language (something Dr. Verghese is sensitive to in the vignette in Chap. 1). In all of the vignettes that we present in this text-anthology an element of anomaly can be discerned, and effective clinical medicine *attends* to such anomalies. Here, we have unpacked the anomalies of the vignette (and we do so also for the literary narrative) partly because this is the beginning of the text-anthology, but most importantly because the setting forth of the strategy of abductive reasoning not only sets forth a formal strategy for diagnosis but also sets forth a formal strategy of attending to anomalies in reading literary texts and the everyday narratives of vignettes and patient stories. In subsequent chapters, such unpacking is left to students and readers.

Relational "Facts"

The anomalies we describe here are particularly apparent in the short story by Dr. Arthur Conan Doyle, "The Resident Patient" (and in detective stories in general). Doyle's creation, Sherlock Holmes, was modeled after one of Doyle's medical school instructors, Dr. Joseph Bell of Edinburgh, who was renowned, like the attending physician in the vignette, for gathering information about patients by means of "inference to the best explanation" (a description of "abduction") that is clearly not logical "deduction," even though Dr. Watson uses this term to describe Holmes's method.

Before turning to the story, let us examine the relations among "deduction," "induction," and "abduction." In logic, deduction describes an inference that is a necessary part of the premise: since a bachelor is an unmarried man, we can deduce if someone is unmarried he is a bachelor. If heavy smoking is a reaction to nervousness, we can deduce, as the police do in "The Resident Patient," that the evidence of heavy smoking (many cigar butts) indicates that the hanged man, Mr. Blessington, was nervous. Induction, on the other hand, describes an inference that is strongly supported but not a "necessary" part of the premise: thus in "The Resident Patient" the police noted the most common time of suicide is 5 am in the morning, and since Blessington was hanged at that time, they concluded, inductively, it was suicide. In his explanation of abduction (1992: 140), Peirce argues that hypothesis formation (abduction) focuses on categories rather than facts; that it focuses on *qualities* rather than quantities: the *difference in kind* between "murder" and "suicide" in "The Resident Patient" (or, in Poe's story "The Murders in the Rue Morgue," the *difference in kind* between "murder" and "killing") is categorical and qualitative rather than simply factual. In both stories, people died, but the qualitative *nature* of their death—murder or suicide or senseless killing—and not the "fact" of death is at issue. In this, what is at issue is the *quality* of understanding, which is not simply a "fact," but the characterization of a fact. (Such qualities are *systematically* ignored or discounted in scientific analysis, which properly aims at systematic reproducible [usually quantitative] explanation; but they are the focus of attention in humanistic analysis, which properly aims at systematic value-judgments [e.g., distinguishing between "murder" vs. "suicide" or "senseless killing"].) As we note in Chap. 1, an important aspect of "relating" facts to one another is to attend to *differences in kind* in empirical data, which is the differences of qualities: the detective in Edgar Allan Poe's story, Auguste Dupin, notices the difference in kind between animal sounds and human language. In "The Resident Patient," the crucial difference in kind is whether Blessington's death is a "suicide" or a "murder." In this chapter's vignette, the attending physician relates the different kinds of fact, that of statement and that of body-language (the resident simply ignores body-language) to one another. In the literary narrative, Holmes notices the differences in kind (literally different *kinds* of cigars) in what the police take to be examples of the single undifferentiated (but enumerated) general category of "cigar." Thus, Holmes is able to *attend to* a surprising fact that the police ignored, namely the number and differences of cigar butts; and he reasons, "abductively," that if A were true (A being the fact that several men were in the room, at least two different men smoking different cigars and cutting them differently), then C (the surprising fact of so many cigar butts) would be a matter of course. From this he formulated an abductive hypothesis, namely that there were four people in the room and that instead of a suicide it was a murder. Again, like the physician in the "Young Mother with Abdominal Pain" vignette in Chap. 5, the hypothesis could have been wrong, but by attending to different *kinds* of fact— different *qualities* of fact—and discovering relations among these "facts," the efficiency of the observation by Holmes was better than that of the police just as the attending physician's observation in the case of hyponatremia was better than the resident's.

Literary Narrative: "The Resident Patient" (1893) by Dr. Arthur Conan Doyle

Author Note: Dr Arthur Conan Doyle (1859–1930) was a physician and writer, best known for his creation of the character of Sherlock Holmes, who appeared in more than 50 short stories and novels. He also wrote many other works focused on fantasy, science fiction, humor, as well as plays, romances, poetry, and non-fiction and historical novels. Doyle served as the clerk to Dr. Joseph Bell at the Edinburgh Royal Infirmary in the 1870s, upon whom, Doyle noted later, he loosely based his character of Sherlock Holmes.

The Resident Patient

In glancing over the somewhat incoherent series of Memoirs with which I have endeavored to illustrate a few of the mental peculiarities of my friend Mr. Sherlock Holmes, I have been struck by the difficulty which I have experienced in picking out examples which shall in every way answer my purpose. For in those cases in which Holmes has performed some tour de force of analytical reasoning and has demonstrated the value of his peculiar methods of investigation, the facts themselves have often been so slight or so commonplace that I could not feel justified in laying them before the public. On the other hand, it has frequently happened that he has been concerned in some research where the facts have been of the most remarkable and dramatic character, but where the share which he has himself taken in determining their causes has been less pronounced than I, as his biographer, could wish. The small matter which I have chronicled under the heading of "A Study in Scarlet," and that other later one connected with the loss of the Gloria Scott, may serve as examples of this Scylla and Charybdis which are forever threatening the historian. It may be that in the business of which I am now about to write the part which my friend played is not sufficiently accentuated; and yet the whole train of circumstances is so remarkable that I cannot bring myself to omit it entirely from this series.

It had been a close, rainy day in October. Our blinds were half-drawn, and Holmes lay curled upon the sofa, reading and re-reading a letter which he had received by the morning post. For myself, my term of service in India had trained me to stand heat better than cold, and a thermometer of ninety was no hardship. But the paper was uninteresting. Parliament had risen. Everybody was out of town, and I yearned for the glades of the New Forest or the shingle of Southsea. A depleted bank account had caused me to postpone my holiday, and as to my companion, neither the country nor the sea presented the slightest attraction to him. He loved to lie in the very centre of five millions of people, with his filaments stretching out and running through them, responsive to every little rumour or suspicion of unsolved crime. Appreciation of nature found no place among his many gifts, and his only change was when he turned his mind from the evildoer of the town to track down his brother of the country.

Finding that Holmes was too absorbed for conversation, I had tossed aside the barren paper, and, leaning back in my chair I fell into a brown study. Suddenly my companion's voice broke in upon my thoughts.

"You are right, Watson," said he. "It does seem a very preposterous way of settling a dispute."

"Most preposterous!" I exclaimed, and then, suddenly realizing how he had echoed the inmost thought of my soul, I sat up in my chair and stared at him in blank amazement.

"What is this, Holmes?" I cried. "This is beyond anything which I could have imagined."

He laughed heartily at my perplexity.

"You remember," said he, "that some little time ago, when I read you the passage in one of Poe's sketches, in which a close reasoner follows the unspoken thoughts of his companion, you were inclined to treat the matter as a mere tour de force of the author. On my remarking that I was constantly in the habit of doing the same thing you expressed incredulity."

"Oh, no!"

"Perhaps not with your tongue, my dear Watson, but certainly with your eyebrows. So when I saw you throw down your paper and enter upon a train of thought, I was very happy to have the opportunity of reading it off, and eventually of breaking into it, as a proof that I had been in rapport with you."

But I was still far from satisfied. "In the example which you read to me," said I, "the reasoner drew his conclusions from the actions of the man whom he observed. If I remember right, he stumbled over a heap of stones, looked up at the stars, and so on. But I have been seated quietly in my chair, and what clues can I have given you?"

"You do yourself an injustice. The features are given to man as the means by which he shall express his emotions, and yours are faithful servants."

"Do you mean to say that you read my train of thoughts from my features?"

"Your features, and especially your eyes. Perhaps you cannot yourself recall how your reverie commenced?"

"No, I cannot."

"Then I will tell you. After throwing down your paper, which was the action which drew my attention to you, you sat for half a minute with a vacant expression. Then your eyes fixed themselves upon your newly framed picture of General Gordon, and I saw by the alteration in your face that a train of thought had been started. But it did not lead very far. Your eyes turned across to the unframed portrait of Henry Ward Beecher, which stands upon the top of your books. You then glanced up at the wall, and of course your meaning was obvious. You were thinking that if the portrait were framed it would just cover that bare space and correspond with Gordon's picture over there."

"You have followed me wonderfully!" I exclaimed.

"So far I could hardly have gone astray. But now your thoughts went back to Beecher, and you looked hard across as if you were studying the character in his features. Then your eyes ceased to pucker, but you continued to look across, and your face was thoughtful. You were recalling the incidents of Beecher's career. I was well aware that you could not do this without thinking of the mission which he undertook on behalf of the North at the time of the Civil War, for I remember you expressing your passionate indignation at the way in which he was received by the more turbulent of our people. You felt so strongly about it that I knew you could not think of Beecher without thinking of that also. When a moment later I saw your eyes wander away from the picture, I suspected that your mind had now turned to the Civil War, and when I observed that your lips set, your eyes sparkled, and your hands clinched, I was positive that you were indeed thinking of the gallantry which was shown by both sides in that desperate struggle. But then, again, your face grew sadder; you shook your head. You were dwelling upon the sadness and horror and useless waste of life. Your hand stole towards your own old wound, and a smile quivered on your lips, which showed me that the ridiculous side of this method of settling international questions had forced itself upon your mind. At this point I agreed with you that it was preposterous, and was glad to find that all my deductions had been correct."

"Absolutely!" said I. "And now that you have explained it, I confess that I am as amazed as before."

"It was very superficial, my dear Watson, I assure you. I should not have intruded it upon your attention had you not shown some incredulity the other day. But the evening has brought a breeze with it. What do you say to a ramble through London?"

I was weary of our little sitting-room and gladly acquiesced. For three hours we strolled about together, watching the everchanging kaleidoscope of life as it ebbs and flows through Fleet Street and the Strand. His characteristic talk, with its keen observance of detail and subtle power of inference, held me amused and enthralled. It was ten o'clock before we reached Baker Street again. A brougham was waiting at our door.

"Hum! A doctor's – general practitioner, I perceive," said Holmes. "Not been long in practice, but has a good deal to do. Come to consult us, I fancy! Lucky we came back!"

I was sufficiently conversant with Holmes's methods to be able to follow his reasoning, and to see that the nature and state of the various medical instruments in the wicker basket which hung in the lamp-light inside the brougham had given him the data for his swift deduction. The light in our window above showed that this late visit was indeed intended for us. With some curiosity as to what could have sent a brother medico to us at such an hour, I followed Holmes into our sanctum.

A pale, taper-faced man with sandy whiskers rose up from a chair by the fire as we entered. His age may not have been more than three or four and thirty, but his haggard expression and unhealthy hue told of a life which had sapped his strength and robbed him of his youth. His manner was nervous and shy, like that of a sensitive gentleman, and the thin white hand which he laid on the mantelpiece as he rose was that of an artist rather than of a surgeon. His dress was quiet and sombre – a black frockcoat, dark trousers, and a touch of colour about his necktie.

"Good-evening, Doctor," said Holmes cheerily. "I am glad to see that you have only been waiting a very few minutes."

"You spoke to my coachman, then?"

"No, it was the candle on the side-table that told me. Pray resume your seat and let me know how I can serve you."

"My name is Dr. Percy Trevelyan," said our visitor, "and I live at 403 Brook Street."

"Are you not the author of a monograph upon obscure nervous lesions?" I asked.

His pale cheeks flushed with pleasure at hearing that his work was known to me.

"I so seldom hear of the work that I thought it was quite dead," said he. "My publishers gave me a most discouraging account of its sale. You are yourself, I presume, a medical man."

"A retired army surgeon."

"My own hobby has always been nervous disease. I should wish to make it an absolute specialty, but of course a man must take what he can get at first. This, however, is beside the question, Mr. Sherlock Holmes, and I quite appreciate how valuable your time is. The fact is that a very singular train of events has occurred recently at my house in Brook Street, and to-night they came to such a head that I felt it was quite impossible for me to wait another hour before asking for your advice and assistance."

Sherlock Holmes sat down and lit his pipe. "You are very welcome to both," said he. "Pray let me have a detailed account of what the circumstances are which have disturbed you."

"One or two of them are so trivial," said Dr. Trevelyan, "that really I am almost ashamed to mention them. But the matter is so inexplicable, and the recent turn which it has taken is so elaborate, that I shall lay it all before you, and you shall judge what is essential and what is not.

"I am compelled, to begin with, to say something of my own college career. I am a London University man, you know, and I am sure that you will not think that I am unduly singing my own praises if I say that my student career was considered by my professors to be a very promising one. After I had graduated I continued to devote myself to research, occupying a minor position in King's College Hospital, and I was fortunate enough to excite considerable interest by my research into the pathology of catalepsy, and finally to win the Bruce Pinkerton prize and medal by the monograph on nervous lesions to which your friend has just alluded. I should not go too far if I were to say that there was a general impression at that time that a distinguished career lay before me.

"But the one great stumbling-block lay in my want of capital. As you will readily understand, a specialist who aims high is compelled to start in one of a dozen streets in the Cavendish Square quarter, all of which entail enormous rents and furnishing expenses. Besides this preliminary outlay, he must be prepared to keep himself for some years, and to hire a presentable carriage and horse. To do this was quite beyond my power, and I could only hope that by economy I might in ten years' time save enough to enable me to put up my plate. Suddenly, however, an unexpected incident opened up quite a new prospect to me.

"This was a visit from a gentleman of the name of Blessington, who was a complete stranger to me. He came up to my room one morning, and plunged into business in an instant.

"'You are the same Percy Trevelyan who has had so distinguished a career and won a great prize lately?' said he.

"I bowed.

"'Answer me frankly,' he continued, 'for you will find it to your interest to do so. You have all the cleverness which makes a successful man. Have you the tact?'

"I could not help smiling at the abruptness of the question.

"'I trust that I have my share,' I said.

"'Any bad habits? Not drawn towards drink, eh?'

"'Really, sir!' I cried.

"'Quite right! That's all right! But I was bound to ask. With all these qualities, why are you not in practice?'

"I shrugged my shoulders.

"'Come, come!' said he, in his bustling way. 'It's the old story. More in your brains than in your pocket, eh? What would you say if I were to start you in Brook Street?'

"I stared at him in astonishment.

"'Oh, it's for my sake, not for yours,' he cried. 'I'll be perfectly frank with you, and if it suits you it will suit me very well. I have a few thousands to invest, d'ye see, and I think I'll sink them in you.'

"'But why?' I gasped.

"'Well, it's just like any other speculation, and safer than most.'

"'What am I to do, then?'

"'I'll tell you. I'll take the house, furnish it, pay the maids, and run the whole place. All you have to do is just to wear out your chair in the consulting-room. I'll let you have pocket-money and everything. Then you hand over to me three quarters of what you earn, and you keep the other quarter for yourself.'

"This was the strange proposal, Mr. Holmes, with which the man Blessington approached me. I won't weary you with the account of how we bargained and negotiated. It ended in my moving into the house next Lady-day, and starting in practice on very much the same conditions as he had suggested. He came himself to live with me in the character of a resident patient. His heart was weak, it appears, and he needed constant medical supervision. He turned the two best rooms of the first floor into a sitting-room and bedroom for himself. He was a man of singular habits, shunning company and very seldom going out. His life was irregular, but in one respect he was regularity itself. Every evening, at the same hour, he walked into the consulting-room, examined the books, put down five and three-pence for every guinea that I had earned, and carried the rest off to the strong-box in his own room.

"I may say with confidence that he never had occasion to regret his speculation. From the first it was a success. A few good cases and the reputation which I had won in the hospital brought me rapidly to the front, and during the last few years I have made him a rich man.

"So much, Mr. Holmes, for my past history and my relations with Mr. Blessington. It only remains for me now to tell you what has occurred to bring me here to-night.

"Some weeks ago Mr. Blessington came down to me in, as it seemed to me, a state of considerable agitation. He spoke of some burglary which, he said, had been committed in the West End, and he appeared, I remember, to be quite unnecessarily excited about it, declaring that a day should not pass before we should add stronger bolts to our windows and doors. For a week he continued to be in a peculiar state of restlessness, peering continually out of the windows, and ceasing to take the short walk which had usually been the prelude to his dinner. From his manner it struck me that he was in mortal dread of something or somebody, but when I questioned him upon the point he became so offensive that I was compelled to drop the subject. Gradually, as time passed, his fears appeared to die away, and he had renewed his former habits, when a fresh event reduced him to the pitiable state of prostration in which he now lies.

"What happened was this. Two days ago I received the letter which I now read to you. Neither address nor date is attached to it.

"'A Russian nobleman who is now resident in England,' it runs, 'would be glad to avail himself of the professional assistance of Dr. Percy Trevelyan. He has been for some years a victim to cataleptic attacks, on which, as is well known, Dr. Trevelyan is an authority. He proposes to call at about quarter past six to-morrow evening, if Dr. Trevelyan will make it convenient to be at home.'

"This letter interested me deeply, because the chief difficulty in the study of catalepsy is the rareness of the disease. You may believe, then, that I was in my consulting-room when, at the appointed hour, the page showed in the patient.

"He was an elderly man, thin, demure, and commonplace—by no means the conception one forms of a Russian nobleman. I was much more struck by the appearance of his companion. This was a tall young man, surprisingly handsome, with a dark, fierce face, and the limbs and chest of a Hercules. He had his hand under the other's arm as they entered, and helped him to a chair with a tenderness which one would hardly have expected from his appearance.

"'You will excuse my coming in, doctor,' said he to me, speaking English with a slight lisp. 'This is my father, and his health is a matter of the most overwhelming importance to me.'

"I was touched by this filial anxiety. 'You would, perhaps, care to remain during the consultation?' said I.

"'Not for the world,' he cried with a gesture of horror. 'It is more painful to me than I can express. If I were to see my father in one of these dreadful seizures I am convinced that I should never survive it. My own nervous system is an exceptionally sensitive one. With your permission, I will remain in the waiting-room while you go into my father's case.'

"To this, of course, I assented, and the young man withdrew. The patient and I then plunged into a discussion of his case, of which I took exhaustive notes. He was not remarkable for intelligence, and his answers were frequently obscure, which I attributed to his limited acquaintance with our language. Suddenly, however, as I sat writing, he ceased to give any answer at all to my inquiries, and on my turning towards him I was shocked to see that he was sitting bolt upright in his chair, staring at me with a perfectly blank and rigid face. He was again in the grip of his mysterious malady.

"My first feeling, as I have just said, was one of pity and horror. My second, I fear, was rather one of professional satisfaction. I made notes of my patient's pulse and temperature, tested the rigidity of his muscles, and examined his reflexes. There was nothing markedly abnormal in any of these conditions, which harmonized with my former experiences. I had obtained good results in such cases by the inhalation of nitrite of amyl, and the present seemed an admirable opportunity of testing its virtues. The bottle was downstairs in my laboratory, so leaving my patient seated in his chair, I ran down to get it. There was some little delay in finding it—five minutes, let us say—and then I returned. Imagine my amazement to find the room empty and the patient gone.

"Of course, my first act was to run into the waiting-room. The son had gone also. The hall door had been closed, but not shut. My page who admits patients is a new boy and by no means quick. He waits downstairs, and runs up to show patients out when I ring the consulting-room bell. He had heard nothing, and the affair remained a complete mystery. Mr. Blessington came in from his walk shortly afterwards, but I did not say anything to him upon the subject, for, to tell the truth, I have got in the way of late of holding as little communication with him as possible.

"Well, I never thought that I should see anything more of the Russian and his son, so you can imagine my amazement when, at the very same hour this evening, they both came marching into my consulting-room, just as they had done before.

"'I feel that I owe you a great many apologies for my abrupt departure yesterday, doctor,' said my patient.

"'I confess that I was very much surprised at it,' said I.

"'Well, the fact is,' he remarked, 'that when I recover from these attacks my mind is always very clouded as to all that has gone before. I woke up in a strange room, as it seemed to me, and made my way out into the street in a sort of dazed way when you were absent.'

"'And I,' said the son, 'seeing my father pass the door of the waiting-room, naturally thought that the consultation had come to an end. It was not until we had reached home that I began to realize the true state of affairs.'

"'Well,' said I, laughing, 'there is no harm done except that you puzzled me terribly; so if you, sir, would kindly step into the waiting-room I shall be happy to continue our consultation which was brought to so abrupt an ending.'

"For half an hour or so I discussed that old gentleman's symptoms with him, and then, having prescribed for him, I saw him go off upon the arm of his son.

"I have told you that Mr. Blessington generally chose this hour of the day for his exercise. He came in shortly afterwards and passed upstairs. An instant later I heard him running down, and he burst into my consulting-room like a man who is mad with panic.

"'Who has been in my room?' he cried.

"'No one,' said I.

"'It's a lie!' He yelled. 'Come up and look!'

"I passed over the grossness of his language, as he seemed half out of his mind with fear. When I went upstairs with him he pointed to several footprints upon the light carpet.

"'D'you mean to say those are mine?' he cried.

"They were certainly very much larger than any which he could have made, and were evidently quite fresh. It rained hard this afternoon, as you know, and my patients were the only people who called. It must have been the case, then, that the man in the waiting-room had, for some reason, while I was busy with the other, ascended to the room of my resident patient. Nothing had been touched or taken, but there were the footprints to prove that the intrusion was an undoubted fact.

"Mr. Blessington seemed more excited over the matter than I should have thought possible, though of course it was enough to disturb anybody's peace of mind. He actually sat crying in an arm-chair, and I could hardly get him to speak coherently. It was his suggestion that I should come round to you, and of course I at once saw the propriety of it, for certainly the incident is a very singular one, though he appears to completely overrate its importance. If you would only come back with me in my brougham, you would at least be able to soothe him, though I can hardly hope that you will be able to explain this remarkable occurrence."

Sherlock Holmes had listened to this long narrative with an intentness which showed me that his interest was keenly aroused. His face was as impassive as ever, but his lids had drooped more heavily over his eyes, and his smoke had curled up more thickly from his pipe to emphasize each curious episode in the doctor's tale. As our visitor concluded, Holmes sprang up without a word, handed me my hat, picked his own from the table, and followed Dr. Trevelyan to the door. Within a quarter of an hour we had been dropped at the door of the physician's residence in Brook Street, one of those sombre, flat-faced houses which one associates with a West End practice. A small page admitted us, and we began at once to ascend the broad, well-carpeted stair.

But a singular interruption brought us to a standstill. The light at the top was suddenly whisked out, and from the darkness came a reedy, quavering voice.

"I have a pistol," it cried. "I give you my word that I'll fire if you come any nearer."

"This really grows outrageous, Mr. Blessington," cried Dr. Trevelyan.

"Oh, then it is you, Doctor." said the voice with a great heave of relief. "But those other gentlemen, are they what they pretend to be?"

We were conscious of a long scrutiny out of the darkness.

"Yes, yes, it's all right," said the voice at last. "You can come up, and I am sorry if my precautions have annoyed you."

He relit the stair gas as he spoke, and we saw before us a singular-looking man, whose appearance, as well as his voice, testified to his jangled nerves. He was very fat, but had apparently at some time been much fatter, so that the skin hung about his face in loose pouches, like the cheeks of a bloodhound. He was of a sickly color, and his thin, sandy hair seemed to bristle up with the intensity of his emotion. In his hand he held a pistol, but he thrust it into his pocket as we advanced.

"Good-evening, Mr. Holmes," said he. "I am sure I am very much obliged to you for coming round. No one ever needed your advice more than I do. I suppose that Dr. Trevelyan has told you of this most unwarrantable intrusion into my rooms."

"Quite so," said Holmes. "Who are these two men, Mr. Blessington, and why do they wish to molest you?"

"Well, well," said the resident patient in a nervous fashion, "of course it is hard to say that. You can hardly expect me to answer that, Mr. Holmes."

"Do you mean that you don't know?"

"Come in here, if you please. Just have the kindness to step in here."

He led the way into his bedroom, which was large and comfortably furnished.

"You see that," said he, pointing to a big black box at the end of his bed. "I have never been a very rich man, Mr. Holmes – never made but one investment in my life, as Dr. Trevelyan would tell you. But I don't believe in bankers. I would never trust a banker, Mr. Holmes. Between ourselves, what little I have is in that box, so you can understand what it means to me when unknown people force themselves into my rooms."

Holmes looked at Blessington in his questioning way and shook his head.

"I cannot possibly advise you if you try to deceive me," said he.

"But I have told you everything."

Holmes turned on his heel with a gesture of disgust. "Good-night, Dr. Trevelyan," said he.

"And no advice for me?" cried Blessington in a breaking voice.

"My advice to you, sir, is to speak the truth."

A minute later we were in the street and walking for home. We had crossed Oxford Street and were halfway down Harley Street before I could get a word from my companion.

"Sorry to bring you out on such a fool's errand, Watson," he said at last. "It is an interesting case, too, at the bottom of it."

"I can make little of it," I confessed.

"Well, it is quite evident that there are two men – more perhaps, but at least two – who are determined for some reason to get at this fellow Blessington. I have no doubt in my mind that both on the first and on the second occasion that young man penetrated to Blessington's room, while his confederate, by an ingenious device, kept the doctor from interfering."

"And the catalepsy?"

"A fraudulent imitation, Watson, though I should hardly dare to hint as much to our specialist. It is a very easy complaint to imitate. I have done it myself."

"And then?"

"By the purest chance Blessington was out on each occasion. Their reason for choosing so unusual an hour for a consultation was obviously to insure that there should be no other patient in the waiting-room. It just happened, however, that this hour coincided with Blessington's constitutional, which seems to show that they were not very well acquainted with his daily routine. Of course, if they had been merely after plunder they would at least have made some attempt to search for it. Besides, I can read in a man's eye when it is his own skin that he is frightened for. It is inconceivable that this fellow could have made two such vindictive enemies as these appear to be without knowing of it. I hold it, therefore, to be certain that he does know who these men are, and that for reasons of his own he suppresses it. It is just possible that to-morrow may find him in a more communicative mood."

"Is there not one alternative," I suggested, "grotesquely improbable, no doubt, but still just conceivable? Might the whole story of the cataleptic Russian and his son be a concoction of Dr. Trevelyan's, who has, for his own purposes, been in Blessington's rooms?"

I saw in the gas-light that Holmes wore an amused smile at this brilliant departure of mine.

"My dear fellow," said he, "it was one of the first solutions which occurred to me, but I was soon able to corroborate the doctor's tale. This young man has left prints upon the stair-carpet which made it quite superfluous for me to ask to see those which he had made in the room. When I tell you that his shoes were square-toed instead of being pointed like Blessington's, and were quite an inch and a third longer than the doctor's, you will acknowledge that there can be no doubt as to his individuality. But we may sleep on it now, for I shall be surprised if we do not hear something further from Brook Street in the morning."

Sherlock Holmes's prophecy was soon fulfilled, and in a dramatic fashion. At half-past seven next morning, in the first dim glimmer of daylight, I found him standing by my bedside in his dressing-gown.

"There's a brougham waiting for us, Watson," said he.

"What's the matter, then?"

"The Brook Street business."

"Any fresh news?"

"Tragic, but ambiguous," said he, pulling up the blind. "Look at this—a sheet from a notebook, with 'For God's sake come at once. P. T.,' scrawled upon it in pencil. Our friend, the doctor, was hard put to it when he wrote this. Come along, my dear fellow, for it's an urgent call."

In a quarter of an hour or so we were back at the physician's house. He came running out to meet us with a face of horror.

"Oh, such a business!" he cried with his hands to his temples.

"What then?"

"Blessington has committed suicide!" Holmes whistled.

"Yes, he hanged himself during the night."

We had entered, and the doctor had preceded us into what was evidently his waiting-room.

"I really hardly know what I am doing," he cried. "The police are already upstairs. It has shaken me most dreadfully."

"When did you find it out?"

"He has a cup of tea taken in to him early every morning. When the maid entered, about seven, there the unfortunate fellow was hanging in the middle of the room. He had tied his cord to the hook on which the heavy lamp used to hang, and he had jumped off from the top of the very box that he showed us yesterday."

Holmes stood for a moment in deep thought.

"With your permission," said he at last, I should like to go upstairs and look into the matter."

We both ascended, followed by the doctor.

It was a dreadful sight which met us as we entered the bedroom door. I have spoken of the impression of flabbiness which this man Blessington conveyed. As he dangled from the hook it was exaggerated and intensified until he was scarce human in his appearance. The neck was drawn out like a plucked chicken's, making the rest of him seem the more obese and unnatural by the contrast. He was clad only in his long nightdress, and his swollen ankles and ungainly feet protruded starkly from beneath it. Beside him stood a smart-looking police-inspector, who was taking notes in a pocketbook.

"Ah, Mr. Holmes," said he heartily as my friend entered, "I am delighted to see you."

"Good-morning, Lanner," answered Holmes, "you won't think me an intruder, I am sure. Have you heard of the events which led up to this affair?"

"Yes, I heard something of them."

"Have you formed any opinion?"

"As far as I can see, the man has been driven out of his senses by fright. The bed has been well slept in, you see. There's his impression, deep enough. It's about five in the morning, you know, that suicides are most common. That would be about his time for hanging himself. It seems to have been a very deliberate affair."

"I should say that he has been dead about three hours, judging by the rigidity of the muscles," said I.

"Noticed anything peculiar about the room?" asked Holmes.

"Found a screw-driver and some screws on the wash-hand stand. Seems to have smoked heavily during the night, too. Here are four cigar-ends that I picked out of the fireplace."

"Hum!" said Holmes, have you got his cigar-holder?"

"No, I have seen none."

"His cigar-case, then?"

"Yes, it was in his coat-pocket." Holmes opened it and smelled the single cigar which it contained.

"Oh, this is a Havana, and these others are cigars of the peculiar sort which are imported by the Dutch from their East Indian colonies. They are usually wrapped in straw, you know, and are thinner for their length than any other brand." He picked up the four ends and examined them with his pocket-lens.

"Two of these have been smoked from a holder and two without," said he. "Two have been cut by a not very sharp knife, and two have had the ends bitten off by a set of excellent teeth. This is no suicide, Mr. Lanner. It is a very deeply planned and cold-blooded murder."

"Impossible!" cried the inspector.

"And why?"

"Why should anyone murder a man in so clumsy a fashion as by hanging him?"

"That is what we have to find out."

"How could they get in?"

"Through the front door."

"It was barred in the morning."

"Then it was barred after them."

"How do you know?"

"I saw their traces. Excuse me a moment, and I may be able to give you some further information about it."

He went over to the door, and turning the lock he examined it in his methodical way. Then he took out the key, which was on the inside, and inspected that also. The bed, the carpet, the chairs, the mantelpiece, the dead body, and the rope were each in turn examined, until at last he professed himself satisfied, and with my aid and that of the inspector cut down the wretched object and laid it reverently under a sheet.

"How about this rope?" he asked.

"It is cut off this," said Dr. Trevelyan, drawing a large coil from under the bed. "He was morbidly nervous of fire, and always kept this beside him, so that he might escape by the window in case the stairs were burning."

"That must have saved them trouble," said Holmes thoughtfully. "Yes, the actual facts are very plain, and I shall be surprised if by the afternoon I cannot give you the reasons for them as well. I will take this photograph of Blessington, which I see upon the mantelpiece, as it may help me in my inquiries."

"But you have told us nothing!" cried the doctor.

"Oh, there can be no doubt as to the sequence of events," said Holmes. "There were three of them in it: the young man, the old man, and a third, to whose identity I have no clue. The first two, I need hardly remark, are the same who masqueraded as the Russian count and his son, so we can give a very full description of them. They were admitted by a confederate inside the house. If I might offer you a word of advice, Inspector, it would be to arrest the page, who, as I understand, has only recently come into your service, Doctor."

"The young imp cannot be found," said Dr. Trevelyan; "the maid and the cook have just been searching for him."

Holmes shrugged his shoulders.

"He has played a not unimportant part in this drama," said he. "The three men having ascended the stairs, which they did on tiptoe, the elder man first, the younger second, and the unknown man in the rear–".

"My dear Holmes!" I ejaculated.

"Oh, there could be no question as to the superimposing of the footmarks. I had the advantage of learning which was which last night. They ascended, then, to Mr. Blessington's room, the door of which they found to be locked. With the help of a wire, however, they forced round the key. Even without the lens you will perceive, by the scratches on this ward, where the pressure was applied.

"On entering the room their first proceeding must have been to gag Mr. Blessington. He may have been asleep, or he may have been so paralyzed with terror as to have been unable to cry out. These walls are thick, and it is conceivable that his shriek, if he had time to utter one, was unheard.

"Having secured him, it is evident to me that a consultation of some sort was held. Probably it was something in the nature of a judicial proceeding. It must have lasted for some time, for it was then that these cigars were smoked. The older man sat in that wicker chair; it was he who used the cigar-holder. The younger man sat over yonder; he knocked his ash off against the chest of drawers. The third follow paced up and down. Blessington, I think, sat upright in the bed, but of that I cannot be absolutely certain.

"Well, it ended by their taking Blessington and hanging him. The matter was so prearranged that it is my belief that they brought with them some sort of block or pulley which might serve as a gallows. That screw-driver and those screws were, as I conceive, for fixing it up. Seeing the hook, however, they naturally saved themselves the trouble. Having finished their work they made off, and the door was barred behind them by their confederate."

We had all listened with the deepest interest to this sketch of the night's doings, which Holmes had deduced from signs so subtle and minute that, even when he had pointed them out to us, we could scarcely follow him in his reasonings. The inspector hurried away on the instant to make inquiries about the page, while Holmes and I returned to Baker Street for breakfast.

"I'll be back by three," said he when we had finished our meal. "Both the inspector and the doctor will meet me here at that hour, and I hope by that time to have cleared up any little obscurity which the case may still present."

Our visitors arrived at the appointed time, but it was a quarter to four before my friend put in an appearance. From his expression as he entered, however, I could see that all had gone well with him.

"Any news, Inspector?"

"We have got the boy, sir."

"Excellent, and I have got the men."

"You have got them!" we cried, all three.

"Well, at least I have got their identity. This so-called Blessington is, as I expected, well known at headquarters, and so are his assailants. Their names are Biddle, Hayward, and Moffat."

"The Worthingdon bank gang," cried the inspector.

"Precisely," said Holmes.

"Then Blessington must have been Sutton."

"Exactly," said Holmes.

"Why, that makes it as clear as crystal," said the inspector.

But Trevelyan and I looked at each other in bewilderment.

"You must surely remember the great Worthingdon bank business," said Holmes. "Five men were in it – these four and a fifth called Cartwright. Tobin, the caretaker, was murdered, and the thieves got away with seven thousand pounds. This was in 1875. They were all five arrested, but the evidence against them was by no means conclusive. This Blessington or Sutton, who was the worst of the gang, turned informer. On his evidence Cartwright was hanged and the other three got fifteen years apiece. When they got out the other day, which was some years before their full term, they set themselves, as you perceive, to hunt down the traitor and to avenge the death of their comrade upon him. Twice they tried to get at him and failed; a third time you see, it came off. Is there anything further which I can explain, Dr. Trevelyan?"

"I think you have made it all remarkably clear," said the doctor. "No doubt the day on which he was so perturbed was the day when he had seen of their release in the newspapers."

"Quite so. His talk about a burglary was the merest blind."

"But why could he not tell you this?"

"Well, my dear sir, knowing the vindictive character of his old associates, he was trying to hide his own identity from everybody as long as he could. His secret was a shameful one and he could not bring himself to divulge it. However, wretch as he was, he was still living under the shield of British law, and I have no doubt, Inspector, that you will see that, though that shield may fail to guard, the sword of justice is still there to avenge."

Such were the singular circumstances in connection with the Resident Patient and the Brook Street Doctor. From that night nothing has been seen of the three murderers by the police, and it is surmised at Scotland Yard that they were among the passengers of the ill-fated steamer Norah Creina, which was lost some years ago with all hands upon the Portuguese coast, some leagues to the north of Oporto. The proceedings against the page broke down for want of evidence, and the Brook Street Mystery, as it was called, has never until now been fully dealt with in any public print.

Twice Told Story

What is remarkable about this story is the manner in which—*twice* in this twice-told story—Holmes is able to develop a full-blown narrative that creates relationships among a number of disparate facts, first in his explanation of Watson's musing, and then in the case he encounters. The beginning of "The Resident Patient" imitates the beginning of Edgar Allan Poe's great detective story (the first written in English), "The Murders in the Rue Morgue," as Holmes explicitly notes. Here, Holmes, watching Dr. Watson, guesses at the narrator's thoughts and explains his guess as a form of deduction. This process is not really deduction but hypothesis generation (abduction) described earlier, just as the process of Auguste Dupin—Poe's detective—is more than simply the empirical observations of induction. Holmes's logic is not deduction in the strictest sense because he focuses on *quality*, namely *differences in kind*, in the evidence he examines. Moreover, his guess turns out to be based on a good hunch and works because Holmes knows Watson so well. This knowledge involves different *kinds* of fact that we noted earlier. When Holmes tells Watson, "The features are given to man as the means by which he shall express his emotions, and yours are faithful servants," he is suggesting what much of the empirical study of the effects of reading we described in Chap. 1 demonstrates, that there are strong relationships between the different categories of cognition and emotion. (We describe this in Part II of the Introduction as the "complexity" of clinical medicine.)

In this opening scene in "The Resident Patient" (which Doyle also uses, almost verbatim, in "The Cardboard Box") Holmes presents the *kind* of reasoning in narrating Watson's chain of thought based on the careful observation of Watson's actions looking around the room. Here, Holmes emphasizes his powers of observation without analyzing the ways in which observations of different *orders of fact* are brought together to generate an explanation. In this narrative Holmes, like an experienced attending physician, is "reasoning backwards": he is making sense of phenomena by supplying their causes that relate them to one another. In this archetypal detective story, the preamble is presented to frame the manner in which Holmes observes details of the death scene and of the circumstances surround the death, which is another version of a "twice-told story." As in "The Murders in the Rue Morgue," the detective solves the crime when the police failed since they erroneously jump to the "obvious" conclusion too early, before all the data are analyzed, classified, and, most of all, *characterized* carefully. (It is such characterization that determines the *kind* of fact.)

That is, the police aim at *classification* rather than *explanation*. In "The Resident Patient" the police assume suicide instead of what Holmes calls "a deeply planned and cold-blooded murder" because they cannot *see* what there is to observe. (In the case of the woman with hyponatremia, the resident assumes the patient is truthfully supplying all the evidence and leaves it at that.) The kind of mistake the police make—the assumption that the goal of diagnosis is classification rather than explanation—is that of clinicians who narrow the differential diagnosis too early, failing to account for all the data, both positive and negative, and failing to attend to the manner or modality of their own observations. When Watson says to Holmes that what the detective saw "was quite invisible to me," Holmes replies: "Not invisible, but unnoticed, Watson. You did not know where to look, and so you missed all that was important." Holmes goes on to tutor Watson: "Never trust to general impressions, my boy," he says, "but concentrate yourself upon details" (Doyle, "A Case of Identity"). That is, narrative comprehension teaches physicians to appreciate detail as used by authors like Chekhov and to understand that diagnosis is a form of explanation not a method of classification. (See Appendix 4 for suggestions for engaging with the vignette and story of this chapter. It includes class hand-outs for discussion.)

Related Poem

For this chapter, the related poem is a sonnet (like that of Yeats in Chap. 10), but here it is what is often called a Shakespearean rather than Italian (or "Petrarchan" [named for the fourteenth-century poet Francesco Petrarch]) sonnet, which has a different form, even though it remains 14 lines. The Shakespearean sonnet has three rhyming quatrains (of four lines) and a two-line rhyming couplet that "sums up" the poem. (This detailed knowledge of a literary form, like Holmes's detailed knowledge of cigars, is the knowledge-based for abductive reasoning.) What "relates" the poem to this chapter is that it is not presented as Shakespeare wrote it, but rather it is presented with its lines mixed up. Students, using the features of literary narrative—especially patterns of sound, narrative syntax (i.e., the overall structure of the poem), narrative semantics (i.e., the overall argument and meaning of the poem)—should relate the facts (i.e., the particular lines) to one another to create a meaningful whole. This is an exercise on attention and systematic hypothesis formation based upon information about syntax, the logic of meaningful order, and a pre-existing knowledge of poetic rhyme-order analogous to Holmes's pre-existing knowledge of cigars.

Poem: Sonnet 73, "That Time of Year Thou Mayst in Me Behold" (1609) by William Shakespeare

Author Note: William Shakespeare (1564–1616) is widely regarded as one of the greatest writers of the English language. Between 1590 and 1613 he wrote at least 37 plays. His sonnets were published in 1609.

That Time of Year Thou Mayst in Me Behold

Which by and by black night doth take away,
Bare ruin'd choirs, where late the sweet birds sang.
In me thou see'st the twilight of such day
That time of year thou mayst in me behold
Consum'd with that which it was nourish'd by.
When yellow leaves, or none, or few, do hang
To love that well which thou must leave ere long.
Death's second self, that seals up all in rest.
In me thou see'st the glowing of such fire
This thou perceiv'st, which makes thy love more strong,
As the death-bed whereon it must expire,
Upon those boughs which shake against the cold,
As after sunset fadeth in the west,
That on the ashes of his youth doth lie,

Lessons for Providers
There is a logic for making a diagnosis—especially the difficult diagnoses. All providers should spend a significant amount of time honing this logical process.

Bibliography

Peirce, Charles Sanders. 1931–1935, 1958. *Collected Papers*, vols. 1–6. Ed. C. Hartshorne and P. Weiss; vols. 7–8. Ed. A. Burks. Cambridge: Harvard University Press.

Peirce, Charles Sanders. 1992. Deduction, Induction, and Hypothesis. In *The Essential Peirce, Volume I (1867–1893)*, ed. Nathan Houser and Christian Kloesel. Bloomington: Indianapolis University Press.

Section III
Professionalism

Literature and Professionalism in Medicine

3

The first two chapters of this book focus on storytelling as part of medicine and the particular "logic" of diagnosis, which is also part of medicine. These aspects of healthcare might be understood as contributing, respectively, to the "art" and "science" of medical practices. Along with these focuses, another global concern of the healthcare professions might be that of "professionalism" itself. In *The Checklist Manifesto*, noted surgeon Dr. Atul Gawande observes that

> All learned occupations have a definition of professionalism, a code of conduct … [consisting of] three common elements.
> First, is an expectation of selflessness … [which] will place the needs and concerns of those who depend on us above our own. Second is an expectation of skill …. Third is an expectation of trust-worthiness: that we will be responsible in our personal behavior toward our charges. (2010: loc 2532)

Appendix 5, which, like this chapter, focuses on Dr. Richard Selzer's short story "Imelda" in its goal of designing a workshop aimed at measuring and promulgating professional behaviors in medicine, sets forth a practical definition of professionalism that aligns nicely with Gawande's general description. "Professionalism," it notes, "is demonstrated through a foundation of clinical competence, communication skills, and ethical understanding, upon which is built the aspiration to and wise application of the principles of professionalism: excellence, humanism, accountability and altruism" (Stern 2006: 19). As such, professionalism is not quite either an "art" or a "science" related to healthcare, but rather a way of understanding the practices of medicine and healthcare as social institutions that participate in a society's overall senses of value and "normal" behavior.

This is perhaps more clear in the vignette discussion in this chapter, which focuses on a passage from Dr. Michael LaCombe's narrative entitled "Playing God," than it is in Selzer's story in this chapter, whose physician has no relationship to temporal authorities. In the Preface to the collection of narratives, in which LaCombe includes this vignette, he notes that when this narrative "was enacted … at the Mayo Clinic, an attorney stood at the back of the theater, [and] reminded me that there is no statute of limitations on felony accessory to murder" (LaCombe 2010: ix). By bringing up the law—the codification of what a society takes to be "normal" (and "abnormal") behavior (we discuss "normative" ethics in Chap. 8)—LaCombe emphasizes, perhaps more explicitly than Selzer in his short story, that canons of behavior in healthcare, as in many other professions, are implicitly and explicitly tied up with standards of behavior of the legal system as well as standards shared by a professional community. Still, the latter—professionalization itself—is a central part of an education in healthcare, as Appendix 5 makes explicitly clear.

A comparison between Selzer's extended narrative exploration in "Imelda," which in its extended form takes up many of the features of narrative we have outlined in Chap. 1, and the single incident (with a single "flashback" memory) of LaCombe's narrative can nicely highlight some of the contributions of including the "art" narrative of short stories within a medical education. Here, we analyze a short clinical narrative vignette, excerpted from a longer (but still quite short) narrative of a physician encountering a suspicious death of the husband of a patient he had treated from childhood and throughout her marriage with an abusive husband. Note that this narrative vignette, unlike Chap. 10, which excerpts the first-person literary narrative of a perpetrator of postmortem spousal abuse, is thoroughly narrated from the point of view of the physician. Selzer's narrative, as we shall see, stands between these two insofar as it is narrative from the point of view of a medical student working with the physician in "Imelda."

Playing God: A Vignette (A Passage from Dr. Michael LaCombe, "Playing God," in *Bedside: The Art of Medicine* [2010])

> *Author Note*: Dr. Michael LaCombe (b. 1942) is a graduate of Harvard Medical School who practiced primary-care general internal medicine and, subsequently, cardiology in Maine. The first author to write fiction for medical journals, he has published over 100 short stories. He has published numerous books, including *Bedside: The Art of Medicine*, from a chapter of which this short vignette-analysis is excerpted. In addition, he has compiled the medical writings of Dr. William Osler, one of the four founding professors of Johns Hopkins Hospital, who created the first residency program and program in bedside clinical training.

> One time back along, I had a bad baby on my hands, a new-born with hydrocephalus and a big cyst at the base of the neck—the crippled-for-life kind of baby you see once in a lifetime. I watched that baby struggle and watched and didn't do a damn thing to save it and apologized to the family afterwards, explaining it was a stillborn, lying to them. That was the one time I played God and it aggravated me, I can tell you. I went home that night and yelled at my wife, kicked the dog, and drank too much—brooded for weeks and never talked about it. It can eat at you. (LaCombe 2010: 25)

LaCombe goes on to note that "there had never been a second time" like this incident with the "bad baby" until the event with Kitty, who is the subject of the longer vignette entitled "Playing God." Kitty was his long-term patient—he had treated her since childhood—and she was reputed to be the "most abused woman in the whole county." The story LaCombe tells here focuses on the time she calls him to make a house call at her husband's seeming suicide. After engaging with Kitty, he notes:

> I grabbed one of the kitchen chairs and a dish towel and went back into the bedroom. In a few minutes I was on the phone and had the dispatcher get a hold of the Chief [of Police]. […]
>
> "Chief," I said, "I'm at Earl Staples' house. He's finally done himself in. He got drunk and shot himself with his deer rifle. […] Clear suicide in my book. I'll be signing it out that way. […]"
>
> I put the phone down and turned back to Kitty. She was staring at the floor. She hadn't moved. I sighed, slapped my thighs, and got up to go.
>
> "Where's your coat, Kitty? I'll drive you over to Kate's, and in the morning," I said, nodding to her left arm, "you come over to the office so I can set that fracture for you one last time." (LaCombe 2010: 26)

These two incidents, like "Berenice" in Chap. 10, present harrowing narratives. But both of them, unlike Edgar Allan Poe's story, set forth the sanctioned power of healthcare providers and physicians—the manner in which members of the healthcare professions have legally binding authority (e.g., the ability to prescribe drugs, create official death reports, etc.) and responsibilities that are part and parcel of healthcare. The physician here—perhaps "playing God," as the title of this vignette suggests—takes upon himself an active role in the social consequences of his professional judgment: he has it within his power, concomitant with his professional attainment, to shape the legal findings and outcome of the nature of the "event" he encounters, here again, as in Arthur Conan Doyle, to distinguish between "murder" and "suicide." In many ways, the moral dilemma suggested by this short narrative—even more than the dilemma suggested by the flashback story of the "crippled-for-life" newborn that he remembers—underlines the *position* of life-and-death decisions that face healthcare providers. Is it the physician's job to make moral/social decisions about the consequences of behaviors he encounters? Is the function of "professionalism"—precisely social-professional modes of behavior designed to govern people who have obtained power over others with their professional attainments—to make "playing God" more unlikely? Another related question for this narrative vignette is whether or not the life expectancy of the "crippled-for-life" newborn makes a difference in assessing the behavior—and even the "professionalism"—of this physician.

Reading "Imelda"

The following story, "Imelda" by Richard Selzer, gains much of its power from the changing judgments of its narrator, a third-year medical student at the beginning of the story, who enacts the ambiguity of the situation in evaluating the "insanely arrogant" behavior of Dr. Franciscus early on and changing his judgment years later in seeing his act as "done, perhaps, to ward off madness" at the end of the story. In this, Selzer creates a "twice-told story." In large part these changing judgments—unlike the two *different* stories in the vignette—makes the evaluation of professionalism a prominent feature of this narrative, the "sensitivity to ambiguity" discussed in Appendix 5.

Literary Narrative: "Imelda" (1982) by Dr. Richard Selzer

Author Note: Dr. Richard Selzer (1928–2016) was a surgeon and professor of surgery at Yale University from 1960 until his retirement in 1985. Beginning in the 1970s, he also pursued a career as a writer, mostly of short stories that focused on medical themes. He won the Pushcart Prize for fiction in 1983 for his stories "Mercy" and "Witness" and was a semifinalist for the Pen/Faulkner Award for Fiction for *The Doctor Stories* (1998). His work has become central to the Medical Humanities.

Imelda

I heard the other day that Hugh Franciscus had died. I knew him once. He was the Chief of Plastic Surgery when I was a medical student at Albany Medical College. Dr. Franciscus was the archetype of the professor of surgery—tall, vigorous, muscular, as precise in his technique as he was impeccable in his dress. Each day a clean lab coat, monkishly starched, that sort of thing. I doubt that he ever read books. One book only, that of the human body, took the place of all others. He never raised his eyes from it. He read it like a printed page as though he knew that in the calligraphy there just beneath the skin were all the secrets of the world. Long before it became visible to anyone else, he could detect the first sign of granulation at the base of a wound, the first blue line of new epithelium at the periphery that would tell him that a wound would heal, or the barest hint of necrosis that presaged failure. This gave him the appearance of a prophet. "This skin graft will take," he would say, and you must believe beyond all cyanosis, exudation, and inflammation that it would.

He had enemies, of course, who said he was arrogant, that he exalted activity for its own sake. Perhaps. But perhaps it was no more than the honesty of one who knows his own worth. Just look at a scalpel, after all. What a feeling of sovereignty, megalomania even, when you know that it is you and you alone who will make certain use of it. It was said, too, that he was a ladies' man. I don't know about that. It was all rumor. Besides, I think he had other things in mind than mere living. Hugh Franciscus was a zealous hunter. Every fall during the season he drove upstate to hunt deer. There was a glass-front case in his office where he showed his guns. How could he shoot a deer? we asked. But he knew better. To us medical students he was someone heroic, someone made up of several gods, beheld at a distance, and always from a lesser height. If he had grown accustomed to his miracles, we had not. He had no close friends on the staff. There was something a little sad in that. As though once long ago he had been flayed by friendship and now the slightest breeze would hurt. Confidences resulted in dishonor. Perhaps the person in whom one confided would scorn him, betray. Even though he spent his days among those less fortunate, weaker than he—the sick, after all—Franciscus seemed aware of an air of personal harshness in his environment to which he reacted by keeping his own counsel, by a certain remoteness. It was what gave him the appearance of being haughty. With the patients he was forthright. All the facts laid out, every question anticipated and answered with specific information. He delivered good news and bad with the same dispassion.

I was a third-year student, just turned onto the wards for the first time, and clerking on Surgery. Everything—the operating room, the morgue, the emergency room, the patients, professors, even the nurses—was terrifying. One picked one's way among the mines and booby traps of the hospital, hoping only to avoid the hemorrhage and perforation of disgrace. The opportunity for humiliation was everywhere.

It all began on ward rounds. Dr. Franciscus was demonstrating a cross-leg flap graft he had constructed to cover a large fleshy defect in the leg of a merchant seaman who had injured himself in a fall. The man was from Spain and spoke no English. There had been a comminuted fracture of the femur, much soft-tissue damage, necrosis. After weeks of debridement and dressings, the wound had been made ready for grafting. Now the patient was in his fifth postoperative day. What we saw was a thick web of pale blue flesh arising from the man's left thigh, and which had been sutured to the open wound on the right thigh. When the surgeon pressed the pedicle with his finger, it blanched; when he let up, there was a slow return of the violaceous color.

"The circulation is good," Franciscus announced. "It will get better." In several weeks, we were told, he would divide the tube of flesh at its site of origin, and tailor it to fit the defect to which, by then, it would have grown more solidly. All at once, the webbed man in the bed reached out, and gripping Franciscus by the arm, began to speak rapidly, pointing to his groin and hip. Franciscus stepped back at once to disengage his arm from the patient's grasp.

"Anyone here know Spanish? I didn't get a word of that."

"The cast is digging into him up above," I said. "The edges of the plaster are rough. When he moves, they hurt."

Without acknowledging my assistance, Dr. Franciscus took a plaster shears from the dressing cart and with several large snips cut away the rough edges of the cast.

"*Gracias, gracias.*" The man in the bed smiled. But Franciscus had already moved on to the next bed. He seemed to me a man of immense strength and ability, yet without affection for the patients. He did not want to be touched by them. It was less kindness that he showed them than a reassurance that he would never give up, that he would bend every effort. If anyone could, he would solve the problems of their flesh.

Ward Rounds had disbanded and I was halfway down the corridor when I heard Dr. Franciscus's voice behind me.

"You speak Spanish." It seemed a command.

"I lived in Spain for two years," I told him.

"I'm taking a surgical team to Honduras next week to operate on the natives down there. I do it every year for three weeks, somewhere. This year, Honduras. I can arrange the time away from your duties here if you'd like to come along. You will act as interpreter. I'll show you how to use the clinical camera. What you'd see would make it worthwhile."

So it was that, a week later, the envy of my classmates, I joined the mobile surgical unit—surgeons, anesthetists, nurses, and equipment—aboard a Military Air Transport plane to spend three weeks performing plastic surgery on people who had been previously selected by an advance team. Honduras. I don't suppose I shall ever see it again. Nor do I especially want to. From the plane it seemed a country made of clay—burnt umber, raw sienna, dry. It had a deadweight quality, as though the ground had no buoyancy, no air sacs through which a breeze might wander. Our destination was Comayagua, a town in the Central Highlands. The town itself was situated on the edge of one of the flatlands that were linked in a network between the granite mountains. Above, all was brown, with only an occasional Spanish cedar tree; below, patches of luxuriant tropical growth. It was a day's bus ride from the airport. For hours, the town kept appearing and disappearing with the convolutions of the road. At last, there it lay before us, panting and exhausted at the bottom of the mountain.

That was all I was to see of the countryside. From then on, there was only the derelict hospital of Comayagua, with the smell of spoiling bananas and the accumulated odors of everyone who had been sick there for the last hundred years. Of the two, I much preferred the frank smell of the sick. The heat of the place was incendiary. So hot that, as we stepped from the bus, our own words did not carry through the air, but hung limply at our lips and chins. Just in front of the hospital was a thirsty courtyard where mobs of waiting people squatted or lay in the meager shade, and where, on dry days, a fine dust rose through which untethered goats shouldered. Against the walls of this courtyard, gaunt, dejected men stood, their faces, like their country, preternaturally solemn, leaden. Here no one looked up at the sky. Every head was bent beneath a wide-brimmed straw hat. In the days that followed, from the doorway of the dispensary I would watch the brown mountains sliding about, drinking the hospital into their shadow as the afternoon grew later and later, flattening us by their very altitude.

The people were mestizos, of mixed Spanish and Indian blood. They had flat, broad, dumb museum feet. At first they seemed to me indistinguishable the one from the other, without animation. All the vitality, the hidden sexuality, was in their black hair. Soon I was to know them by the fissures with which each face was graven. But, even so, compared to us, they were masked, shut away. My job was to follow Dr. Franciscus around, photograph the patients before and after surgery, interpret and generally act as aide-de-camp. It was exhilarating. Within days I had decided that I was not just useful, but essential. Despite that we spent all day in each other's company, there were no overtures of friendship from Dr. Franciscus. He knew my place, and I knew it, too. In the afternoon he examined the patients scheduled for the next day's surgery. I would call out a name from the doorway to the examining room. In the courtyard someone would rise. I would usher the patient in, and nudge him to the examining table where Franciscus stood, always, I thought, on the verge of irritability. I would read aloud the case history, then wait while he carried out his examination. While I took the "before" photographs, Dr. Franciscus would dictate into a tape recorder:

"Ulcerating basal-cell carcinoma of the right orbit—six by eight centimeters—involving the right eye and extending into the floor of the orbit. Operative plan: wide excision with enucleation of the eye. Later, bone and skin grafting." The next morning we would be in the operating room where the procedure would be carried out.

We were more than two weeks into our tour of duty—a few days to go—when it happened. Earlier in the day I had caught sight of her through the window of the dispensary. A thin, dark Indian girl about fourteen years old. A figurine, orange-brown, terra-cotta, and still attached to the unshaped clay from which she had been carved. An older, sun-weathered woman stood behind and somewhat to the left of the girl. The mother was short and dumpy. She wore a broad-brimmed hat with a high crown, and a shapeless dress like a cassock. The girl had long, loose black hair. There were tiny gold hoops in her ears. The dress she wore could have been her mother's. Far too big, it hung from her thin shoulders at some risk of slipping down her arms. Even with her in it, the dress was empty, something hanging on the back of a door. Her breasts made only the smallest imprint in the cloth, her hips none at all. All the while, she pressed to her mouth a filthy, pink, balled-up rag as though to stanch a flow or buttress against pain. I knew that what she had come to show us, what we were there to see, was hidden beneath that pink cloth. As I watched, the woman handed down to her a gourd from which the girl drank, lapping like a dog. She was the last patient of the day. They had been waiting in the courtyard for hours.

"Imelda Valdez," I called out. Slowly she rose to her feet, the cloth never leaving her mouth, and followed her mother to the examining-room door. I shooed them in.

"You sit up there on the table," I told her. "Mother, you stand over there, please." I read from the chart:

"This is a fourteen-year-old girl with a complete, unilateral, left-sided cleft lip and cleft palate. No other diseases or congenital defects. Laboratory tests, chest X-ray—negative."

"Tell her to take the rag away," said Dr. Franciscus. I did, and the girl shrank back, pressing the cloth all the more firmly.

"Listen, this is silly," said Franciscus. "Tell her I've got to see it. Either she behaves, or send her away."

"Please give me the cloth," I said to the girl as gently as possible. She did not. She could not. Just then, Franciscus reached up and, taking the hand that held the rag, pulled it away with a hard jerk. For an instant the girl's head followed the cloth as it left her face, one arm still upflung against showing. Against all hope, she would hide herself. A moment later, she relaxed and sat still. She seemed to me then like an animal that looks outward at the infinite, at death, without fear, with recognition only.

Set as it was in the center of the girl's face, the defect was utterly hideous—a nude rubbery insect that had fastened there. The upper lip was widely split all the way to the nose. One white tooth perched upon the protruding upper jaw projected through the hole. Some of the bone seemed to have been gnawed away as well. Above the thing, clear almond eyes and long black hair reflected the light. Below, a slender neck where the pulse trilled visibly.

Under our gaze the girl's eyes fell to her lap where her hands lay palms upward, half open. She was a beautiful bird with a crushed beak. And tense with the expectation of more shame.

"Open your mouth," said the surgeon. I translated. She did so, and the surgeon tipped back her head to see inside.

"The palate, too. Complete," he said. There was a long silence. At last he spoke.

"What is your name?" The margins of the wound melted until she herself was being sucked into it.

"Imelda." The syllables leaked through the hole with a slosh and a whistle.

"Tomorrow," said the surgeon, "I will fix your lip. *Mañana*."

It seemed to me that Hugh Franciscus, in spite of his years of experience, in spite of all the dreadful things he had seen, must have been awed by the sight of this girl. I could see it flit across his face for an instant. Perhaps it was her small act of concealment, that he had had to demand that she show him the lip, that he had had to force her to show it to him. Perhaps it was her resistance that intensified the disfigurement. Had she brought her mouth to him willingly, without shame, she would have been for him neither more nor less than any other patient.

He measured the defect with calipers, studied it from different angles, turning her head with a finger at her chin.

"How can it ever be put back together?" I asked.

"Take her picture," he said. And to her, "Look straight ahead." Through the eye of the camera she seemed more pitiful than ever, her humiliation more complete.

"Wait!" The surgeon stopped me. I lowered the camera. A strand of her hair had fallen across her face and found its way to her mouth, becoming stuck there by saliva. He removed the hair and secured it behind her ear.

"Go ahead," he ordered. There was the click of the camera. The girl winced.

"Take three more, just in case."

When the girl and her mother had left, he took paper and pen and with a few lines drew a remarkable likeness of the girl's face.

"Look," he said. "If this dot is A, and this one B, this, C and this, D, the incisions are made A to B, then C to D. CD must equal AB. It is all equilateral triangles." All well and good, but then came X and Y and rotation flaps and the rest.

"Do you see?" he asked.

"It is confusing," I told him.

"It is simply a matter of dropping the upper lip into a normal position, then crossing the gap with two triangular flaps. It is geometry," he said.

"Yes," I said. "Geometry." And relinquished all hope of becoming a plastic surgeon.

* * *

In the operating room the next morning, the anesthesia had already been administered when we arrived from ward rounds. The tube emerging from the girl's mouth was pressed against her lower lip to be kept out of the field of surgery. Already, a nurse was scrubbing the face which swam in a reddish brown lather. The tiny gold earrings were included in the scrub. Now and then, one of them gave a brave flash. The face was washed for the last time, and dried. Green towels were placed over the face to hide everything but the mouth and nose. The drapes were applied.

"Calipers!" The surgeon measured, locating the peak of the distorted Cupid's bow.

"Marking pen!" He placed the first blue dot at the apex of the bow. The nasal sills were dotted; next, the inferior philtral dimple, the vermilion line. The A flap and the B flap were outlined. On he worked, peppering the lip and nose, making sense out of chaos, realizing the lip that lay waiting in that deep essential pink, that only he could see. The last dot and line were placed. He was ready.

"Scalpel!" He held the knife above the girl's mouth.
"Okay to go ahead?" he asked the anesthetist.
"Yes."
He lowered the knife.
"No! Wait!" The anesthetist's voice was tense, staccato. "Hold it!"
The surgeon's hand was motionless.
"What's the matter?"
"Something's wrong. I'm not sure. God, she's hot as a pistol. Blood pressure is way up. Pulse one-eighty. Get a rectal temperature." A nurse fumbled beneath the drapes. We waited. The nurse retrieved the thermometer.
"One hundred seven ... no ... eight." There was disbelief in her voice.
"Malignant hyperthermia," said the anesthetist. "Ice! Ice! Get lots of ice!" I raced out the door, accosted the first nurse I saw.
"Ice!" I shouted. "*Hielo!* Quickly! *Hielo!*" The woman's expression was blank. I ran to another. "*Hielo! Hielo!* For the love of God, ice!"
"*Hielo?*" She shrugged. "*Nada.*" I ran back to the operating room.
"There isn't any ice," I reported.
Dr. Franciscus had ripped off his rubber gloves and was feeling the skin of the girl's abdomen. Above the mask his eyes were the eyes of a horse in battle.
"The EKG is wild ...."
"I can't get a pulse ...."
"What the hell ...."
The surgeon reached for the girl's groin. No femoral pulse.
"EKG flat. My God! She's dead!"
"She can't be."
"She is."
The surgeon's fingers pressed the groin where there was no pulse to be felt, only his own pulse hammering at the girl's flesh to be let in.

* * *

It was noon, four hours later, when we left the operating room. It was a day so hot and humid I felt steamed-open like an envelope. The woman was sitting on a bench in the courtyard in her dress like a cassock. In one hand she held the piece of cloth the girl had used to conceal her mouth. As we watched, she folded it once neatly, and then again, smoothing it, cleaning the cloth which might have been the head of the girl in her lap that she stroked and consoled.
"I'll do the talking here," he said. He would tell her himself, in whatever Spanish he could find. Only if she did not understand was I to speak for him. I watched him brace himself, set his shoulders. How could he tell her? I wondered. What? But I knew he would tell her everything, exactly as it had happened. As much for himself as for her, he needed to explain. But suppose she screamed, fell to the ground, attacked him, even? All that hope of love ... gone. Even in his discomfort I knew that he was teaching me. The way to do it was professionally. Now he was standing above her. When the woman saw that he did not speak, she lifted her eyes and saw what he held crammed in his mouth to tell her. She knew, and rose to her feet.
"*Señora,*" he began, "I am sorry." All at once he seemed to me shorter than he was, scarcely taller than she. There was a place at the crown of his head where the hair had grown thin. His lips were stones. He could hardly move them. The voice dry, dusty.
"No one could have known. Some bad reaction to the medicine for sleeping. It poisoned her. High fever. She did not wake up." The last, a whisper. The woman studied his lips as though she were deaf. He tried, but could not control a twitching at the corner of his mouth. He raised a thumb and forefinger to press something back into his eyes.

"*Muerte*," the woman announced to herself. Her eyes were human, deadly.

"*Sí, muerte.*" At that moment he was like someone cast, still alive, as an effigy for his own tomb. He closed his eyes. Nor did he open them until he felt the touch of the woman's hand on his arm, a touch from which he did not withdraw. Then he looked and saw the grief corroding her face, breaking it down, melting the features so that eyes, nose, mouth ran together in a distortion, like the girl's. For a long time they stood in silence. It seemed to me that minutes passed. At last her face cleared, the features rearranged themselves. She spoke, the words coming slowly to make certain that he understood her. She would go home now. The next day her sons would come for the girl, to take her home for burial. The doctor must not be sad. God has decided. And she was happy now that the harelip had been fixed so that her daughter might go to Heaven without it. Her bare feet retreating were the felted pads of a great bereft animal.

* * *

The next morning I did not go to the wards, but stood at the gate leading from the courtyard to the road outside. Two young men in striped ponchos lifted the girl's body wrapped in a straw mat onto the back of a wooden cart. A donkey waited. I had been drawn to this place as one is drawn, inexplicably, to certain scenes of desolation—executions, battlefields. All at once, the woman looked up and saw me. She had taken off her hat. The heavy-hanging coil of her hair made her head seem larger, darker, noble. I pressed some money into her hand.

"For flowers," I said. "A priest." Her cheeks shook as though minutes ago a stone had been dropped into her navel and the ripples were just now reaching her head. I regretted having come to that place.

"*Sí, Sí*," the woman said. Her own face was stitched with flies. "The doctor is one of the angels. He has finished the work of God. My daughter is beautiful."

What could she mean! The lip had not been fixed. The girl had died before he would have done it.

"Only a fine line that God will erase in time," she said.

I reached into the cart and lifted a corner of the mat in which the girl had been rolled. Where the cleft had been there was now a fresh line of tiny sutures. The Cupid's bow was delicately shaped, the vermilion border aligned. The flattened nostril had now the same rounded shape as the other one. I let the mat fall over the face of the dead girl, but not before I had seen the touching place where the finest black hairs sprang from the temple.

"*Adiós, adiós*…" And the cart creaked away to the sound of hooves, a tinkling bell.

* * *

There are events in a doctor's life that seem to mark the boundary between youth and age, seeing and perceiving. Like certain dreams, they illuminate a whole lifetime of past behavior. After such an event, a doctor is not the same as he was before. It had seemed to me then to have been the act of someone demented, or at least insanely arrogant. An attempt to reorder events. Her death had come to him out of order. It should have come after the lip had been repaired, not before. He could have told the mother that, no, the lip had not been fixed. But he did not. He said nothing. It had been an act of omission, one of those strange lapses to which all of us are subject and which we live to regret. It must have been then, at that moment, that the knowledge of what he would do appeared to him. The words of the mother had not consoled him; they had hunted him down. He had not done it for her. The dire necessity was his. He would not accept that Imelda had died before he could repair her lip. People who do such things break free from society. They follow their own lonely path. They have a secret which they can never reveal. I must never let on that I knew.

* * *

How often I have imagined it. Ten o'clock at night. The hospital of Comayagua is all but dark. Here and there lanterns tilt and skitter up and down the corridors. One of these lamps breaks free from the others and descends the stone steps to the underground room that is the morgue of the hospital. This room wears the expression as if it had waited all night for someone to come. No silence so deep as this place with its cargo of newly dead. Only the slow drip of water over stone. The door closes gassily and clicks shut. The lock is turned. There are four tables, each with a body encased in a paper shroud. There is no mistaking her. She is the smallest. The surgeon takes a knife from his pocket and slits open the paper shroud, that part in which the girl's head is enclosed. The wound seems to be living on long after she has died. Waves of heat emanate from it, blurring his vision. All at once, he turns to peer over his shoulder. He sees nothing, only a wooden crucifix on the wall.

* * *

He removes a package of instruments from a satchel and arranges them on a tray. Scalpel, scissors, forceps, needle holder. Sutures and gauze sponges are produced. Stealthy, hunched, engaged, he begins. The dots of blue dye are still there upon her mouth. He raises the scalpel, pauses. A second glance into the darkness. From the wall a small lizard watches and accepts. The first cut is made. A sluggish flow of dark blood appears. He wipes it away with a sponge. No new blood comes to take its place. Again and again he cuts, connecting each of the blue dots until the whole of the zigzag slice is made, first on one side of the cleft, then on the other. Now the edges of the cleft are lined with fresh tissue. He sets down the scalpel and takes up scissors and forceps, undermining the little flaps until each triangle is attached only at one side. He rotates each flap into its new position. He must be certain that they can be swung without tension. They can. He is ready to suture. He fits the tiny curved needle into the jaws of the needle holder. Each suture is placed precisely the same number of millimeters from the cut edge, and the same distance apart. He ties each knot down until the edges are apposed. Not too tightly. These are the most meticulous sutures of his life. He cuts each thread close to the knot. It goes well. The vermilion border with its white skin roll is exactly aligned. One more stitch and the Cupid's bow appears as if by magic. The man's face shines with moisture. Now the nostril is incised around the margin, released, and sutured into a round shape to match its mate. He wipes the blood from the face of the girl with gauze that he has dipped in water. Crumbs of light are scattered on the girl's face. The shroud is folded once more about her. The instruments are handed into the satchel. In a moment the morgue is dark and a lone lantern ascends the stairs and is extinguished.

* * *

Six weeks later I was in the darkened amphitheater of the Medical School. Tiers of seats rose in a semicircle above the small stage where Hugh Franciscus stood presenting the case material he had encountered in Honduras. It was the highlight of the year. The hall was filled. The night before, he had arranged the slides in the order in which they were shown. I was at the controls of the slide projector.

"Next slide!" he would order from time to time in that military voice which had called forth blind obedience from generations of medical students, interns, residents, and patients.

"This is a fifty-seven-year-old man with a severe burn contracture of the neck. You will notice the rigid webbing that has fused the chin to the presternal tissues. No motion of the head on the torso is possible. ... Next slide!"

Click, went the projector.

"Here he is after the excision of the scar tissue and with the head in full extension for the first time. The defect was then covered ... Next slide!"

Click.

"... with full-thickness drums of skin taken from the abdomen with the Padgett dermatome. Next slide!"

Click.

And suddenly there she was, extracted from the shadows, suspended above and beyond all of us like a resurrection. There was the oval face, the long black hair unbraided, the tiny gold hoops in her ears. And that luminous gnawed mouth. The whole of her life seemed to have been summed up in this photograph. A long silence followed that was the surgeon's alone to break. Almost at once, like the anesthetist in the operating room in Comayagua, I knew that something was wrong. It was not that the man would not speak as that he could not. The audience of doctors, nurses, and students seemed to have been infected by the black, limitless silence. My own pulse doubled. It was hard to breathe. Why did he not call out for the next slide? Why did he not save himself? Why had he not removed this slide from the ones to be shown? All at once I knew that he had used his camera on her again. I could see the long black shadows of her hair flowing into the darker shadows of the morgue. The sudden blinding flash.... The next slide would be the one taken in the morgue. He would be exposed.

In the dim light reflected from the slide, I saw him gazing up at her, seeing not the colored photograph, I thought, but the negative of it where the ghost of the girl was. For me, the amphitheater had become Honduras. I saw again that courtyard littered with patients. I could see the dust in the beam of light from the projector. It was then that I knew that she was his measure of perfection and pain—the one lost, the other gained. He, too, had heard the click of the camera, had seen her wince and felt his mercy enlarge. At last he spoke.

"Imelda." It was the one word he had heard her say. At the sound of his voice I removed the next slide from the projector. *Click* ... and she was gone. *Click* again, and in her place the man with the orbital cancer. For a long moment Franciscus looked up in my direction, on his face an expression that I have given up trying to interpret. Gratitude? Sorrow? It made me think of the gaze of the girl when at last she understood that she must hand over to him the evidence of her body.

"This is a sixty-two-year-old man with a basal-cell carcinoma of the temple eroding into the bony orbit ..." he began, as though nothing had happened.

At the end of the hour, even before the lights went on, there was loud applause. I hurried to find him among the departing crowd. I could not. Some weeks went by before I caught sight of him. He seemed vaguely convalescent, as though a fever had taken its toll before burning out.

Hugh Franciscus continued to teach for fifteen years, although he operated a good deal less, then gave it up entirely. It was as though he had grown tired of blood, of always having to be involved with blood, of having to draw it, spill it, wipe it away, stanch it. He was a quieter, softer man, I heard, the ferocity diminished. There were no more expeditions to Honduras or anywhere else.

I, too, have not been entirely free of her. Now and then, in the years that have passed, I see that donkey-cart cortege, or his face bent over hers in the morgue. I would like to have told him what I now know, that his unrealistic act was one of goodness, one of those small, persevering acts done, perhaps, to ward off madness. Like lighting a lamp, boiling water for tea, washing a shirt. But, of course, it's too late now.

Workshop or Reading

Given that Appendix 5 offers a practical design for discussing "Imelda" in relation to professionalism, one could approach this story by staging a professionalization workshop in class or by simply reading the story and attending to the ways it promotes vicarious experience and an ethical discussion of professionalism. Alternatively, one could combine these approaches and discuss the ways in which literary features—the provocation of empathy and vicarious experience, the uses of language (such as the similes in the story's last paragraph), the strategies of narrative presentation—clarify issues that arise in considering the "professionalism" of healthcare that supplement its "art" and "science."

Related Poem

For this chapter, the related poem is a touching poem about a mother who leaves her home and son before dawn to take up her job as a physician. The poem presents and quietly questions the balancing of personal and professional life for healthcare providers. That balance is difficult because the *profession* of healthcare overlaps in profound ways with the personal *caring* that is part and parcel of our private and domestic lives. The poem's very title—the beginning of the workweek after the seeming respite of the weekend—touches on issues of professionalism in profound ways.

Poem: "Monday Morning" (1992) by Dr. Audrey Shafer

> *Author Note*: Dr. Audrey Shafer (b. 1956) is a Professor of Anesthesiology, Perioperative, and Pain Medicine at Stanford University School of Medicine/Palo Alto Veterans Affairs Health Care System and a member of the Stanford Center for Biomedical Ethics at Stanford University. In addition, she directs the Stanford Medicine and the Muse: Medical Humanities and the Arts Program. In 2007 she won the Henry J. Kaiser Foundation Award for Outstanding & Innovative Contributions to Medicine Education at Stanford. She is the author of *Sleep Talker: Poems by a Doctor/Mother*, from which "Monday Morning" is taken (it first appeared in *Annals of Internal Medicine* in 1992), and *The Mailbox* (young adult novel).

Monday Morning

> In the prelight
> A heavy sound from upstairs
> I turn from the front door
>   to investigate.
>
> My three-year-old son stands
>   naked
>   in the soft penumbra of dimmed hallway light
> Clutching his favorite blanket
>   picture book and well-rubbed panther
>   to his chest.
> His toes curl on the wooden floor.
>
> I am dressed and beepered –
> No snuggling in the warm water bed this morning
>   floating back to sleep till sunlight wakens.

Instead, we hug.
I kiss
      his thin neck.
I feel his small breaths.

His bedroom door stands closed,
    heavy in shadows.

At the operating suite,
The residents still at lecture
The patient not yet here,
     I enjoy the rote motions –
follow green snake tubing to the ceiling
barbotage dissolving drugs into syringes
snap open the laryngoscope.

Around me all is bright   pristine   ordered
Primed.
Sterile instruments attend in precise, metallic rows.

I try to recall his just awakened warmth
   in the brief moment
         before

The patient arrives
Naked under hospital issue
Ready to sleep.

Lessons for Providers

1. In its vignettes and short story, this chapter raises the complex question of "truth-telling" as a professional issue. It is too simple to develop a rule that must always be applied in medicine—such as is done in normative (or "deontological") ethics. As providers, we must struggle with when and how truth-telling—as opposed to silence rather than outright lying—is used in our practice.
2. This chapter, in conjunction with Appendix 5, also calls attention to the manner in which truth-telling is an element of a complex set of attributes of professionalism in healthcare.
3. Finally, the poem of this chapter raises the important question of the ways professional healthcare providers can and should attend to the balance between professional responsibilities and family and personal lives.

Bibliography

Gawande, Atul. 2010. *The Checklist Manifesto: How to Get Things Right*. Amazon Kindle ed. New York: Metropolitan Books.
LaCombe, Michael. 2010. *Bedside: The Art of Medicine*. Orono: University of Maine Press.
Stern, D.T., ed. 2006. *Measuring Medical Professionalism*. New York: Oxford Press.

Section IV

Building the Patient-Provider Relationship

Rapport and Empathy in Medicine

The four chapters of this section offer literary narratives that allow healthcare providers to more fully understand the complexity of clinical medicine by understanding the cognitive/affective nature of the patient-provider relationship, to discover strategies of listening, and to see that "patient" and "healthcare provider" are social roles, which people assume without discarding other aspects of their lives. As we have already mentioned, for almost two decades the writers/editors of this book team-taught a course that brings together the humanities and medicine with the goal of widening the understanding and experience of people who have committed themselves to engaging with and caring for others who face crises of health and well-being in their lives. This course came about because of an experience Dr. Vannatta had in his practice, which was related to Toni Morrison's great novel *Beloved*. The novel, one of the most important in American literature, traces a family of African American slaves in their lives right before the American Civil War. It narrates the life and experience of a slave woman, Sethe, who finds life as a slave so hateful that she kills her infant daughter to prevent her from living a life of slavery, which includes her child becoming a motherless child. What is so powerful about the novel is the way it makes us empathize with this woman who commits such a terrible crime. It is an important book because it allows its readers to feel and *experience* the life of slavery. Before reading this book, both of us—like many fellow Americans—only had the most abstract sense of the terrible condition of chattel slavery in our country. Morrison's book makes this terrible time in our history terribly *real* for those who read it. It also allows us to understand much more fully the associated poem in Chap. 7, the Slave Spiritual "Sometimes I Feel Like a Motherless Child." The experience of this novel changed Dr. Vannatta's practice. Here is his vignette.

An Elderly African American Patient: A Vignette
(Excerpt from *The Chief Concern of Medicine*)

I came to this whole interest in narrative, literature, and the practice of medicine through an experience I had in my own practice. I'm a general internist, and I had an elderly African American woman who came back to the office for an office visit after having been in the hospital. I didn't get to know her real well in the hospital because she was cared for primarily by the residents and the medical students on my service, but when she came back for an office visit, I was providing the care. And she rapidly told me that she was having trouble getting her medications. As I was interacting with her, there was just really no connection being made. That makes me so uncomfortable when I'm really not connecting with the patient, so, as I usually do when I'm not connecting well, I backed up and sort of took a psychosocial history. I basically just said, "Tell me about your life."

She began to tell me a story about having grown up in east Texas on a sharecropping farm where her father was a sharecropper, and he, when she was fifteen, made her marry a man who was twenty-one. It really wasn't the man she wanted to marry; she was in love with a sixteen year-old, but he made her marry the twenty-one year-old because he could provide for a living. In fact, she said to me during the story that "he wasn't very good at making a living, but he was sure good at making babies," and she had seventeen of them. And I thought at the time she said that, "My goodness, that could have rolled right out of a wonderful novel or short story." She went on to say that she, oftentimes to make ends meet, walked two miles to a white man's house and two miles back to do domestic work. And she told me that sometimes the white man would give her a dozen eggs, and sometimes he would give her a two-gallon pail of milk to carry back to the family. And then she looked at me and said, "Doctor, have you ever carried a two-gallon pail of milk two miles?" And, in fact, I did grow up on a farm, and I can remember carrying those galvanized pails of water around the farm to the chickens and whatnot, and I could just see that wire handle just burying itself and cutting into her hand.

But more importantly, I was thinking that I was seeing her carrying this pail of milk on a dusty, sort of rocky road, probably with not very good shoes. And as I was thinking about her feet, making this journey back, I began to think of this novel, Toni Morrison's novel, *Beloved*, which I had just read a few months earlier, at that time, the most remarkable novel I had ever read, a very disturbing story about slavery in America. And the protagonist, Sethe, is running from slavery. She's pregnant, she's trying to escape, and she's tired and she's about to deliver a baby, and she's hiding up under a bush and a little white girl finds her. One of the things that's striking about that scene is her swollen, bleeding and pussy, infected feet. And that image of those feet came back to me just in a flood, and the emotions that I had felt, I think, when I read the novel were seemingly stored in memory. And along with the image of the feet, these emotions came flooding back to me. And the remarkable thing that happened in the room was that those emotions were available to me to be able to connect with this lady, not that she was a slave, but in some way she was telling me a story about her economic enslavement and somehow they connected. I don't know how that works, but nonetheless, it happened. It was an experience that was dramatic for me, and from that point on, we began to make a more meaningful connection, and we rapidly sort of problem solved her ability to buy her medications and get them so that she could take them. And at the end of the interaction, we stood up to leave and a remarkable thing happened, which usually doesn't happen in my practice, which is we embraced. And she knew that a wonderful relationship had begun, and so did I.

In this narrative of his experience with his patient, Dr. Vannatta notes that he experienced a "flood" of emotion that allowed him "to connect" to his patient, although, as he says, "I don't know how that works." In the last generation, the work in cognitive psychology we describe in Chap. 1 attempts to understand how this works, which is to say how a literary narrative such as Morrison's *Beloved* and even an everyday narrative, such as that of his patient, give rise to experience, knowledge, and social connection. Even the patient's observation, in the midst of an anxious and fretful meeting with her physician, that her husband "wasn't very good at making a living, but he was sure good at making babies," draws our attention to the formal patterns of language—here, the play on the word "making" used figuratively in terms of "making a living" and literally in terms of "making babies"—suggesting that the experience narrative provokes can be analyzed in terms of linguistic strategies (here the opposition between literal and figurative meanings discussed in Chap. 1 above).

In this chapter, we turn to another "literary" physician, Dr. Anton Chekhov, who revolutionized both the short story and drama in the early twentieth century. Like Grace Paley's father, who pursued careers as a physician and an artist, Chekhov too exhibits great interest "in details, craft, technique." His "technique," like that of Morrison, focuses on provoking empathy through his craft.

Literary Narrative: "A Doctor's Visit" (1898) by Dr. Anton Chekhov

Author Note: Dr. Anton Chekhov (1860–1904), grandson of a Russian serf, was a physician, short story writer, and playwright. He is considered to be among the greatest writers of short fiction, and his innovations in drama in the early twentieth century are also considered to have transformed the theater. Throughout his writing career he practiced medicine. He once said that "medicine is my lawful wife, and literature is my mistress."

A Doctor's Visit

The Professor received a telegram from the Lyalikovs' factory; he was asked to come as quickly as possible. The daughter of some Madame Lyalikov, apparently the owner of the factory, was ill, and that was all that one could make out of the long, incoherent telegram. And the Professor did not go himself, but sent instead his assistant, Korolyov.

It was two stations from Moscow, and there was a drive of three miles from the station. A carriage with three horses had been sent to the station to meet Korolyov; the coachman wore a hat with a peacock's feather on it, and answered every question in a loud voice like a soldier: "No, sir!" "Certainly, sir!"

It was Saturday evening; the sun was setting, the workpeople were coming in crowds from the factory to the station, and they bowed to the carriage in which Korolyov was driving. And he was charmed with the evening, the farmhouses and villas on the road, and the birch-trees, and the quiet atmosphere all around, when the fields and woods and the sun seemed preparing, like the workpeople now on the eve of the holiday, to rest, and perhaps to pray….

He was born and had grown up in Moscow; he did not know the country, and he had never taken any interest in factories, or been inside one, but he had happened to read about factories, and had been in the houses of manufacturers and had talked to them; and whenever he saw a factory far or near, he always thought how quiet and peaceable it was outside, but within there was always sure to be impenetrable ignorance and dull egoism on the side of the owners, wearisome, unhealthy toil on the side of the workpeople, squabbling, vermin, vodka. And now when the workpeople timidly and respectfully made way for the carriage, in their faces, their caps, their walk, he read physical impurity, drunkenness, nervous exhaustion, bewilderment.

They drove in at the factory gates. On each side he caught glimpses of the little houses of workpeople, of the faces of women, of quilts and linen on the railings. "Look out!" shouted the coachman, not pulling up the horses. It was a wide courtyard without grass, with five immense blocks of buildings with tall chimneys a little distance one from another, warehouses and barracks, and over everything a sort of grey powder as though from dust. Here and there, like oases in the desert, there were pitiful gardens, and the green and red roofs of the houses in which the managers and clerks lived. The coachman suddenly pulled up the horses, and the carriage stopped at the house, which had been newly painted grey; here was a flower garden, with a lilac bush covered with dust, and on the yellow steps at the front door there was a strong smell of paint.

"Please come in, doctor," said women's voices in the passage and the entry, and at the same time he heard sighs and whisperings. "Pray walk in …. We've been expecting you so long … we're in real trouble. Here, this way."

Madame Lyalikov– a stout elderly lady wearing a black silk dress with fashionable sleeves, but, judging from her face, a simple uneducated woman – looked at the doctor in a flutter, and could not bring herself to hold out her hand to him; she did not dare. Beside her stood a personage with short hair and a pince-nez; she was wearing a blouse of many colours, and was very thin and no longer young. The servants called her Christina Dmitryevna, and Korolyov guessed that this was the governess. Probably, as the person of most education in the house, she had been charged to meet and receive the doctor, for she began immediately, in great haste, stating the causes of the illness, giving trivial and tiresome details, but without saying who was ill or what was the matter.

The doctor and the governess were sitting talking while the lady of the house stood motionless at the door, waiting. From the conversation Korolyov learned that the patient was Madame Lyalikov's only daughter and heiress, a girl of twenty, called Liza; she had been ill for a long time, and had consulted various doctors, and the previous night she had suffered till morning from such violent palpitations of the heart, that no one in the house had slept, and they had been afraid she might die.

"She has been, one may say, ailing from a child," said Christina Dmitryevna in a sing-song voice, continually wiping her lips with her hand. "The doctors say it is nerves; when she was a little girl she was scrofulous, and the doctors drove it inwards, so I think it may be due to that."

They went to see the invalid. Fully grown up, big and tall, but ugly like her mother, with the same little eyes and disproportionate breadth of the lower part of the face, lying with her hair in disorder, muffled up to the chin, she made upon Korolyov at the first minute the impression of a poor, destitute creature, sheltered and cared for her out of charity, and he could hardly believe that this was the heiress of the five huge buildings.

"I am the doctor come to see you," said Korolyov. "Good evening."

He mentioned his name and pressed her hand, a large, cold, ugly hand; she sat up, and, evidently accustomed to doctors, let herself be sounded, without showing the least concern that her shoulders and chest were uncovered.

"I have palpitations of the heart," she said, "It was so awful all night…. I almost died of fright! Do give me something."

"I will, I will; don't worry yourself."

Korolyov examined her and shrugged his shoulders.

"The heart is all right," he said; "it's all going on satisfactorily; everything is in good order. Your nerves must have been playing pranks a little, but that's so common. The attack is over by now, one must suppose; lie down and go to sleep."

At that moment a lamp was brought into the bed-room. The patient screwed up her eyes at the light, then suddenly put her hands to her head and broke into sobs. And the impression of a destitute, ugly creature vanished, and Korolyov no longer noticed the little eyes or the heavy development of the lower part of the face. He saw a soft, suffering expression which was intelligent and touching: she seemed to him altogether graceful, feminine, and simple; and he longed to soothe her, not with drugs, not with advice, but with simple, kindly words. Her mother put her arms round her head and hugged her. What despair, what grief was in the old woman's face! She, her mother, had reared her and brought her up, spared nothing, and devoted her whole life to having her daughter taught French, dancing, music: had engaged a dozen teachers for her; had consulted the best doctors, kept a governess. And now she could not make out the reason of these tears, why there was all this misery, she could not understand, and was bewildered; and she had a guilty, agitated, despairing expression, as though she had omitted something very important, had left something undone, had neglected to call in somebody—and whom, she did not know.

"Lizanka, you are crying again … again," she said, hugging her daughter to her. "My own, my darling, my child, tell me what it is! Have pity on me! Tell me."

Both wept bitterly. Korolyov sat down on the side of the bed and took Liza's hand.

"Come, give over; it's no use crying," he said kindly. "Why, there is nothing in the world that is worth those tears. Come, we won't cry; that's no good …."

And inwardly he thought:

"It's high time she was married …."

"Our doctor at the factory gave her kalibromati," said the governess, "but I notice it only makes her worse. I should have thought that if she is given anything for the heart it ought to be drops …. I forget the name …. Convallaria, isn't it?"

And there followed all sorts of details. She interrupted the doctor, preventing his speaking, and there was a look of effort on her face, as though she supposed that, as the woman of most education in the house, she was duty bound to keep up a conversation with the doctor, and on no other subject but medicine.

Korolyov felt bored.

"I find nothing special the matter," he said, addressing the mother as he went out of the bedroom. "If your daughter is being attended by the factory doctor, let him go on attending her. The treatment so far has been perfectly correct, and I see no reason for changing your doctor. Why change? It's such an ordinary trouble; there's nothing seriously wrong."

He spoke deliberately as he put on his gloves, while Madame Lyalikov stood without moving, and looked at him with her tearful eyes.

"I have half an hour to catch the ten o'clock train," he said. "I hope I am not too late."

"And can't you stay?" she asked, and tears trickled down her cheeks again. "I am ashamed to trouble you, but if you would be so good…. For God's sake," she went on in an undertone, glancing towards the door, "do stay to-night with us! She is all I have … my only daughter …. She frightened me last night; I can't get over it…. Don't go away, for goodness' sake! …"

He wanted to tell her that he had a great deal of work in Moscow, that his family were expecting him home; it was disagreeable to him to spend the evening and the whole night in a strange house quite needlessly; but he looked at her face, heaved a sigh, and began taking off his gloves without a word.

All the lamps and candles were lighted in his honour in the drawing-room and the dining-room. He sat down at the piano and began turning over the music. Then he looked at the pictures on the walls, at the portraits. The pictures, oil-paintings in gold frames, were views of the Crimea – a stormy sea with a ship, a Catholic monk with a wineglass; they were all dull, smooth daubs, with no trace of talent in them. There was not a single good-looking face among the portraits, nothing but broad cheekbones and astonished-looking eyes. Lyalikov, Liza's father, had a low forehead and a self-satisfied expression; his uniform sat like a sack on his bulky plebeian figure; on his breast was a medal and a Red Cross Badge. There was little sign of culture, and the luxury was senseless and haphazard, and was as ill fitting as that uniform. The floors irritated him with their brilliant polish, the lustres on the chandelier irritated him, and he was reminded for some reason of the story of the merchant who used to go to the baths with a medal on his neck ….

He heard a whispering in the entry; some one was softly snoring. And suddenly from outside came harsh, abrupt, metallic sounds, such as Korolyov had never heard before, and which he did not understand now; they roused strange, unpleasant echoes in his soul.

"I believe nothing would induce me to remain here to live ..." he thought, and went back to the music-books again.

"Doctor, please come to supper!" the governess called him in a low voice.

He went into supper. The table was large and laid with a vast number of dishes and wines, but there were only two to supper: himself and Christina Dmitryevna. She drank Madeira, ate rapidly, and talked, looking at him through her pince-nez:

"Our workpeople are very contented. We have performances at the factory every winter; the workpeople act themselves. They have lectures with a magic lantern, a splendid tea-room, and everything they want. They are very much attached to us, and when they heard that Lizanka was worse they had a service sung for her. Though they have no education, they have their feelings, too."

"It looks as though you have no man in the house at all," said Korolyov.

"Not one. Pyotr Nikanoritch died a year and a half ago, and left us alone. And so there are the three of us. In the summer we live here, and in winter we live in Moscow, in Polianka. I have been living with them for eleven years—as one of the family."

At supper they served sterlet, chicken rissoles, and stewed fruit; the wines were expensive French wines.

"Please don't stand on ceremony, doctor," said Christina Dmitryevna, eating and wiping her mouth with her fist, and it was evident she found her life here exceedingly pleasant. "Please have some more."

After supper the doctor was shown to his room, where a bed had been made up for him, but he did not feel sleepy. The room was stuffy and it smelt of paint; he put on his coat and went out.

It was cool in the open air; there was already a glimmer of dawn, and all the five blocks of buildings, with their tall chimneys, barracks, and warehouses, were distinctly outlined against the damp air. As it was a holiday, they were not working, and the windows were dark, and in only one of the buildings was there a furnace burning; two windows were crimson, and fire mixed with smoke came from time to time from the chimney. Far away beyond the yard the frogs were croaking and the nightingales singing.

Looking at the factory buildings and the barracks, where the workpeople were asleep, he thought again what he always thought when he saw a factory. They may have performances for the workpeople, magic lanterns, factory doctors, and improvements of all sorts, but, all the same, the workpeople he had met that day on his way from the station did not look in any way different from those he had known long ago in his childhood, before there were factory performances and improvements. As a doctor accustomed to judging correctly of chronic complaints, the radical cause of which was incomprehensible and incurable, he looked upon factories as something baffling, the cause of which also was obscure and not removable, and all the improvements in the life of the factory hands he looked upon not as superfluous, but as comparable with the treatment of incurable illnesses.

"There is something baffling in it, of course ..." he thought, looking at the crimson windows. "Fifteen hundred or two thousand workpeople are working without rest in unhealthy surroundings, making bad cotton goods, living on the verge of starvation, and only waking from this nightmare at rare intervals in the tavern; a hundred people act as overseers, and the whole life of that hundred is spent in imposing fines, in abuse, in injustice, and only two or three so-called owners enjoy the profits, though they don't work at all, and despise the wretched cotton. But what are the profits, and how do they enjoy them? Madame Lyalikov and her daughter are unhappy—it makes one wretched to look at them; the only one who enjoys her life is Christina Dmitryevna, a stupid, middle-aged maiden lady in pince-nez. And so it appears that all these five blocks of buildings are at work, and inferior cotton is sold in the Eastern markets, simply that Christina Dmitryevna may eat sterlet and drink Madeira."

Suddenly there came a strange noise, the same sound Korolyov had heard before supper. Some one was striking on a sheet of metal near one of the buildings; he struck a note, and then at once checked the vibrations, so that short, abrupt, discordant sounds were produced, rather like "Dair ... dair ... dair...." Then there was half a minute of stillness, and from another building there came sounds equally abrupt and unpleasant, lower bass notes: "Drin ... drin ... drin..." eleven times. Evidently it was the watchman striking the hour. Near the third building he heard: "Zhuk ... zhuk ... zhuk...." And so near all the buildings, and then behind the barracks and beyond the gates. And in the stillness of the night it seemed as though these sounds were uttered by a monster with crimson eyes—the devil himself, who controlled the owners and the work-people alike, and was deceiving both.

Korolyov went out of the yard into the open country.

"Who goes there?" some one called to him at the gates in an abrupt voice.

"It's just like being in prison," he thought, and made no answer.

Here the nightingales and the frogs could be heard more distinctly, and one could feel it was a night in May. From the station came the noise of a train; somewhere in the distance drowsy cocks were crowing; but, all the same, the night was still, the world was sleeping tranquilly. In a field not far from the factory there could be seen the framework of a house and heaps of building material:

Korolyov sat down on the planks and went on thinking.

"The only person who feels happy here is the governess, and the factory hands are working for her gratification. But that's only apparent: she is only the figurehead. The real person, for whom everything is being done, is the devil."

And he thought about the devil, in whom he did not believe, and he looked round at the two windows where the fires were gleaming. It seemed to him that out of those crimson eyes the devil himself was looking at him – that unknown force that had created the mutual relation of the strong and the weak, that coarse blunder which one could never correct. The strong must hinder the weak from living – such was the law of Nature; but only in a newspaper article or in a school book was that intelligible and easily accepted. In the hotchpotch which was everyday life, in the tangle of trivialities out of which human relations were woven, it was no longer a law, but a logical absurdity, when the strong and the weak were both equally victims of their mutual relations, unwillingly submitting to some directing force, unknown, standing outside life, apart from man.

So thought Korolyov, sitting on the planks, and little by little he was possessed by a feeling that this unknown and mysterious force was really close by and looking at him. Meanwhile the east was growing paler, time passed rapidly; when there was not a soul anywhere near, as though everything were dead, the five buildings and their chimneys against the grey background of the dawn had a peculiar look – not the same as by day; one forgot altogether that inside there were steam motors, electricity, telephones, and kept thinking of lake-dwellings, of the Stone Age, feeling the presence of a crude, unconscious force ....

And again there came the sound: "Dair ... dair ... dair ... dair ..." twelve times. Then there was stillness, stillness for half a minute, and at the other end of the yard there rang out.

"Drin ... drin ... drin...."

"Horribly disagreeable," thought Korolyov.

"Zhuk ... zhuk ..." there resounded from a third place, abruptly, sharply, as though with annoyance—"Zhuk ... zhuk...."

And it took four minutes to strike twelve. Then there was a hush; and again it seemed as though everything were dead.

Korolyov sat a little longer, then went to the house, but sat up for a good while longer. In the adjoining rooms there was whispering, there was a sound of shuffling slippers and bare feet.

"Is she having another attack?" thought Korolyov.

He went out to have a look at the patient. By now it was quite light in the rooms, and a faint glimmer of sunlight, piercing through the morning mist, quivered on the floor and on the wall of the drawing-room. The door of Liza's room was open, and she was sitting in a low chair beside her bed, with her hair down, wearing a dressing-gown and wrapped in a shawl. The blinds were down on the windows.

"How do you feel?" asked Korolyov.

"Well, thank you."

He touched her pulse, then straightened her hair, that had fallen over her forehead.

"You are not asleep," he said. "It's beautiful weather outside. It's spring. The nightingales are singing, and you sit in the dark and think of something."

She listened and looked into his face; her eyes were sorrowful and intelligent, and it was evident she wanted to say something to him.

"Does this happen to you often?" he said.

She moved her lips, and answered:

"Often, I feel wretched almost every night."

At that moment the watchman in the yard began striking two o'clock. They heard: "Dair … dair …" and she shuddered.

"Do those knockings worry you?" he asked.

"I don't know. Everything here worries me," she answered, and pondered. "Everything worries me. I hear sympathy in your voice; it seemed to me as soon as I saw you that I could tell you all about it."

"Tell me, I beg you."

"I want to tell you of my opinion. It seems to me that I have no illness, but that I am weary and frightened, because it is bound to be so and cannot be otherwise. Even the healthiest person can't help being uneasy if, for instance, a robber is moving about under his window. I am constantly being doctored," she went on, looking at her knees, and she gave a shy smile. "I am very grateful, of course, and I do not deny that the treatment is a benefit; but I should like to talk, not with a doctor, but with some intimate friend who would understand me and would convince me that I was right or wrong."

"Have you no friends?" asked Korolyov.

"I am lonely. I have a mother; I love her, but, all the same, I am lonely. That's how it happens to be …. Lonely people read a great deal, but say little and hear little. Life for them is mysterious; they are mystics and often see the devil where he is not. Lermontov's Tamara was lonely and she saw the devil."

"Do you read a great deal?"

"Yes. You see, my whole time is free from morning till night. I read by day, and by night my head is empty; instead of thoughts there are shadows in it."

"Do you see anything at night?" asked Korolyov.

"No, but I feel…."

She smiled again, raised her eyes to the doctor, and looked at him so sorrowfully, so intelligently; and it seemed to him that she trusted him, and that she wanted to speak frankly to him, and that she thought the same as he did. But she was silent, perhaps waiting for him to speak.

And he knew what to say to her. It was clear to him that she needed as quickly as possible to give up the five buildings and the million if she had it—to leave that devil that looked out at night; it was clear to him, too, that she thought so herself, and was only waiting for some one she trusted to confirm her.

But he did not know how to say it. How? One is shy of asking men under sentence what they have been sentenced for; and in the same way it is awkward to ask very rich people what they want so much money for, why they make such a poor use of their wealth, why they don't give it up, even when they see in it their unhappiness; and if they begin a conversation about it themselves, it is usually embarrassing, awkward, and long.

"How is one to say it?" Korolyov wondered. "And is it necessary to speak?"

And he said what he meant in a roundabout way:

"You in the position of a factory owner and a wealthy heiress are dissatisfied; you don't believe in your right to it; and here now you can't sleep. That, of course, is better than if you were satisfied, slept soundly, and thought everything was satisfactory. Your sleeplessness does you credit; in any case, it is a good sign. In reality, such a conversation as this between us now would have been unthinkable for our parents. At night they did not talk, but slept sound; we, our generation, sleep badly, are restless, but talk a great deal, and are always trying to settle whether we are right or not. For our children or grandchildren that question – whether they are right or not – will have been settled. Things will be clearer for them than for us. Life will be good in fifty years' time; it's only a pity we shall not last out till then. It would be interesting to have a peep at it."

"What will our children and grandchildren do?" asked Liza.

"I don't know…. I suppose they will throw it all up and go away."

"Go where?"

"Where? … Why, where they like," said Korolyov; and he laughed. "There are lots of places a good, intelligent person can go to."

He glanced at his watch.

"The sun has risen, though," he said. "It is time you were asleep. Undress and sleep soundly. Very glad to have made your acquaintance," he went on, pressing her hand. "You are a good, interesting woman. Good-night!"

He went to his room and went to bed.

In the morning, when the carriage was brought round they all came out on to the steps to see him off. Liza, pale and exhausted, was in a white dress as though for a holiday, with a flower in her hair; she looked at him, as yesterday, sorrowfully and intelligently, smiled and talked, and all with an expression as though she wanted to tell him something special, important – him alone. They could hear the larks trilling and the church bells pealing. The windows in the factory buildings were sparkling gaily, and, driving across the yard and afterwards along the road to the station, Korolyov thought neither of the workpeople nor of lake dwellings, nor of the devil, but thought of the time, perhaps close at hand, when life would be as bright and joyous as that still Sunday morning; and he thought how pleasant it was on such a morning in the spring to drive with three horses in a good carriage, and to bask in the sunshine.

—translated by Constance Garnett

Empathy

"A Doctor's Visit" is a wonderful example of a twice-told story: it begins with the odd detail that after the Professor-Physician received a "long, incoherent telegram" from someone out in the country, he decided to send his assistant rather than go himself. Moreover, that assistant, Korolyov, has never been inside a factory, hardly knew the country, but he possessed a stereotype of factory-life so that "whenever he saw a factory far or near, he always thought how … within there was always sure to be impenetrable ignorance and dull egoism on the side of the owners, wearisome, unhealthy toil on the side of the workpeople, squabbling, vermin, vodka." Thus, from the very beginning, he encounters the stereotypes he brings to his experience: of Madame Lyalikov, "judging from her face [he took her to be] a simple uneducated woman"; of the governess, he found her "a stupid, middle-aged maiden lady in a pince-nez"; and on first meeting Liza, the invalid herself, he found her "ugly like her mother" and her ailment only warranting a shrug. But when she begins to sob, "the impression of a destitute, ugly creature vanished," and "she seemed to him altogether graceful, feminine, and simple; and he longed to sooth her, not with drugs, not with advice, but with simple, kindly words."

Even so, he remains bored and tells her "there's nothing seriously wrong." Still, against his own wishes, he allows himself to be persuaded to spend the night at this house so that his experience itself is "twice-told," and being there he finds the factory and factory hands "as comparable with the treatment of incurable illnesses," and even finds himself encountering "the devil, in whom [as a sophisticated Muscovite,] he did not believe." This second experience—of something he had no knowledge, as we, thank goodness, have no personal knowledge of chattel slavery—gives him insight into his patient Liza, and in his nighttime conversation with her, he feels the need to convey his understanding "in a roundabout way." The great question of this story is why is a "roundabout" conversation necessary? And to what degree do the details of a narrative—the felt presence of supernatural force in Chekhov's story, the striking image of pus-filled feet in Morrison's narrative—give rise to empathetic feelings, which themselves give rise to focused action?

Related Poem

For this chapter, the related poem written by Dr. John Stone, "He Makes a House Call," is not quite a twice-told story, but it does situate two different stories, one in which the physician was in charge "Six, seven years ago"—the doctor cannot clearly remember when—and another in which the patient is "in charge—of figs, beans, / tomatoes, life."

Poem: "He Makes a House Call" (1980) by Dr. John Stone

Author Note: Dr. John Stone (1936–2008) was a poet and physician (a cardiologist) and four times the writer of the year for the state of Georgia in the United States. He wrote six volumes of poetry and numerous essays. He was also the co-editor (with Richard Reynolds—and others in later editions) of *On Doctoring: Stories, Poems, Essays*, an anthology that, for many years, was given to every first-year medical student in the United States by the Johnson Foundation.

He Makes a House Call

Six, seven years ago
When you began to begin to faint
I painted your leg with iodine

threaded the artery
with the needle and then the tube
pumped your heart with dye enough

to see the valve
almost closed with stone.
We were both under pressure.

Today, in your garden,
kneeling under the sticky fig tree
for tomatoes

I keep remembering your blood.
Seven it was. I was just
beginning to learn the heart

inside out.
Afterward, your surgery
and the precise valve of steel

and plastic that still pops and clicks
inside like a ping-pong ball.
I should try

chewing tobacco sometimes
if only to see how it tastes.
There is a trace of it at the corner

of your leathery smile
which insists that I see inside
the house: someone named Bill I'm supposed

to know; the royal plastic soldier
whose body fills with whiskey
and marches on a music box

How Dry I Am;
the illuminated 3-D Christ who turns
into Mary from different angles;

the watery basement,
the pills you take, the ivy
that may grow around the ceiling

if it must. Here, you
are in charge – of figs, beans,
tomatoes, life.

At the hospital, a thousand times
I have heard your heart valve open, close.
I know how clumsy it is.

But health is whatever works
and for as long. I keep thinking
of seven years without a faint

on my way to the car
loaded with vegetables
I keep thinking of seven years ago

when you bled in my hands like a saint.

"He Makes a House Call" tells a story of a cardiologist, visiting his patient seven years after a heart operation. The patient welcomes him into the garden, the living room, and the basement, and, as part of the visit, she gives the doctor some vegetables. Meanwhile, the doctor keeps thinking of the operation those many years ago during his encounter with the patient who, here in her house, is the person in charge. Together, the visit and the memory teach the doctor a new definition of health—"whatever works / and for as long."

Dr. Stone himself has spoken movingly about this poem. "In the writing of that poem," he notes,

> I discovered at least two things about this encounter and about medicine in general. The first is a definition of health, which I can still defend and would gladly defend today. Health is whatever works and for as long. A utilitarian view, probably befits the internist. And the second one is an emphasis on the sacred relationship between doctor and patient, emphasized in the last line, "When you bled in my hands like a saint." The common dousing in the blood of the patient is a very important part of the practice of medicine. It's usually a metaphorical dousing in the blood of the patient, but an involvement in his or her life to a marked degree. It's the most privileged encounter in the professions, as privileged as the theologians among us. (Vannatta et al. 2005: Chap. 2, screen 18 [video])

What is striking about this poem is that it helps define the patient-provider relationship, where the authority does not solely reside in the healthcare provider but moves backward and forward between provider and patient. Moreover, it might be that such movement is closely related to Stone's pragmatic definition of health as measured in relation to the life-activities of the patient. A striking formal feature of this poem is its sole rhyme, in the last lines, "faint"/"saint."

Lessons for Providers
1. Vicarious experience gained through reading literary narrative has many potential pay-offs for providers of health care. Among them, demonstrated in these readings, is empathy for characters in the story which is transferable to patients.
2. This process of empathy development—practiced through close reading of literary narrative—provides a portal through which the provider can engage meaningfully and therefore "connect" with the patient.

Bibliography

Vannatta, Jerry, Ronald Schleifer, and Sheila Crow. 2005. *Medicine and Humanistic Understanding: The Significance of Narrative in Medical Practices.* Philadelphia: University of Pennsylvania Press. A DVD-Rom publication.

Listening to Patients

In their Introduction to a special issue of the journal *Literature and Medicine* focused on narrative, pain, and suffering, Dr. Rita Charon and Maura Spiegel argue that

> narratives that emerge from suffering differ from those born elsewhere …. Not restricted to the linear, the orderly, the emplotted, or the clean, these narratives that come from the ill contain unruly fragments, silences, bodily processes rendered in code. The language is deputized to point to things not ordinarily admitted into prose or poetry or text of other kinds – shameful, painful, prelingual limitations, absences, breath-taking fears. (2005: vi)

Understanding and engaging with such disorienting stories requires special attention to time sequences, and such understanding often entails the reordering the events and the recreating of the story in the interlocutor's mind in order to clearly understand the concern, plots, and consequences of the actions of the characters. That is, in many patient narratives—as often in literary narratives—there is an element of the "unsaid" that can be discerned and acted upon. Patient stories are almost always told in retrospect, given that the symptoms have already been experienced before the patient tells his or her story to the doctor. Often, as Charon and Spiegel note, patients do not present their stories chronologically—they often get the sequence of events wrong—because of faulty memory, heightened emotional state at the time of the telling, anxiety or pain, or for other reasons. It is the task of the listener to question the patient, to review carefully what he/she has heard, and to ensure that the message sent was the message received. This task requires careful listening that encompasses anticipation that the narrator may make mistakes, may misremember the details, and on occasion, the narrator himself is disoriented in time. One important aspect of engaging with literary texts in a systematic manner is the way that such engagement trains readers/listeners to discern what is unsaid and the implications of narrative to the end of recovering a "whole" narrative from a fragmented story.

Young Mother with Abdominal Pain: A Vignette
(Excerpt from *The Chief Concern of Medicine*)

Here is a description of a clinical encounter with a patient, who was notably unresponsive, complaining about vague abdominal pain. In the face of her unresponsiveness, the attending physician (narrating this vignette) asks her about her family.

> A twenty-year-old mother of a three-month-old baby was admitted to the hospital through the emergency room at eleven o'clock at night. The admitting resident, intern, and medical student were tired and not too empathic this late at night. The woman's husband left her and took the baby before the intern completed her evaluation and was therefore not available for an interview.
>
> The following morning, at check-in rounds as attending physician, I met the patient for the first time. The intern formally presented her case history in the following manner:
>
> *This young lady was brought to the emergency room for abdominal pain last night. The pain is described as "all over" and diffuse. She is completely unable to give any more detail about the pain. She has never had this pain before. She has had no surgeries, and denies vomiting, diarrhea, and constipation. She has had no fever, takes no medications, and denies being pregnant. She is vague about her family history, but it does not sound as if the pain is in any way familial. This lady and her husband are transients and in town looking for work. They have little money and sound as if they live from paycheck to paycheck. Her examination is totally normal except for some very poorly localized abdominal tenderness. The abdominal X-Rays are normal as is the admitting laboratory examination. She has no evidence of infection, including in her bladder. Our assessment is that this lady is a "crock," and for some reason would rather be in the hospital than at home. We do, however, want to rule out cancer by doing a barium enema, and a CT scan of her abdomen. If those tests are normal, we want to discharge her to home.*
>
> Upon entering the room I noticed a young woman with straight, oily hair, who buried her face in the pillow even when spoken to. The intern introduced me, but the young woman did not look up. She turned over when addressed and responded to a few questions. She allowed me to examine her abdomen. However, she always looked down when answering the questions and never smiled or engaged in a meaningful conversation. She spoke with a very quiet voice, and appeared full of shame. The history was as the intern had presented, as was the examination of her abdomen. In fact, her history was unremarkable and her examination essentially normal. She was in the hospital because her husband had left her in the emergency room and there was no phone or any other way to contact him.
>
> The intern and resident wanted to discharge her as soon as they performed a diagnostic test called a barium enema, which is a radiological procedure, and an expensive CT scan of her abdomen. Because there were no good indications to do these procedures, I declined to allow them to be done. I instead instructed them to interview the husband when he came to visit to see if any additional information could be obtained. The next day the resident reported that the husband had not come to the hospital all day, and that she did not know where he was. The pain had persisted, and the examination had remained the same. They again wanted to discharge her but there was nowhere for her to go. I again would not allow them to do expensive and time consuming tests for which there was no indication. Late in the evening on the second hospital day I went to her room to interview her in private. Again I found a young woman lying in the dark, with her face toward the wall. She finally turned over and began to talk to me. At first she was very shy and quiet but eventually she began to open up and talk more freely.
>
> The story she told me of her abdominal pain fit no diagnostic category, just as the intern had reported. Because I was getting nowhere pursuing a biomedical diagnosis, I changed direction, and explored the psychosocial aspect of her history. I said: "Tell me about your

family. Where did you grow up?" I heard a story about her mother, her sister and two brothers in North Carolina. She shared the story of her leaving town after high school graduation in search of adventure. After losing several jobs, she found herself in a homeless shelter in Indianapolis, where she met her future husband. She described how he convinced her to jump a freight train with him and go find work, and soon they found themselves in Oklahoma City, still out of work, hungry, and caring for a three-month-old baby. At the pause in her story, I said, "You told me of your family, but you didn't mention your father." She looked down, frowned and stated in a barely audible voice, "He wasn't very nice to me." After a long silence, she continued. "He hurt me and was dirty with me."
"Did he hurt you physically?"
"Yes, with a belt, many times."
"Did he sexually hurt you too?"
Looking down and in a very quiet voice, she replied, "Yeah."
"I am sorry that happened to you; I can imagine that it is very difficult to live with."
"Yes, but I think things are better now."
I waited for a few moments. "Tell me about your husband. Does he hurt you?"
At first she denied that her husband was anything but perfect, but after some facilitation, she related that he was emotionally abusive, and that she was afraid that soon he was going to begin physically abusing her. She related that every night when he was preparing for bed, he would take off his leather belt and fold it in two. She would shiver with fright, remembering her father's belt and its hateful pain. (For a fuller account, see *The Chief Concern:* 9–14)

In this narrative, the intern solicits—or at least focuses on her retelling—only the "facts" of the case, only what is *said*, without attending to the patient's story, which combines the said and the unsaid. It is probable that the intern faced an encounter that took the form of monosyllabic responses to the questions she asked the patient, probably spoken as quietly and shamefully as the attending physician reports in his first interview with this patient. But when a narrative is solicited, the patient presents her doctors with a not-yet-completed story: it is "fragmented" insofar as it doesn't possess a complete set of characters found in narrative (in this case, it is missing an antagonist, or what we describe in Appendix 5 as the hero's "opponent"; see Section B, "Roles of Narrative," pp. 266–267). That is, in this narrative, what is missing—what is *unsaid*—is a particular "role" or character-actor that is also part of an understanding of narrative that engagements with literary narratives trains one to notice. Narratives present a small number of general roles, and when one of them is missing or unarticulated, an understanding of what they are allows us to infer and suggest how they might be filled—what "character" might fill this role—and even allows us to infer *why* they were missing. Appendix 5 sets forth the parallel between general narrative roles and parts of speech in a sentence in order to allow readers and listeners one strategy to *notice* what is *not there* in a particular narrative.

The phenomenon of "foregrounding" in literature examined and tested in empirical studies, as we mentioned in Chap. 1, includes intentional "stylistic variations," and reading literary texts calls upon readers to become attentive to these features and their implications in understanding stories. (The fragmentation, silence, encrypted codes in patient stories that Charon and Spiegel describe are everyday versions of such "stylistic variations.") In the vignette describing the unresponsive patient, when the physician asks about her father, he discovers a character-actor which was missing from her story, namely the "opponent" who, in narrative, always struggles with the hero. In this clinical narrative, the opponent turns out to be the patient's father, and by getting his patient to "complete" her story, the physician is able to discover a "fact" in his patient's story beyond and probably underlying the biomedical fact of her complaint. Awareness of the elements or features of narrative, the roles that inhabit narrative, the implications of "encrypted" meaning, the fact that silence—like the body language of the patient in the Chap. 2 vignette, "The Woman with Hyponatremia"—can allow healthcare professionals to more fully engage with, learn from, and empathetically respond to their patients. Moreover, the physician is able to "facilitate" his patient's further story about her husband—again, a "twice-told story" in which she finds patterns of repetition in her experience with her father and her husband—which allows the physician to discern an immediate context for her seeking "asylum" in the hospital.

One final note: as we suggest in discussing "The Red Wheelbarrow" in this chapter and as we note more fully in Chap. 2 in the discussion of "The Logic of Making a Diagnosis," the physician's "educated guess" about the role of his patient's father is *provisional* and could be mistaken. There are multiple possible reasons for the absence of the father in the patient's story: her parents were divorced, he travels a lot and was relatively absent in his children's lives, he didn't exhibit the warmth and closeness of her other family members. But a focus on what is unsaid in narrative—whether a patient's or a literary author's—creates a form of attention that promotes engagement and discernment that enhance understanding, emotional identification, and action.

Literary Narrative: "Araby" (1914) by James Joyce

Author Note: James Joyce (1882–1941) was an Irish novelist and short story writer. He was a major writer in English in the twentieth century: his novel *Ulysses*, published in 1922, is considered one of the great masterpieces in English in the twentieth century. Its style, which is considered difficult and rewarding, helped shape the vision of many "modernist" literary texts. His collection of short stories, *Dubliners*—from which "Araby" is taken—appeared in 1914 and depicts the lives of lower middle-class Dubliners under the rule of late British colonialism. These stories are noted for their distinctive and understated style—Joyce himself described it as a style of "scrupulous meanness," which places great emphasis on the "unsaid."

Araby

North Richmond Street, being blind, was a quiet street except at the hour when the Christian Brothers' School set the boys free. An uninhabited house of two storeys stood at the blind end, detached from its neighbours in a square ground. The other houses of the street, conscious of decent lives within them, gazed at one another with brown imperturbable faces.

The former tenant of our house, a priest, had died in the back drawing-room. Air, musty from having been long enclosed, hung in all the rooms, and the waste room behind the kitchen was littered with old useless papers. Among these I found a few paper-covered books, the pages of which were curled and damp: *The Abbot*, by Walter Scott, *The Devout Communicant* and *The Memoirs of Vidocq*. I liked the last best because its leaves were yellow. The wild garden behind the house contained a central apple-tree and a few straggling bushes under one of which I found the late tenant's rusty bicycle-pump. He had been a very charitable priest; in his will he had left all his money to institutions and the furniture of his house to his sister.

When the short days of winter came dusk fell before we had well eaten our dinners. When we met in the street the houses had grown sombre. The space of sky above us was the colour of ever-changing violet and towards it the lamps of the street lifted their feeble lanterns. The cold air stung us and we played till our bodies glowed. Our shouts echoed in the silent street. The career of our play brought us through the dark muddy lanes behind the houses where we ran the gauntlet of the rough tribes from the cottages, to the back doors of the dark dripping gardens where odours arose from the ashpits, to the dark odorous stables where a coachman smoothed and combed the horse or shook music from the buckled harness. When we returned to the street light from the kitchen windows had filled the areas. If my uncle was seen turning the corner we hid in the shadow until we had seen him safely housed. Or if Mangan's sister came out on the doorstep to call her brother in to his tea we watched her from our shadow peer up and down the street. We waited to see whether she would remain or go in and, if she remained, we left our shadow and walked up to Mangan's steps resignedly. She was waiting for us, her figure defined by the light from the half-opened door. Her brother always teased her before he obeyed and I stood by the railings looking at her. Her dress swung as she moved her body and the soft rope of her hair tossed from side to side.

Every morning I lay on the floor in the front parlour watching her door. The blind was pulled down to within an inch of the sash so that I could not be seen. When she came out on the doorstep my heart leaped. I ran to the hall, seized my books and followed her. I kept her brown figure always in my eye and, when we came near the point at which our ways diverged, I quickened my pace and passed her. This happened morning after morning. I had never spoken to her, except for a few casual words, and yet her name was like a summons to all my foolish blood.

Her image accompanied me even in places the most hostile to romance. On Saturday evenings when my aunt went marketing I had to go to carry some of the parcels. We walked through the flaring streets, jostled by drunken men and bargaining women, amid the curses of labourers, the shrill litanies of shop-boys who stood on guard by the barrels of pigs' cheeks, the nasal chanting of street-singers, who sang a *come-all-you* about O'Donovan Rossa, or a ballad about the troubles in our native land. These noises converged in a single sensation of life for me: I imagined that I bore my chalice safely through a throng of foes. Her name sprang to my lips at moments in strange prayers and praises which I myself did not understand. My eyes were often full of tears (I could not tell why) and at times a flood from my heart seemed to pour itself out into my bosom. I thought little of the future. I did not know whether I would ever speak to her or not or, if I spoke to her, how I could tell her of my confused adoration. But my body was like a harp and her words and gestures were like fingers running upon the wires.

One evening I went into the back drawing-room in which the priest had died. It was a dark rainy evening and there was no sound in the house. Through one of the broken panes I heard the rain impinge upon the earth, the fine incessant needles of water playing in the sodden beds. Some distant lamp or lighted window gleamed below me. I was thankful that I could see so little. All my senses seemed to desire to veil themselves and, feeling that I was about to slip from them, I pressed the palms of my hands together until they trembled, murmuring: "O love! O love!" many times.

At last she spoke to me. When she addressed the first words to me I was so confused that I did not know what to answer. She asked me was I going to *Araby*. I forgot whether I answered yes or no. It would be a splendid bazaar, she said; she would love to go.

"And why can't you?" I asked.

While she spoke she turned a silver bracelet round and round her wrist. She could not go, she said, because there would be a retreat that week in her convent. Her brother and two other boys were fighting for their caps and I was alone at the railings. She held one of the spikes, bowing her head towards me. The light from the lamp opposite our door caught the white curve of her neck, lit up her hair that rested there and, falling, lit up the hand upon the railing. It fell over one side of her dress and caught the white border of a petticoat, just visible as she stood at ease.

"It's well for you," she said.

"If I go," I said, "I will bring you something."

What innumerable follies laid waste my waking and sleeping thoughts after that evening! I wished to annihilate the tedious intervening days. I chafed against the work of school. At night in my bedroom and by day in the classroom her image came between me and the page I strove to read. The syllables of the word *Araby* were called to me through the silence in which my soul luxuriated and cast an Eastern enchantment over me. I asked for leave to go to the bazaar on Saturday night. My aunt was surprised and hoped it was not some Freemason affair. I answered few questions in class. I watched my master's face pass from amiability to sternness; he hoped I was not beginning to idle. I could not call my wandering thoughts together. I had hardly any patience with the serious work of life which, now that it stood between me and my desire, seemed to me child's play, ugly monotonous child's play.

On Saturday morning I reminded my uncle that I wished to go to the bazaar in the evening. He was fussing at the hallstand, looking for the hat-brush, and answered me curtly:

"Yes, boy, I know."

As he was in the hall I could not go into the front parlour and lie at the window. I left the house in bad humour and walked slowly towards the school. The air was pitilessly raw and already my heart misgave me.

When I came home to dinner my uncle had not yet been home. Still it was early. I sat staring at the clock for some time and, when its ticking began to irritate me, I left the room. I mounted the staircase and gained the upper part of the house. The high cold empty gloomy rooms liberated me and I went from room to room singing. From the front window I saw my companions playing below in the street. Their cries reached me weakened and indistinct and, leaning my forehead against the cool glass, I looked over at the dark house where she lived. I may have stood there for an hour, seeing nothing but the brown-clad figure cast by my imagination, touched discreetly by the lamplight at the curved neck, at the hand upon the railings and at the border below the dress.

When I came downstairs again I found Mrs Mercer sitting at the fire. She was an old garrulous woman, a pawnbroker's widow, who collected used stamps for some pious purpose. I had to endure the gossip of the tea-table. The meal was prolonged beyond an hour and still my uncle did not come. Mrs Mercer stood up to go: she was sorry she couldn't wait any longer, but it was after eight o'clock and she did not like to be out late as the night air was bad for her. When she had gone I began to walk up and down the room, clenching my fists. My aunt said:

"I'm afraid you may put off your bazaar for this night of Our Lord."

At nine o'clock I heard my uncle's latchkey in the halldoor. I heard him talking to himself and heard the hallstand rocking when it had received the weight of his overcoat. I could interpret these signs. When he was midway through his dinner I asked him to give me the money to go to the bazaar. He had forgotten.

"The people are in bed and after their first sleep now," he said.

I did not smile. My aunt said to him energetically:

"Can't you give him the money and let him go? You've kept him late enough as it is."

My uncle said he was very sorry he had forgotten. He said he believed in the old saying: "All work and no play makes Jack a dull boy." He asked me where I was going and, when I had told him a second time he asked me did I know *The Arab's Farewell to his Steed*. When I left the kitchen he was about to recite the opening lines of the piece to my aunt.

I held a florin tightly in my hand as I strode down Buckingham Street towards the station. The sight of the streets thronged with buyers and glaring with gas recalled to me the purpose of my journey. I took my seat in a third-class carriage of a deserted train. After an intolerable delay the train moved out of the station slowly. It crept onward among ruinous houses and over the twinkling river. At Westland Row Station a crowd of people pressed to the carriage doors; but the porters moved them back, saying that it was a special train for the bazaar. I remained alone in the bare carriage. In a few minutes the train drew up beside an improvised wooden platform. I passed out on to the road and saw by the lighted dial of a clock that it was ten minutes to ten. In front of me was a large building which displayed the magical name.

I could not find any sixpenny entrance and, fearing that the bazaar would be closed, I passed in quickly through a turnstile, handing a shilling to a weary-looking man. I found myself in a big hall girdled at half its height by a gallery. Nearly all the stalls were closed and the greater part of the hall was in darkness. I recognised a silence like that which pervades a church after a service. I walked into the centre of the bazaar timidly. A few people were gathered about the stalls which were still open. Before a curtain, over which the words *Café Chantant* were written in coloured lamps, two men were counting money on a salver. I listened to the fall of the coins.

Remembering with difficulty why I had come I went over to one of the stalls and examined porcelain vases and flowered tea-sets. At the door of the stall a young lady was talking and laughing with two young gentlemen. I remarked their English accents and listened vaguely to their conversation.

"O, I never said such a thing!"

"O, but you did!"

"O, but I didn't!"

"Didn't she say that?"

"Yes. I heard her."

"O, there's a ... fib!"

Observing me the young lady came over and asked me did I wish to buy anything. The tone of her voice was not encouraging; she seemed to have spoken to me out of a sense of duty. I looked humbly at the great jars that stood like eastern guards at either side of the dark entrance to the stall and murmured:

"No, thank you."

The young lady changed the position of one of the vases and went back to the two young men. They began to talk of the same subject. Once or twice the young lady glanced at me over her shoulder.

I lingered before her stall, though I knew my stay was useless, to make my interest in her wares seem the more real. Then I turned away slowly and walked down the middle of the bazaar. I allowed the two pennies to fall against the sixpence in my pocket. I heard a voice call from one end of the gallery that the light was out. The upper part of the hall was now completely dark.

Gazing up into the darkness I saw myself as a creature driven and derided by vanity; and my eyes burned with anguish and anger.

The Unsaid

We can ask of James Joyce's story the same *kind* of questions that healthcare professionals can bring to the patient. That is, we can inquire about the social and familiar situation by asking: who is the small boy in this story? How old is he? What emotions does he exhibit in what he says and in what is said of him? We can even ask, to use the language of the medical interview, what is his chief complaint? And we can ask, as we note in Part II of the Introduction, what is his chief concern. Like so many of Joyce's narratives—and like so many patient narratives—this story offers no information beyond the experience of the young protagonist. That is, the story offers events without reflective commentary that makes explicit what is going on in a vocabulary outside the events themselves. The job of understanding the story entails understanding the unspoken context for its events.

In fact, none of the background of the story is explicitly stated (the boy does not have to think about it), and it is the job of the listener to figure out and piece together a narrative context from the small details. Why does the boy describe his feelings as carrying his "chalice safely through a throng of foes"? Why does he describe them as "confused adoration"? What are his feelings anyway? Why does the story end the way it does? What is the "overall meaning" of this story? That is, why would Joyce tell it? What ends are served by its "cryptic" language? In other words, can a reader retell this story to account for what is not said in it and to figure out what is going on. Even the minor detail of the interactions of the clerks at the bazaar calls for an explanation not supplied by the story.

In interesting ways, this story could be paired with Ernest Hemingway's famous (and widely available) story "Indian Camp," which depicts a crude cesarean birth, performed without anesthetics, through the eyes of a young narrator. Hemingway, like Joyce, is famous for not telling his readers facts his characters do not have to talk about. Appendix 6 describes our experience in teaching medical students to be attentive to the "unsaid" in narratives—literary narratives and patient narratives—by engaging with Hemingway's story.

Related Poem

For this chapter, the related poem, written by Dr. William Carlos Williams, is quite famous for its own cryptic presentation of a scene that requires its readers to comprehend what is not said.

Poem: "The Red Wheelbarrow" (1923) by Dr. William Carlos Williams

> *Author Note*: William Carlos Williams (1883–1963) was a physician and poet, short-story writer, and essayist whose literary work—especially his poetry—was closely associated with modernism in the early twentieth century. He was born in New Jersey to an English father and a Puerto Rican mother, and studied in Europe as well as in the United States. He delivered more than 3000 babies in an ethnically mixed neighborhood in New Jersey and won the Bollingen Prize, the National Book Award, and the Pulitzer Prize (posthumously).

The Red Wheelbarrow

so much depends
upon

a red wheel
barrow

glazed with rain
water

beside the white
chickens

Williams' famous poem "The Red Wheelbarrow" is useful for physicians because, although it does not present an explicit narrative, still it can help us learn to recognize and recover narrative knowledge. As Charon and Spiegel note, quite often, the meaning of the patient's story is not apparent. The story presents itself as a series of disparate facts, emotions, anecdotes that suggest without explicitly articulating the patient's condition and the agenda motivating the visit, as well as the emotion he brings to the encounter with his physician. In a poem like "The Red Wheelbarrow" the elements of narrative and significance—some of the features of narrative—need to be gathered together to make it meaningful.

Moreover, this poem suggests that even a short engagement with the interpretation of poetry is particularly useful in developing the competence of healthcare professionals in recovering the information and meaning of a patient's story. Thus, in this famous poem, Williams presents a single sentence that simply observes details in the environment and *asserts value* in the phrase "so much depends." This arouses several questions: what is the antecedent of "much"? Why does Williams split up compound words (e.g., "wheelbarrow") with his poetic lines? What is the importance, if any, of the single use of figurative language (i.e., "glazed")? The same questions we ask of the cryptic language of Williams' poem can be asked of the patient's story. In fact, Dr. Verghese asks (or thinks about asking) his patient in the vignette to Chap. 1: how do the details of appearance add up? What does Gordon's manner of speaking suggest? What is the overall meaning of how he presents himself?

Lessons for Providers

1. Listening to patients is the most important skill to be developed by healthcare providers.
2. Attending to the patient's "chief concern" while simultaneously attending to the "chief complaint" will lead to better diagnosis, better patient satisfaction, and less provider burnout.

Bibliography

Charon, Rita, and Maura Spiegel. 2005. On Conveying Pain/On Conferring Form. *Literature and Medicine* 24: vi–ix.

Schleifer, Ronald, and Jerry Vannatta. 2013. *The Chief Concern of Medicine: The Integration of the Medical Humanities and Narrative Knowledge into Medical Practices*. Ann Arbor: University of Michigan Press.

The Patient

6

So far in Section IV, we have explored the healthcare provider's feelings and actions in relation to patients—rapport and empathy—and strategies for the healthcare provider to more carefully listen to and engage with the patient. In this chapter, we examine the qualities of the *position* of being a patient, the *role* of a patient as the culmination of a nexus of social interactions, with healthcare providers, family, and simply the understanding of illness within a cultural setting. Of course, this role is represented throughout *Literature and Medicine*: the relationship between Grace Paley and her father in our first chapter is a relationship that is struggling to understand the nature of her father's "patient-hood"; and Tolstoy's novella in our last chapter is an extended narrative of Ivan Ilych's experience as a patient. In between—in Arthur Conan Doyle's resident patient, in Richard Selzer's "Imelda," in Chekhov, Melville, even Flaubert—men, women, and children find themselves in the *position* of patient, a role conferred by illness or some kind of somatic or psychic dysfunction. The texts in the present chapter, however, focus on the aspects of the role of a patient: the vignette describing a woman's response to breast cancer as both a personal and social experience, which is very different from the illness experienced from the point of view of the physician in Chaps. 4 and 5; the short story by Charlotte Perkins Gilman focuses on the descent into illness and madness of a woman who is desperately alone; and the poem by Rafael Campo understands the role of patient as encompassing family beyond the ill person himself.

In other words, these narrative and literary texts allow us to comprehend the manner in which the seeming simple and "self-evident" nature of being a patient—being sick and seeking help from others—is, in fact, a *complex* social situation, a "role" with aspects and functions that can be isolated, grasped, and understood. Such understanding can allow people training to work in healthcare or already pursuing a career that is committed to engaging with people who call upon them—suffering, fearful, and seeking aid and solace—to more fully understand that the care in healthcare necessarily goes beyond strict biomedical strategies. Suffering from cancer and amputation, Audre Lorde suggests—and, in counterpoint, Anatole

Broyard repeatedly notes—that there is something primitive, demonic, and seemingly supernatural in serious illness (Broyard 1992), and she hopes to discover that in the pain, despair, rage, and sadness of terrible illness, she might also find life and "whatever strength [that] can lie at the core of this experience."

Introduction to *The Cancer Journals*: A Vignette (Excerpt from Audre Lorde, *The Cancer Journals* [1980])

Author Note: Audre Lorde (1934–1992) was a writer, poet, feminist, and civil rights activist from New York City. In 1973 she was nominated for the National Book Award for poetry for her volume *From a Land Where Other People Live*; and *Coal*, published by Norton in 1976, helped establish her as an influential voice in the Black Arts Movement. Her autobiographical meditation on her breast cancer and mastectomy, *The Cancer Journals*, was published in 1980. She died in 1992 of liver cancer.

Introduction

1

Each woman responds to the crisis that breast cancer brings to her life out of a whole pattern, which is the design of who she is and how her life has been lived. The weave of her every day existence is the training ground for how she handles crisis. Some women obscure their painful feelings surrounding mastectomy with a blanket of business-as-usual, thus keeping those feelings forever under cover, but expressed elsewhere. For some women, in a valiant effort not to be seen as merely victims, this means an insistence that no such feelings exist and that nothing much has occurred. For some women it means the warrior's painstaking examination of yet another weapon, unwanted but useful.

I am a post-mastectomy woman who believes our feelings need voice in order to be recognized, respected, and of use.

I do not wish my anger and pain and fear about cancer to fossilize into yet another silence, nor to rob me of whatever strength can lie at the core of this experience, openly acknowledged and examined. For other women of all ages, colors, and sexual identities who recognize that imposed silence about any area of our lives is a tool for separation and powerlessness, and for myself, I have tried to voice some of my feelings and thoughts about the travesty of prosthesis, the pain of amputation, the function of cancer in a profit economy, my confrontation with mortality, the strength of women loving, and the power and rewards of self-conscious living.

Breast cancer and mastectomy are not unique experiences, but ones shared by thousands of american women. Each of these women has a particular voice to be raised in what must become a female outcry against all preventable cancers, as well as against the secret fears that allow those cancers to flourish. May these words serve as encouragement for other women to speak and to act out of our experiences with cancer and with other threats of death, for silence has never brought us anything of worth. Most of all, may these words underline the possibilities of self-healing and the richness of living for all women.

There is a commonality of isolation and painful reassessment which is shared by all women with breast cancer, whether this commonality is recognized or not. It is not my intention to judge the woman who has chosen the path of prosthesis, of silence and invisibility, the woman who wishes to be 'the same as before.' She has survived on another kind of courage, and she is not alone. Each of us struggles daily with the pressure of conformity and the loneliness of difference from which those choices seem to offer escape. I only know that those choices do not work for me, nor for other woman who, not without fear, have survived cancer by scrutinizing its meaning our lives, and by attempting to integrate this crises into useful strengths for change.

2

These selected journal entries, which begin 6 months after my modified radical mastectomy for breast cancer and extend beyond the completion of the essays in this book, exemplify the process of integrating this crisis into my life.

January 26, 1979
I'm not feeling very hopeful these days, about selfhood or anything else. I handle the outward motions of each day while pain fills me like a puspocket and every touch threatens to breech the taut membrane that keeps it from flowing through and poisoning my whole existence. Sometimes despair sweeps across my consciousness like luna winds across a barren moonscape. Ironshod horses rage back and forth over every nerve. Oh Seboulisa ma, help me remember what I have paid so much to learn. I could die of difference, or live – myriad selves.

February 5, 1979
The terrible thing is that nothing goes past me these days, nothing. Each horror remains like a steel vise in my flesh, another magnet to the flame. Buster has joined the rolecall of useless wasteful deaths of young Black people; in the gallery today everywhere ugly images of women offering up distorted bodies for whatever fantasy passes in the name of male art. Gargoyles of pleasure. Beautiful laughing Buster, shot down in a hallway for ninety cents. Shall I unlearn that tongue in which my curse is written?

March 1, 1979
It is such an effort to find decent food in this place, not to just give up and eat the old poison. But I must tend my body with at least as much care as I tend the compost, particularly now when it seems so beside the point. Is this pain and despair that surround me a result of cancer, or has it just been released by cancer? I feel so unequal to what I always handled before, the abominations outside that echo the pain within. And yes I am completely self-referenced right now because it is the only translation I can trust, and I do believe not until every woman traces her weave back strand by bloody self-referenced strand, will we begin to alter the whole pattern.

April 16, 1979
The enormity of our task, to turn the world around. It feels like turning my life around, inside out. If I can look directly at my life and my death without flinching I know there is nothing they can ever do to me again. I must be content to see how really little I can do and still do it with an open heart. I can never accept this, like I can't accept that turning my life around is so hard, eating differently, sleeping differently, moving differently, being differently. Like Martha said, I want the old me, bad as before.

April 22, 1979
I must let this pain flow through me and pass on. If I resist or try to stop it, it will detonate inside me, shatter me, splatter my pieces against every wall and person that I touch

May 1, 1979
Spring comes, and still I feel despair like a pale cloud waiting to consume me, engulf me like another cancer, swallow me into immobility, metabolize me into cells of itself; my body, a barometer. I need to remind myself of the joy, the lightness, the laughter so vital to my living and my health. Otherwise, the other will always be waiting to eat me up into despair again. And that means destruction. I don't know how, but it does.

September, 1979
There is no room around me in which to be still, to examine and explore what pain is mine alone – no device to separate my struggle within from my fury at the outside world's viciousness, the stupid brutal lack of consciousness or concern that passes for the way things are. The arrogant blindness of comfortable white women. What is this work all for? What does it matter whether I ever speak again or not? I try. The blood of black women sloshes from coast to coast and Daly says race is of no concern to women. So that means we are either immortal or born to die and no note taken, un-women.

October 3, 1979
I don't feel like being strong, but do I have a choice? It hurts when even my sisters look at me in the street with cold and silent eyes. I am defined as other in every group I'm a part of. The outsider, both strength and weakness. Yet without community there is certainly no liberation, no future, only the most vulnerable and temporary, armistice between me and my oppression.

November 19, 1979
I want to write rage but all that comes is sadness. We have been sad long enough to make this earth either weep or grow fertile. I am an anachronism, a sport, like the bee that was never meant to fly. Science said so. I am not supposed to exist. I carry death around in my body like a condemnation. But I do live. The bee flies. There must be some way to integrate death into living, neither ignoring it nor giving in to it.

January 1, 1980
Faith is the last day of Kwanza, and the name of the war against despair, the battle I fight daily. I become better at it. I want to write about that battle, the skirmishes, the losses, the small yet so important victories that make the sweetness of my life.

January 20, 1980
The novel is finished at last. It has been a lifeline. I do not have to win in order to know my dreams are valid, I only have to believe in a process of which I am a part. My work kept me alive this past year, my work and the love of women. They are inseparable from each other. In the recognition of the existence of love lies the answer to despair. Work is that recognition given voice and name.

February 18, 1980
I am 46 years living today and very pleased to be alive, very glad and very happy. Fear and pain and despair do not disappear. They only become slowly less and less important. Although sometimes I still long for a simple orderly life with a hunger sharp as that sudden vegetarian hunger for meat.

April 6, 1980
Somedays, if bitterness were a whetstone, I could be sharp as grief.

May 30, 1980
Last spring was another piece of the fall and winter before, a progression from all the pain and sadness of that time, ruminated over. But somehow this summer which is almost upon me feels like a part of my future. Like a brand new time, and I'm pleased to know it, wherever it leads. I feel like another woman, de-chrysalised and become a broader, stretched-out me, strong and excited, a muscle flexed and honed for action.

June 20, 1980
I do not forget cancer for very long, ever. That keeps me armed and on my toes, but also with a slight background noise of fear. Carl Simonton's book, Getting Well Again, *has been really helpful to me, even though his smugness infuriates me sometimes. The visualizations and deep relaxing techniques that I learned from it help make me a less anxious person, which seems strange, because in other ways, I live with the constant fear of recurrence of another cancer. But fear and anxiety are not the same at all. One is an appropriate response*

to a real situation which I can accept and learn to work through just as I work through semi-blindness. But the other, anxiety, is an immobilizing yield to things that go bump in the night, a surrender to namelessness, formlessness, voicelessness, and silence.

July 10, 1980
I dreamt I had begun training to change my life, with a teacher who is very shadowy. I was not attending classes, but I was going to learn how to change my whole life, live differently, do everything in a new and different way. I didn't really understand, but I trusted this shadowy teacher. Another young woman who was there told me she was taking a course in 'language crazure,' the opposite of discrazure (the cracking and wearing away of rock). I thought it would be very exciting to study the formation and crack and composure of words, so I told my teacher I wanted to take that course. My teacher said okay, but it wasn't going to help me any because I had to learn something else, and I wouldn't get anything new from that class. I replied maybe not, but even though I knew all about rocks, for instance, I still liked studying their composition, and giving a name to the different ingredients of which they were made. It's very exciting to think of me being all the people in this dream.

3

I have learned much in the 18 months since my mastectomy. My visions of a future I can create have been honed by the lessons of my limitations. Now I wish to give form with honesty and precision to the pain faith labor and loving which this period of my life has translated into strength for me.

Sometimes fear stalks me like another malignancy, sapping energy and power and attention from my work. A cold becomes sinister; a cough, lung cancer; a bruise, leukemia. Those fears are most powerful when they are not given voice, and close upon their heels come the fury that I cannot shake them. I am learning to live beyond fear by living through it, and in the process learning to turn fury at my own limitations into some more creative energy. I realize that if I wait until I am no longer afraid to act, write, speak, be, I'll be sending messages on a ouija board, cryptic complaints from the other side. When I dare to be powerful, to use my strength in the service of my vision, then it becomes less important whether or not I am unafraid.

As women we were raised to fear. If I cannot banish fear completely, I can learn to count with it less. For then fear becomes not a tyrant against which I waste my energy fighting, but a companion, not particularly desirable, yet one whose knowledge can be useful.

I write so much here about fear because in shaping this introduction to *The Cancer Journals*, I found fear laid across my hands like a steel bar. When I tried to reexamine the 18 months since my mastectomy, some of what I touched was molten despair and waves of mourning – for my lost breast, for time, for the luxury of false power. Not only were these emotions difficult and painful to relive, but they were entwined with the terror that if I opened myself once again to scrutiny, to feeling the pain of loss, of despair, of victories too minor in my eyes to rejoice over, then I might also open myself again to disease. I had to remind myself that I had lived through it all, already. I had known the pain, and survived it. It only remained for me to give it voice, to share it for use, that the pain not be wasted.

Living a self-conscious life, under the pressure of time, I work with the consciousness of death at my shoulder, not constantly, but often enough to leave a mark upon all of my life's decisions and actions. And it does not matter whether this death comes next week or thirty years from now; this consciousness gives my life another breadth. It helps shape the words I speak, the ways I love, my politic of action, the strength of my vision and purpose, the depth of my appreciation of living.

I would lie if I did not also speak of loss. Any amputation is a physical and psychic reality that must be integrated into a new sense of self. The absence of my breast is a recurrent sadness, but certainly not one that dominates my life. I miss it, sometimes piercingly. When other one-breasted women hide behind the mask of prosthesis or the dangerous fantasy of reconstruction, I find little support in the broader female environment for my rejection of what feels like a cosmetic sham. But I believe that socially sanctioned prosthesis is merely another way of keeping women with breast cancer silent and separate from each other. For instance, what would happen if an army of one-breasted women descended upon Congress and demanded that the use of carcinogenic, fat-stored hormones in beef-feed be outlawed?

The lessons of the past 18 months have been many: How do I provide myself with the best physical and psychic nourishment to repair past, and minimize future damage to my body? How do I give voice to my quests so that other women can take what they need from my experiences? How do my experiences with cancer fit into the larger tapestry of my work as a Black woman, into the history of all women? And most of all, how do I fight the despair born of fear and anger and powerlessness which is my greatest internal enemy?

I have found that battling despair does not mean closing my eyes to the enormity of the tasks of effecting change, nor ignoring the strength and the barbarity of the forces aligned against us. It means teaching, surviving and fighting with the most important resource I have, myself, and taking joy in that battle. It means, for me, recognizing the enemy outside and the enemy within, and knowing that my work is part of a continuum of women's work, of reclaiming this earth and our power, and knowing that this work did not begin with my birth nor will it end with my death. And it means knowing that within this continuum, my life and my love and my work has particular power and meaning relative to others.

It means trout fishing on the Missisquoi River at dawn and tasting the green silence, and knowing that this beauty too is mine forever.

August 29, 1980

In this powerful meditation of what it means to be critically ill and be subject to amputation, Lorde offers a side of medicine and healthcare that training in these fields necessarily discount: namely, the feeling by patients of being "subjected" by forces outside the control of the patient, forces that from time immemorial have seemed overwhelming. Of course, it is the goal of scientific healthcare to understand illness as natural phenomena susceptible to the systematic understanding of biomedical science. But sometimes—or better, "at times" —part of the "condition" of the seriously ill is the rage, sadness, and loneliness that Lorde describes in marking the enormous incongruity between what the patient takes to be *the* great crisis of her life and the fact that such crises, for healthcare providers, often seem a part of the everyday routine of their professions. Such incongruity is easily overlooked, not only in the everyday routines of healthcare, but as a way the healthcare professional—like many breast cancer patients Lorde describes—protects herself with "neutrality" and "silence." Lorde's account of her cancer—and the powerlessness it reinforced, which she had felt as an "outsider," a lesbian, a black person, a woman—nicely parallels another account of the *situation* of a patient, Anatole Broyard's widely available "Doctor Talk to Me" (1992), which more closely focuses on the patient–provider relationship from the point of view of the patient. (Lorde focuses on the patient experience, while Broyard explores the relationship between patient and provider.) Broyard explicitly, and Lorde implicitly, emphasize the incongruity between the personal experience of illness (no matter how much, as Lorde suggests, it is embedded in social formations) and the professional experience of those who care for the ill.

Such incongruity lies at the heart of Charlotte Perkins Gilman's famous short story, "The Yellow Wallpaper." In this story, the patient is a young woman who has recently given birth to her child and is suffering from postpartum depression that seems to be aggravated into postpartum psychosis. Her husband is a physician, and in the face of her ailment, prescribes a "rest cure"—a cure developed by the Civil War physician Silas Weir Mitchell. The great incongruity—like that between the debilitating pain felt by suffers of chronic pain and the great difficulty for anyone (including physicians) to recognize such pain in another person described by Lous Heshusius in the patient-centered vignette in Chap. 11—is that between the experience of a suffering patient and the understanding of those who encounter her. In Gilman's narrative, this is exacerbated by the patent sexism displayed by the sufferer's husband—and, implicitly, by the members of his profession and the more general discounting of women's pain and suffering that persists in our culture even today, more than 125 years after "The Yellow Wallpaper" was published. (For an account of contemporary dismissals of women's pain, see Schleifer 2014: 107–12.)

Literary Narrative: "The Yellow Wallpaper" (1892) by Charlotte Perkins Gilman

> *Author Note*: Charlotte Perkins Gilman (1860–1935) was a prominent American feminist, fiction writer, sociologist, and proponent of social reform. Perhaps she is best remembered for her short story "The Yellow Wallpaper," but her *Women and Economics* (1898) is a powerful argument for the inclusion of women in the public sphere outside the home. Another important book is *The Home: Its Work and Influence* (1903).

The Yellow Wallpaper

> It is very seldom that mere ordinary people like John and myself secure ancestral halls for the summer.
> A colonial mansion, a hereditary estate, I would say a haunted house, and reach the height of romantic felicity—but that would be asking too much of fate!
> Still I will proudly declare that there is something queer about it.
> Else, why should it be let so cheaply? And why have stood so long untenanted?
> John laughs at me, of course, but one expects that in marriage.
> John is practical in the extreme. He has no patience with faith, an intense horror of superstition, and he scoffs openly at any talk of things not to be felt and seen and put down in figures.
> John is a physician, and *perhaps*—(I would not say it to a living soul, of course, but this is dead paper and a great relief to my mind)—*perhaps* that is one reason I do not get well faster.
> You see he does not believe I am sick!
> And what can one do?
> If a physician of high standing, and one's own husband, assures friends and relatives that there is really nothing the matter with one but temporary nervous depression—a slight hysterical tendency—what is one to do?
> My brother is also a physician, and also of high standing, and he says the same thing.
> So I take phosphates or phosphites—whichever it is, and tonics, and journeys, and air, and exercise, and am absolutely forbidden to "work" until I am well again.
> Personally, I disagree with their ideas.

Personally, I believe that congenial work, with excitement and change, would do me good. But what is one to do?

I did write for a while in spite of them; but it does exhaust me a good deal—having to be so sly about it, or else meet with heavy opposition.

I sometimes fancy that in my condition if I had less opposition and more society and stimulus—but John says the very worst thing I can do is to think about my condition, and I confess it always makes me feel bad.

So I will let it alone and talk about the house.

The most beautiful place! It is quite alone standing well back from the road, quite three miles from the village. It makes me think of English places that you read about, for there are hedges and walls and gates that lock, and lots of separate little houses for the gardeners and people.

There is a *delicious* garden! I never saw such a garden—large and shady, full of boxbordered paths, and lined with long grape-covered arbors with seats under them.

There were greenhouses, too, but they are all broken now.

There was some legal trouble, I believe, something about the heirs and coheirs; anyhow, the place has been empty for years.

That spoils my ghostliness, I am afraid, but I don't care—there is something strange about the house—I can feel it.

I even said so to John one moonlight evening but he said what I felt was a *draught*, and shut the window.

I get unreasonably angry with John sometimes I'm sure I never used to be so sensitive. I think it is due to this nervous condition.

But John says if I feel so, I shall neglect proper self-control; so I take pains to control myself—before him, at least, and that makes me very tired.

I don't like our room a bit. I wanted one downstairs that opened on the piazza and had roses all over the window, and such pretty old-fashioned chintz hangings! but John would not hear of it.

He said there was only one window and not room for two beds, and no near room for him if he took another.

He is very careful and loving, and hardly lets me stir without special direction.

I have a schedule prescription for each hour in the day; he takes all care from me, and so I feel basely ungrateful not to value it more.

He said we came here solely on my account, that I was to have perfect rest and all the air I could get. "Your exercise depends on your strength, my dear," said he, "and your food somewhat on your appetite; but air you can absorb all the time." So we took the nursery at the top of the house.

It is a big, airy room, the whole floor nearly, with windows that look all ways, and air and sunshine galore. It was nursery first and then playroom and gymnasium, I should judge; for the windows are barred for little children, and there are rings and things in the walls.

The paint and paper look as if a boys' school had used it. It is stripped off—the paper in great patches all around the head of my bed, about as far as I can reach, and in a great place on the other side of the room low down. I never saw a worse paper in my life. One of those sprawling flamboyant patterns committing every artistic sin.

It is dull enough to confuse the eye in following, pronounced enough to constantly irritate and provoke study, and when you follow the lame uncertain curves for a little distance they suddenly commit suicide—plunge off at outrageous angles, destroy themselves in unheard of contradictions.

The color is repellent, almost revolting; a smoldering unclean yellow, strangely faded by the slow-turning sunlight.

It is a dull yet lurid orange in some places, a sickly sulphur tint in others.

No wonder the children hated it! I should hate it myself if I had to live in this room long.

There comes John, and I must put this away—he hates to have me write a word.

***

We have been here two weeks, and I haven't felt like writing before, since that first day. I am sitting by the window now, up in this atrocious nursery, and there is nothing to hinder my writing as much as I please, save lack of strength.

John is away all day, and even some nights when his cases are serious.

I am glad my case is not serious!

But these nervous troubles are dreadfully depressing.

John does not know how much I really suffer. He knows there is no reason to suffer, and that satisfies him.

Of course it is only nervousness. It does weigh on me so not to do my duty in any way! I meant to be such a help to John, such a real rest and comfort, and here I am a comparative burden already!

Nobody would believe what an effort it is to do what little I am able—to dress and entertain, and order things.

It is fortunate Mary is so good with the baby. Such a dear baby!

And yet I cannot be with him, it makes me so nervous.

I suppose John never was nervous in his life. He laughs at me so about this wallpaper! At first he meant to repaper the room, but afterwards he said that I was letting it get the better of me, and that nothing was worse for a nervous patient than to give way to such fancies. He said that after the wallpaper was changed it would be the heavy bedstead, and then the barred windows, and then that gate at the head of the stairs, and so on.

"You know the place is doing you good," he said, "and really, dear, I don't care to renovate the house just for a three months' rental."

"Then do let us go downstairs," I said, "there are such pretty rooms there."

Then he took me in his arms and called me a blessed little goose, and said he would go down to the cellar, if I wished, and have it whitewashed into the bargain.

But he is right enough about the beds and windows and things.

It is an airy and comfortable room as any one need wish, and, of course, I would not be so silly as to make him uncomfortable just for a whim.

I'm really getting quite fond of the big room, all but that horrid paper.

Out of one window I can see the garden, those mysterious deep shaded arbors, the riotous old-fashioned flowers, and bushes and gnarly trees.

Out of another I get a lovely view of the bay and a little private wharf belonging to the estate. There is a beautiful shaded lane that runs down there from the house. I always fancy I see people walking in these numerous paths and arbors, but John has cautioned me not to give way to fancy in the least. He says that with my imaginative power and habit of story-making, a nervous weakness like mine is sure to lead to all manner of excited fancies, and that I ought to use my will and good sense to check the tendency. So I try.

I think sometimes that if I were only well enough to write a little it would relieve the press of ideas and rest me.

But I find I get pretty tired when I try.

It is so discouraging not to have any advice and companionship about my work. When I get really well, John says we will ask Cousin Henry and Julia down for a long visit; but he says he would as soon put fireworks in my pillow-case as to let me have those stimulating people about now.

I wish I could get well faster.

But I must not think about that. This paper looks to me as if it knew what a vicious influence it had!

There is a recurrent spot where the pattern lolls like a broken neck and two bulbous eyes stare at you upside down.

I get positively angry with the impertinence of it and the everlastingness. Up and down and sideways they crawl, and those absurd, unblinking eyes are everywhere. There is one place where two breaths didn't match, and the eyes go all up and down the line, one a little higher than the other.

I never saw so much expression in an inanimate thing before, and we all know how much expression they have! I used to lie awake as a child and get more entertainment and terror out of blank walls and plain furniture than most children could find in a toy-store.

I remember what a kindly wink the knobs of our big, old bureau used to have, and there was one chair that always seemed like a strong friend.

I used to feel that if any of the other things looked too fierce I could always hop into that chair and be safe.

The furniture in this room is no worse than inharmonious, however, for we had to bring it all from downstairs. I suppose when this was used as a playroom they had to take the nursery things out, and no wonder! I never saw such ravages as the children have made here.

The wallpaper, as I said before, is torn off in spots, and it sticketh closer than a brother—they must have had perseverance as well as hatred.

Then the floor is scratched and gouged and splintered, the plaster itself is dug out here and there, and this great heavy bed which is all we found in the room, looks as if it had been through the wars.

But I don't mind it a bit—only the paper.

There comes John's sister. Such a dear girl as she is, and so careful of me! I must not let her find me writing.

She is a perfect and enthusiastic housekeeper, and hopes for no better profession. I verily believe she thinks it is the writing which made me sick!

But I can write when she is out, and see her a long way off from these windows.

There is one that commands the road, a lovely shaded winding road, and one that just looks off over the country. A lovely country, too, full of great elms and velvet meadows.

This wall-paper has a kind of sub-pattern in a, different shade, a particularly irritating one, for you can only see it in certain lights, and not clearly then.

But in the places where it isn't faded and where the sun is just so—I can see a strange, provoking, formless sort of figure, that seems to skulk about behind that silly and conspicuous front design.

There's sister on the stairs!

***

Well, the Fourth of July is over! The people are all gone and I am tired out. John thought it might do me good to see a little company, so we just had mother and Nellie and the children down for a week.

Of course I didn't do a thing. Jennie sees to everything now.

But it tired me all the same.

John says if I don't pick up faster he shall send me to Weir Mitchell in the fall.

But I don't want to go there at all. I had a friend who was in his hands once, and she says he is just like John and my brother, only more so! Besides, it is such an undertaking to go so far.

I don't feel as if it was worthwhile to turn my hand over for anything, and I'm getting dreadfully fretful and querulous.

I cry at nothing, and cry most of the time.

Of course I don't when John is here, or anybody else, but when I am alone.

And I am alone a good deal just now. John is kept in town very often by serious cases, and Jennie is good and lets me alone when I want her to.

So I walk a little in the garden or down that lovely lane, sit on the porch under the roses, and lie down up here a good deal.

I'm getting really fond of the room in spite of the wallpaper. Perhaps *because* of the wallpaper.

It dwells in my mind so!

I lie here on this great immovable bed—it is nailed down, I believe—and follow that pattern about by the hour. It is as good as gymnastics, I assure you. I start, we'll say, at the bottom, down in the corner over there where it has not been touched, and I determine for the thousandth time that I *will* follow that pointless pattern to some sort of a conclusion.

I know a little of the principle of design, and I know this thing was not arranged on any laws of radiation, or alternation, or repetition, or symmetry, or anything else that I ever heard of.

It is repeated, of course, by the breadths, but not otherwise.

Looked at in one way each breadth stands alone, the bloated curves and flourishes—a kind of "debased Romanesque" with *delirium tremens*—go waddling up and down in isolated columns of fatuity.

But, on the other hand, they connect diagonally, and the sprawling outlines run off in great slanting waves of optic horror, like a lot of wallowing seaweeds in full chase.

The whole thing goes horizontally, too, at least it seems so, and I exhaust myself in trying to distinguish the order of its going in that direction.

They have used a horizontal breadth for a frieze, and that adds wonderfully to the confusion.

There is one end of the room where it is almost intact, and there, when the crosslights fade and the low sun shines directly upon it, I can almost fancy radiation after all,—the interminable grotesques seem to form around a common centre and rush off in headlong plunges of equal distraction.

It makes me tired to follow it. I will take a nap I guess.

***

I don't know why I should write this.

I don't want to.

I don't feel able. And I know John would think it absurd. But I must say what I feel and think in some way—it is such a relief!

But the effort is getting to be greater than the relief.

Half the time now I am awfully lazy, and lie down ever so much.

John says I mustn't lose my strength, and has me take cod liver oil and lots of tonics and things, to say nothing of ale and wine and rare meat.

Dear John! He loves me very dearly, and hates to have me sick. I tried to have a real earnest reasonable talk with him the other day, and tell him how I wish he would let me go and make a visit to Cousin Henry and Julia.

But he said I wasn't able to go, nor able to stand it after I got there; and I did not make out a very good case for myself, for I was crying before I had finished.

It is getting to be a great effort for me to think straight. Just this nervous weakness I suppose.

And dear John gathered me up in his arms, and just carried me upstairs and laid me on the bed, and sat by me and read to me till it tired my head.

He said I was his darling and his comfort and all he had, and that I must take care of myself for his sake, and keep well.

He says no one but myself can help me out of it, that I must use my will and self-control and not let any silly fancies run away with me.

There's one comfort, the baby is well and happy, and does not have to occupy this nursery with the horrid wallpaper.

If we had not used it, that blessed child would have! What a fortunate escape! Why, I wouldn't have a child of mine, an impressionable little thing, live in such a room for worlds. I never thought of it before, but it is lucky that John kept me here after all, I can stand it so much easier than a baby, you see.

Of course I never mention it to them any more—I am too wise,—but I keep watch of it all the same.

There are things in that paper that nobody knows but me, or ever will.

Behind that outside pattern the dim shapes get clearer every day.

It is always the same shape, only very numerous.

And it is like a woman stooping down and creeping about behind that pattern. I don't like it a bit. I wonder—I begin to think—I wish John would take me away from here!

***

It is so hard to talk with John about my case, because he is so wise, and because he loves me so.

But I tried it last night.

It was moonlight. The moon shines in all around just as the sun does.

I hate to see it sometimes, it creeps so slowly, and always comes in by one window or another.

John was asleep and I hated to waken him, so I kept still and watched the moonlight on that undulating wallpaper till I felt creepy.

The faint figure behind seemed to shake the pattern, just as if she wanted to get out. I got up softly and went to feel and see if the paper *did* move, and when I came back John was awake.

"What is it, little girl?" he said. "Don't go walking about like that—you'll get cold."

I thought it was a good time to talk, so I told him that I really was not gaining here, and that I wished he would take me away.

"Why darling!" said he, "our lease will be up in three weeks, and I can't see how to leave before.

"The repairs are not done at home, and I cannot possibly leave town just now. Of course if you were in any danger, I could and would, but you really are better, dear, whether you can see it or not. I am a doctor, dear, and I know. You are gaining flesh and color, your appetite is better, I feel really much easier about you."

"I don't weigh a bit more," said I, "nor as much; and my appetite may be better in the evening when you are here, but it is worse in the morning when you are away!"

"Bless her little heart!" said he with a big hug, "she shall be as sick as she pleases! But now let's improve the shining hours by going to sleep, and talk about it in the morning!"

"And you won't go away?" I asked gloomily.

"Why, how can 1, dear? It is only three weeks more and then we will take a nice little trip of a few days while Jennie is getting the house ready. Really dear you are better!"

"Better in body perhaps—" I began, and stopped short, for he sat up straight and looked at me with such a stern, reproachful look that I could not say another word.

"My darling," said he, "I beg of you, for my sake and for our child's sake, as well as for your own, that you will never for one instant let that idea enter your mind! There is nothing so dangerous, so fascinating, to a temperament like yours. It is a false and foolish fancy. Can you not trust me as a physician when I tell you so?"

So of course I said no more on that score, and we went to sleep before long. He thought I was asleep first, but I wasn't, and lay there for hours trying to decide whether that front pattern and the back pattern really did move together or separately.

***

On a pattern like this, by daylight, there is a lack of sequence, a defiance of law, that is a constant irritant to a normal mind.

The color is hideous enough, and unreliable enough, and infuriating enough, but the pattern is torturing.

You think you have mastered it, but just as you get well underway in following, it turns a back somersault and there you are. It slaps you in the face, knocks you down, and tramples upon you. It is like a bad dream.

The outside pattern is a florid arabesque, reminding one of a fungus. If you can imagine a toadstool in joints, an interminable string of toadstools, budding and sprouting in endless convolutions—why, that is something like it.

That is, sometimes!

There is one marked peculiarity about this paper, a thing nobody seems to notice but myself, and that is that it changes as the light changes.

When the sun shoots in through the east window—I always watch for that first long, straight ray—it changes so quickly that I never can quite believe it.

That is why I watch it always.

By moonlight—the moon shines in all night when there is a moon—I wouldn't know it was the same paper.

At night in any kind of light, in twilight, candlelight, lamplight, and worst of all by moonlight, it becomes bars! The outside pattern I mean, and the woman behind it is as plain as can be.

I didn't realize for a long time what the thing was that showed behind, that dim sub-pattern, but now I am quite sure it is a woman.

By daylight she is subdued, quiet. I fancy it is the pattern that keeps her so still. It is so puzzling. It keeps me quiet by the hour.

I lie down ever so much now. John says it is good for me, and to sleep all I can.

Indeed he started the habit by making me lie down for an hour after each meal.

It is a very bad habit I am convinced, for you see I don't sleep.

And that cultivates deceit, for I don't tell them I'm awake—O no!

The fact is I am getting a little afraid of John.

He seems very queer sometimes, and even Jennie has an inexplicable look.

It strikes me occasionally, just as a scientific hypothesis—that perhaps it is the paper!

I have watched John when he did not know I was looking, and come into the room suddenly on the most innocent excuses, and I've caught him several times *looking at the paper*! And Jennie too. I caught Jennie with her hand on it once.

She didn't know I was in the room, and when I asked her in a quiet, a very quiet voice, with the most restrained manner possible, what she was doing with the paper—she turned around as if she had been caught stealing, and looked quite angry—asked me why I should frighten her so!

Then she said that the paper stained everything it touched, that she had found yellow smooches on all my clothes and John's, and she wished we would be more careful!

Did not that sound innocent? But I know she was studying that pattern, and I am determined that nobody shall find it out but myself!

***

Life is very much more exciting now than it used to be. You see I have something more to expect, to look forward to, to watch. I really do eat better, and am more quiet than I was.

John is so pleased to see me improve! He laughed a little the other day, and said I seemed to be flourishing in spite of my wall-paper.

I turned it off with a laugh. I had no intention of telling him it was because of the wall-paper—he would make fun of me. He might even want to take me away.

I don't want to leave now until I have found it out. There is a week more, and I think that will be enough.

***

I'm feeling ever so much better! I don't sleep much at night, for it is so interesting to watch developments; but I sleep a good deal in the daytime.

In the daytime it is tiresome and perplexing.

There are always new shoots on the fungus, and new shades of yellow all over it. I cannot keep count of them, though I have tried conscientiously.

It is the strangest yellow, that wallpaper! It makes me think of all the yellow things I ever saw—not beautiful ones like buttercups, but old foul, bad yellow things.

But there is something else about that paper— the smell! I noticed it the moment we came into the room, but with so much air and sun it was not bad. Now we have had a week of fog and rain, and whether the windows are open or not, the smell is here.

It creeps all over the house.

I find it hovering in the dining room, skulking in the parlor, hiding in the hall, lying in wait for me on the stairs.

It gets into my hair.

Even when I go to ride, if I turn my head suddenly and surprise it—there is that smell! Such a peculiar odor, too! I have spent hours in trying to analyze it, to find what it smelled like.

It is not bad—at first, and very gentle, but quite the subtlest, most enduring odor I ever met.

In this damp weather it is awful, I wake up in the night and find it hanging over me.

It used to disturb me at first. I thought seriously of burning the house—to reach the smell.

But now I am used to it. The only thing I can think of that it is like is the *color* of the paper! A yellow smell.

There is a very funny mark on this wall, low down, near the mopboard. A streak that runs round the room. It goes behind every piece of furniture, except the bed, a long, straight, even *smooch*, as if it had been rubbed over and over.

I wonder how it was done and who did it, and what they did it for. Round and round and round—round and round and round—it makes me dizzy!

***

I really have discovered something at last.

Through watching so much at night, when it changes so, I have finally found out.

The front pattern does move—and no wonder! The woman behind shakes it!

Sometimes I think there are a great many women behind, and sometimes only one, and she crawls around fast, and her crawling shakes it all over.

Then in the very bright spots she keeps still, and in the very shady spots she just takes hold of the bars and shakes them hard.

And she is all the time trying to climb through. But nobody could climb through that pattern—it strangles so; I think that is why it has so many heads.

They get through, and then the pattern strangles them off and turns them upside down, and makes their eyes white!

If those heads were covered or taken off it would not be half so bad.

***

I think that woman gets out in the daytime!

And I'll tell you why—privately—I've seen her!

I can see her out of every one of my windows!

It is the same woman, I know, for she is always creeping, and most women do not creep by daylight.

I see her on that long road under the trees, creeping along, and when a carriage comes she hides under the blackberry vines.

I don't blame her a bit. It must be very humiliating to be caught creeping by daylight! I always lock the door when I creep by daylight. I can't do it at night, for I know John would suspect something at once.

And John is so queer now, that I don't want to irritate him. I wish he would take another room! Besides, I don't want anybody to get that woman out at night but myself.

I often wonder if I could see her out of all the windows at once.

But, turn as fast as I can, I can only see out of one at one time.
And though I always see her, she may be able to creep faster than I can turn!
I have watched her sometimes away off in the open country, creeping as fast as a cloud shadow in a high wind.

***

If only that top pattern could be gotten off from the under one! I mean to try it, little by little.
I have found out another funny thing, but I shan't tell it this time! It does not do to trust people too much.
There are only two more days to get this paper off, and I believe John is beginning to notice.
I don't like the look in his eyes.
And I heard him ask Jennie a lot of professional questions about me. She had a very good report to give.
She said I slept a good deal in the daytime.
John knows I don't sleep very well at night, for all I'm so quiet!
He asked me all sorts of questions, too, and pretended to be very loving and kind.
As if I couldn't see through him!
Still, I don't wonder he acts so, sleeping under this paper for three months.
It only interests me, but I feel sure John and Jennie are secretly affected by it.

***

Hurrah! This is the last day, but it is enough. John to stay in town over night, and won't be out until this evening.
Jennie wanted to sleep with me—the sly thing! but I told her I should undoubtedly rest better for a night all alone.
That was clever, for really I wasn't alone a bit! As soon as it was moonlight and that poor thing began to crawl and shake the pattern, I got up and ran to help her.
I pulled and she shook, I shook and she pulled, and before morning we had peeled off yards of that paper.
A strip about as high as my head and half around the room.
And then when the sun came and that awful pattern began to laugh at me, I declared I would finish it to-day!
We go away to-morrow, and they are moving all my furniture down again to leave things as they were before.
Jennie looked at the wall in amazement, but I told her merrily that I did it out of pure spite at the vicious thing.
She laughed and said she wouldn't mind doing it herself, but I must not get tired.
How she betrayed herself that time!
But I am here, and no person touches this paper but me—not alive!
She tried to get me out of the room—it was too patent! But I said it was so quiet and empty and clean now that I believed I would lie down again and sleep all I could; and not to wake me even for dinner—I would call when I woke.
So now she is gone, and the servants are gone, and the things are gone, and there is nothing left but that great bedstead nailed down, with the canvas mattress we found on it.
We shall sleep downstairs to-night, and take the boat home to-morrow.
I quite enjoy the room, now it is bare again.
How those children did tear about here!
This bedstead is fairly gnawed!
But I must get to work.

I have locked the door and thrown the key down into the front path.

I don't want to go out, and I don't want to have anybody come in, till John comes.

I want to astonish him.

I've got a rope up here that even Jennie did not find. If that woman does get out, and tries to get away, I can tie her!

But I forgot I could not reach far without anything to stand on!

This bed will not move!

I tried to lift and push it until I was lame, and then I got so angry I bit off a little piece at one corner—but it hurt my teeth.

Then I peeled off all the paper I could reach standing on the floor. It sticks horribly and the pattern just enjoys it! All those strangled heads and bulbous eyes and waddling fungus growths just shriek with derision!

I am getting angry enough to do something desperate. To jump out of the window would be admirable exercise, but the bars are too strong even to try.

Besides I wouldn't do it. Of course not. I know well enough that a step like that is improper and might be misconstrued.

I don't like to *look* out of the windows even—there are so many of those creeping women, and they creep so fast.

I wonder if they all come out of that wall-paper as I did?

But I am securely fastened now by my well-hidden rope—you don't get me out in the road there!

I suppose I shall have to get back behind the pattern when it comes night, and that is hard!

It is so pleasant to be out in this great room and creep around as I please!

I don't want to go outside. I won't, even if Jennie asks me to.

For outside you have to creep on the ground, and everything is green instead of yellow. But here I can creep smoothly on the floor, and my shoulder just fits in that long smooch around the wall, so I cannot lose my way.

Why there's John at the door!

It is no use, young man, you can't open it!

How he does call and pound!

Now he's crying for an axe.

It would be a shame to break down that beautiful door!

"John dear!" said I in the gentlest voice, "the key is down by the front steps, under a plantain leaf!"

That silenced him for a few moments.

Then he said—very quietly indeed, "Open the door, my darling!"

"I can't," said I. "The key is down by the front door under a plantain leaf!"

And then I said it again, several times, very gently and slowly, and said it so often that he had to go and see, and he got it of course, and came in. He stopped short by the door. "What is the matter?" he cried. "For God's sake, what are you doing!" I kept on creeping just the same, but I looked at him over my shoulder.

"I've got out at last," said I, "in spite of you and Jane. And I've pulled off most of the paper, so you can't put me back!"

Now why should that man have fainted? But he did, and right across my path by the wall, so that I had to creep over him every time!

Expression and Comprehension

What is most striking about Gilman's narrative is the vast incongruity between what her narrator expresses and what she comprehends. While "The Yellow Wallpaper" narrates the protagonist's descent into psychosis—beginning with her observations about the new house she and her husband share and ending with out-and-out delusions about the room she is forced to inhabit—throughout the narrator rarely understands her words in the ways that her readers do. Thus, early on, she notes "I get unreasonably angry with John sometimes. I'm sure I never used to be so sensitive. I think it is due to this nervous condition." In fact this incongruity—between her expressed feelings and emotions, and her lack of understanding of those feelings—is one that is often presented by patients, albeit hardly in the extreme form that Gilman presents. This story, in fact, offers a powerful occasion for its readers to understand and experience the *affective engagement* that we note in Part II of the Introduction is a pronounced—although sometimes not systematically taught—aspect of clinical healthcare.

Related Poem

For this chapter, the related poem, written by Dr. Rafael Campo, supplements the understanding of the patient as a lucid individual such as Audre Lorde or as an oppressed and deluded individual such as the narrator of "The Yellow Wallpaper" with an understanding of the "patient" extending to those related to an ailing individual. What is striking about this poem is the manner in which Dr. Campo comes to understand that his patient is the couple of the poem's title, and how the poem itself comes to understand that illness often needs to be understood in relation to the social network in which the ailing individual is situated. (This is also consistently true of pediatric patients, especially the very young.)

Poem: "The Couple" (2002) by Dr. Rafael Campo

> *Author Note*: Rafael Campo (b. 1964) practices internal medicine and teaches at Harvard Medical School. He is also a prominent poet and essayist, who focuses on healthcare issues, as well as justice and equality for LGBTQ people, Hispanic people, and poor people. His book, *The Desire to Heal: A Doctor's Education in Empathy, Identity, and Poetry* offers moving discussions of the nature of poetry and its relation to healthcare.

The Couple

> Releasing his determined grip, he lets
> her take the spoon; the cube of cherry Jell-O
> teeters on it, about to drop as if
> no precipice were any steeper, no
> oblivion more final. Earlier
> today, he hemorrhaged, the blood so fast
> a torrent that it splattered onto her.
> She washed herself, unwillingly it seemed,

perhaps not wanting to remove what was
his ending life from where it stained her skin.
I watch them now, the way they love across
the gap between them that their bodies make:
how cruel our life-long separation seems.
The ward keeps narrowing itself to that
bright point outside his door—the muffled screams
along a hallway to the absolute—
and as I turn away from them it's not
their privacy, or even my beginning shame
I wish I could escape. It is the light,
the awful light of what we know must come.

In talking about this poem, Dr. Campo has noted the ways in which "illness is almost never an isolated experience or individual experience, that this is a shared experience in the poem between two people who are in love, and that the end of life, I think, which is visible in the poem, is something that both people present in the poem must confront." He goes on to add that it is "a poem about the mystery of human suffering and how suffering, in a sense, is, perhaps, made more visible by the presence of another" (in Vannatta et al. 2005: Chap. 2, screen 53 [video]). In noting "the way they love across / the gap between them that their bodies make," he poignantly describes the opposition between physical lives and relationships and suggests, as does "The Yellow Wallpaper" in a very different register, the manner in which "good health" is a function of social relationships as well as somatic well-being.

Lessons for Providers
Attending to narratives that describe how patients see their condition and their providers is important feedback—providing us the opportunity for improvement.

Bibliography

Broyard, Anatole. 1992. Doctor Talk to Me. *New York Times*. https://www.nytimes.com/1990/08/26/magazine/doctor-talk-to-me.html

Schleifer, Ronald. 2014. *Pain and Suffering*. New York: Routledge.

Vannatta, Jerry, Ronald Schleifer, and Sheila Crow. 2005. *Medicine and Humanistic Understanding: The Significance of Narrative in Medical Practices*. Philadelphia: University of Pennsylvania Press. A DVD-Rom publication.

The Doctor

7

Just as Chap. 6 focused on the patient as a social role that is assumed by particular people, so this chapter focuses on the "person" of a physician or healthcare provider—who can be pre-judged by her race, gender, or other seeming social distinctions—in relation to the social role of caretaker. Such prejudice is notably what Dr. Damon Tweedy describes as "implicit" prejudice in the different behaviors toward his African American patient that the white physician exhibits before and after he knows the patient is also a physician in the vignette in this chapter excerpted from Dr. Tweedy's *Black Man in a White Coat*. In his book, Tweedy mentions a question he had been asking himself since medical school, "one that I heard many doctors, in frustrated moments, bring up: How much impact can we really have on patients' lives when their own behavior influences their health to such a large extent?" (2015: 203). Another influence on healthcare, he also notes, is the ordinary, everyday behavior of healthcare providers themselves. (Throughout his book he also offers repeated instances of the everyday racist behavior of patients as well.) Later Tweedy returns to the behavior of healthcare providers in describing two patients: one who negatively affected his health with his smoking and alcohol consumption, resulting in a stroke, and another patient who positively enhanced his health with exercise and diet. Considering these outcomes, Tweedy thought: "maybe ... at least part of the difference lay with me" (2015: 216), since he had experienced his second patient's struggles with diet and exercise but never smoked or used alcohol as did his first patient. This issue of the role of "everyday" behavior—tied up with the representation of "The Doctor" in the vignettes of this chapter and out-and-out racism in its literary narrative—is something to which we return in Chap. 8 in the discussion of everyday ethics in healthcare practices.

The racism and sexism described in this chapter are not only important for an understanding of the occasional incongruity between person and role of a healthcare worker, but also for the understanding of the *felt differences* that patients might sometimes feel between themselves and their physicians and vice versa: not only differences of race, culture, class, education, manners, and assumptions about the world—themes to which we return in Section VI of this book, which focuses on the ways literary texts provoke vicarious experience—but also the differences that arise out of the emotional state of patients, which almost always differs from the emotional state of healthcare providers (though in such stories as Dr. William Carlos Williams' "The Use of Force" and Dr. Richard Selzer's "Brute" physicians shares their patients' anger). Although Tweedy focuses on these differences in relation to white America and black America, he also notes that such differences arise between physicians trained outside the United States and their patients. (About 25 percent of practicing physicians in the US were born and trained in other countries.) Abraham Verghese touches on this in *My Own Country* (though not in the passage analyzed in Chap. 1). In *The Chief Concern of Medicine*, we examine such differences of under the category of "story filters" (a term Tweedy uses in the vignette below). These include emotional filters (anger, fear, sadness) and cultural filters (cultural background, gender, age, social class [see *The Chief Concern*: 196–210]). Emotional filters loom large in Appendix 6, which examines the importance of engaging patients' emotions in the patient-provider relationship.

When Doctors Discriminate: A Vignette (Excerpt from Dr. Damon Tweedy *Black, Man in a White Coat: A Doctor's Reflections on Race and Medicine* [2015])

> *Author Note*: Dr. Damon Tweedy (b. 1974) is a graduate of Duke University School of Medicine. He is Associate Professor of Psychiatry at Duke University Medical Center and staff physician at the Durham Veteran Affairs Medical Center. He has published articles about race and medicine in *The New York Times*, *The Washington Post*, as well as in various medical journals.

[This excerpt follows from Tweedy's description of how his medical colleagues had added a "psychiatric diagnosis" of Obsessive-Compulsive Personality Disorder to their black patient Gary because he preferred to try diet and exercise rather than follow their advice for blood-pressure medication. They judged him to be a "difficult" patient, and responded with a negative official (mis)diagnosis on his chart. Dr. Tweedy is an African American physician.]

Surely, I had no reason to think of these doctors as racist in any classic sense. I'd had lunch with Bruce and we'd discussed in depth our internship experiences and future ambitions; he'd given me advice prior to one of my rotations that proved helpful. I'd talked about pro football and college basketball with Carl, who'd gone to a Big Ten school, and he'd invited me for drinks with some of his friends. Dr. Rhodes had mentored a few black students and residents in the past and was always friendly with me. As far as I could tell, all three doctors regarded me as a genuine peer, as one of them, in contrast to the way it seemed they saw Gary.

But at that moment, I didn't feel like I was really one of them. Nor was I like Gary, who reminded me of a past that I could never reclaim. I had a foot in both worlds, but didn't have two feet in either.

My suspicion that, if confronted, these doctors would have vociferously denied that Gary's race influenced their psychiatric diagnosis is supported by the Kaiser Family Foundation's 2002 national survey of physicians, published not long before our encounter with Gary. It found that an overwhelming 75 percent of white physicians said race and ethnicity do not affect the treatment of patients, while 77 percent of black doctors said that

When Doctors Discriminate: A Vignette (Excerpt from Dr. Damon Tweedy *Black, Man*...)

race and ethnicity do impact how patients are treated. Smart people from two groups were seeing entirely different realities.

It was clear that my colleagues did not see their actions toward Gary as racially biased, or else they would not have been so brazen in my presence. But I avoided approaching them about what they had done. Once again, personal ambition and comfort trumped racial solidarity. Learning to be a doctor was hard enough without my trying to change the whole system too. Further, I didn't want to deal with possibly being mislabeled as racially paranoid, especially considering how deeply most educated white people take offense to being accused of racial bias. But was I selling myself, and my race, short in the process?

In the end, I pretended that nothing had happened. We went about our usual business. Life went on. Gary probably never learned how his doctors had callously mislabeled him.

***

Several years later, I had an experience similar to Gary's. In my mid-thirties, my knees were paying the price for many years of playing basketball. I'd grown up spending hours upon hours running, jumping, and cutting on the unforgiving blacktop. […].

The cumulative effect was gimpy knees. I had torn my right ACL many years before, but I'd recovered fairly well with surgery and physical therapy. Now my left knee was the one bothering me. Recently when I had played tennis, it had buckled slightly as I rushed to the net to retrieve a drop shot. When the swelling didn't go down after several days, I decided to get it checked out. It didn't seem serious enough to justify a visit to the emergency room, but I didn't want to wait another week to see my primary care physician or several weeks to see an orthopedic surgeon. An urgent care clinic, part of the same health care system where my primary doctor worked, had recently opened. This seemed the best option.

Within a few minutes of arriving, an energetic nurse called my name. She gave a warm, friendly smile. "Good morning," she said. We shook hands. "Follow me."

I limped behind her down the hall into an exam room. […]

"Dr. Parker should be in shortly," she said. "It's pretty slow here today."

True to her words, the doctor opened the door moments later. He gave me a weak handshake as his eyes scanned me from head to toe. It was only then that I realized how casually I was dressed. In contrast to the usual shirt, tie, and slacks I wore to work, I had on a fleece pullover and sweatpants. In haste, I'd put on white socks that were slightly mismatched. I didn't look homeless, but I didn't look like I had taken much care with my appearance.

With virtually no eye contact, his eyes fixed on the computer, Dr. Parker verified the information the nurse had obtained. He then had me pull up my sweatpants so he could look at both my knees. Next, he asked me to stand. My knees creaked, like a door hinge in need of lubricant. The pain made me grimace.

"You're fine," he said. "Probably just a bruise or sprain. Just take it easy for a while."

That's it? All he had done was look at my leg. He had not touched it to feel if my knee was unusually warm or cold, or whether it had accumulated excess fluid. Nor had he moved my knee through the various ranges of motion. He'd offered no explanation of what part of the knee was bruised or sprained. There'd been no mention of pain meds, ointments, or other analgesia. He did not offer nor suggest any type of knee bracing, just rest. But what if I had a job that required me to move around? He was all set to leave. I knew I had to say something.

"I really just want to make sure there's nothing serious," I said, hurrying to stop him from walking out the door. "Last summer I walked around with a sore hand for three days before I got an X-ray that showed a left third metacarpal fracture."

He looked up and established eye contact for the first time. "Are you a medical person?"

"Yes," I said.

"Are you an X-ray tech?"

"No, I'm a physician."

His eyes widened with surprise and what seemed to me admiration, as if the last thing he had expected was for us to be in the same profession. We traded a few words about the challenges of internship and residency training and the adjustment to life afterward. "Let me take a closer look at your knee," he said.

He went through a detailed physical exam—the kind I had expected from the beginning.

"Everything seems okay," he said. "But I think it would be good to get an X-ray too."

The nurse returned and escorted me to the basement for X-rays. Dr. Parker then came down and reviewed the film with me. It showed some early knee arthritis, but no other problems. He assured me that when the radiologist gave the official reading, he would call me himself. In the meantime, he recommended a brace, and offered me crutches just to have on hand. He also offered me a prescription for pain medication. Based on the X-ray, I told him that the brace would suffice; I didn't need crutches and would take over-the-counter ibuprofen.

As promised, Dr. Parker called me the next day. The radiologist's report had confirmed his preliminary review. He told me that he'd gone over to the nearby orthopedic surgery office and gotten me a better brace than the one they had in the urgent care clinic, free of charge. In the end, Dr. Parker's initial impression was correct; I had a mild-moderate knee sprain. With a few more weeks away from basketball and tennis courts, the pain and swelling receded.

But I couldn't get out of my mind how I'd been treated as two entirely different patients. Damon Tweedy, the unknown black man, dressed like he was about to mow the lawn, couldn't get the doctor to look him in the eye or touch him; Damon Tweedy, M.D., was worthy of personal, first-class service. While it's widely known that doctors get special treatment from their colleagues, this went far beyond the usual professional courtesy of an earlier or more convenient appointment. Receiving a physical exam, an X-ray, medication, and a brace, when you otherwise would not, wasn't just better service: it was different medical care altogether.

Was Dr. Parker aware that his initial lack of attention had been unfair and insulting, leading him to overcompensate with his subsequent actions? Perhaps, but I was more interested in the reasons for his initial approach to me. He evidently saw me through a mental filter, and his assumptions were not positive. Several authors have written about the negative stereotypes that many doctors associate with black patients: poorly educated, drug abusing, not likely to comply with treatments; in short, the kind of person most doctors don't want to treat. […]

***

Doctors, like all other people, are capable of prejudice and discrimination. While bias can be a problem in any profession, in medicine, the stakes are much greater. Missing a blood clot in a patient's painful leg because the doctor thinks that black people in a given clinic or hospital are likely to be drug addicts seeking their next fix is a far more dangerous kind of insult than a salesperson assuming that a black customer can't afford a Brooks Brothers suit or Rolex watch. These high stakes make it vitally important for doctors to understand their capacity for prejudice.

The "implicit" discrimination Tweedy describes here is even more pronounced in the "offhand" paragraph we present in a second narrative-vignette for this chapter, a short passage from Dr. Michael LaCombe's narrative of the experience of a woman physician. This paragraph, describing almost unconscious everyday acts of "chauvinism" and sexism is hardly pertinent to the larger theme of LaCombe's narrative concerning the time, focus, and energy required in a difficult diagnosis, yet it is—in its very offhandedness from LaCombe's main theme—a good example of "implicit" bias.

Offhand Paragraph: A Vignette-Discussion (Excerpt from Dr. Michael LaCombe "Diagnosis," in *Bedside: The Art of Medicine* [2010])

Author Note: Dr. Michael LaCombe (b. 1942) is a graduate of Harvard Medical School who practiced primary-care general internal medicine and, subsequently, cardiology in Maine. The first author to write fiction for medical journals, he has published over 100 short stories. He has published numerous books, including *Bedside: The Art of Medicine*, from a chapter of which this short vignette-analysis is excerpted. In addition, he has compiled the medical writings of Dr. William Osler, one of the four founding professors of Johns Hopkins Hospital, who created the first residency program and program in bedside clinical training.

The resident presented the next case to her. She listened absently. She could feel her uneasiness, a certain apprehension. Reviewing her day, she analyzed the feelings: the department meeting that morning and her Chief's chauvinism, the sexual joke and mumbled halfhearted apology, the condescension, the intern staring at her chest—all old hat in this man's world—the sharp words with the macho surgeon, his snickering resident, their supreme arrogance—this no longer nettled her for very long. The image of her last patient filled her mind: his simple honesty, his absolute trust in her, his blind acceptance of what little she had to give him. His sincerity. His headaches. The uneasiness welled up within her. (LaCombe 2010: 98).

We should also note that the two global manifestations of bias narrated in these vignettes—namely racism and sexism—are present in the literary narrative in this chapter, Paul Laurence Dunbar's "The Lynching of Jube Benson." The story is narrated by a white physician who serves both the white and black communities, yet whose language—despite his regret for participating in the lynching-murder of his African American friend—participates in a profound racism and, in its depiction of the heroine of the story, Annie Daly, a condescending sexism.

Everyday Behavior

Both of these vignettes describe prejudice and discrimination under the category of what we are calling "everyday behavior." The second short vignette—really a paragraph from a narrative that is about the difficulties and qualities of a medical diagnosis that happens to be formed by a woman physician in relation to her patient—demonstrates the "offhand" or unreflected-upon nature of everyday bias and prejudgment, what is sometimes described as "unconscious (implicit) bias" (Tweedy 2015: 270). Tweedy's longer vignette does so as well. In fact, it is really two vignettes: one concerning the experience of a physician who *does* reflect upon prejudicial action in which he is complicit (the excerpt does not narrate the fact that Tweedy signs Gary's written discharge order, which includes the misdiagnosis of Obsessive-Compulsive Personality Disorder); and a second in which Tweedy is the object of a physician's prejudicial action. As he notes in the vignette—and as LaCombe suggests in describing how the female doctor in his narrative regularly encounters unreflected-upon slights and insults—such almost non-intentional, *habitual* behavior may not be "racist in any classic sense" (such as the outright racist epithets he describes hearing from some of his patients in other chapters of *Black Man in a White Coat* or the meditation of "blackness" by the white physician in Dunbar's short story). But this, in large part, is just the point: by including the results from the Kaiser Family Foundation's 2002 national survey that white physicians are unaware of any impact of race and ethnicity on their behavior, whereas black physicians are aware of such impact, he suggests that many white physicians are simply not sensitive to and easily overlook some of the everyday behaviors both he and LaCombe describe in these vignettes. The sense of the everyday nature of prejudice manifesting themselves in small gestures and seeming habitual actions (i.e., unreflected-upon behavior), as opposed to explicit racist appellations and even physical violence of one sort or another, is the most difficult aspect of discrimination to understand and recognize in oneself, precisely because it hardly feels like "prejudice" to those who manifest it.

For this reason, the literary text we bring together with these vignettes does not focus on medicine or healthcare in any direct way, even if its narrator is a physician. Rather, it examines the great complexity of prejudice: not only the overt verbal and violent racism it describes, but also the subtleties of the condescending representations of Jube seeking approval, the participation in cultural norms (what Dunbar's narrator calls "tradition" and philosophers call "habits of thought" shared by a community), the inability to imagine a world different from "this man's world" the female physician finds herself in, which LaCombe describes in passing. Paul Laurence Dunbar's short story "The Lynching of Jube Benson" combines explicit and implicit racism in describing post-Civil War America. Like Ralph Ellison's powerful depiction of implicit racism in the central idea of "invisibility" in his great novel, *Invisible Man*, the very fact of the discussion of lynching as a spectacle and leisure activity by the white smokers at the beginning of the story makes the violence of racism almost "invisible" so that white people simply, unreflectingly "look through" the suffering of the falsely accused black person in a manner similar to the way that Tweedy describes himself as "the unknown black man, dressed like he was about to mow the lawn," a patient, seemingly not quite there, who couldn't get the doctor to look him in the eye or touch him. In this narrative, Dunbar also sets forth explicit racism in the physical violence, the dehumanization, and explicit racist/chauvinist appellations, even, perhaps, in the dialect-language for which he was justly famous. (Among other things, Dunbar wrote the lyrics for *In Dahomey*, the first musical written and performed entirely by African Americans on Broadway in 1903.)

Literary Narrative: "The Lynching of Jube Benson" (1904) by Paul Laurence Dunbar

> *Author Note*: Paul Laurence Dunbar (1872–1906), born in Ohio to parents who had been enslaved in Kentucky before the Civil War, began writing at an early age and published his first poem in a Dayton newspaper when he was 16. His work was praised by William Dean Howells, and Dunbar was one of the first African American writers to establish an international reputation. He was a prolific writer, who produced a dozen books of poetry, four novels, and four books of short stories in his short lifetime. Maya Angelou entitled her autobiography *I Know Why the Caged Bird Sings* from a line in Dunbar's poem "Sympathy."

The Lynching of Jube Benson

Gordon Fairfax's library held but three men, but the air was dense with clouds of smoke. The talk had drifted from one topic to another much as the smoke wreaths had puffed, floated, and thinned away. Then Handon Gay, who was an ambitious young reporter, spoke of a lynching story in a recent magazine, and the matter of punishment without trial put new life into the conversation.

"I should like to see a real lynching," said Gay rather callously.

"Well, I should hardly express it that way," said Fairfax, "but if a real, live lynching were to come my way, I should not avoid it."

"I should," spoke the other from the depths of his chair, where he had been puffing in moody silence. Judged by his hair, which was freely sprinkled with gray, the speaker might have been a man of forty-five or fifty, but his face, though lined and serious, was youthful, the face of a man hardly past thirty.

"What, you, Dr. Melville? Why, I thought that you physicians wouldn't weaken at anything."

"I have seen one such affair," said the doctor gravely, "in fact, I took a prominent part in it."

"Tell us about it," said the reporter, feeling for his pencil and notebook, which he was, nevertheless, careful to hide from the speaker.

The men drew their chairs eagerly up to the doctor's, but for a minute he did not seem to see them, but sat gazing abstractedly into the fire, then he took a long draw upon his cigar and began:

"I can see it all very vividly now. It was in the summer time and about seven years ago. I was practicing at the time down in the little town of Bradford. It was a small and primitive place, just the location for an impecunious medical man, recently out of college.

"In lieu of a regular office, I attended to business in the first of two rooms which I rented from Hiram Daly, one of the more prosperous of the townsmen. Here I boarded and here also came my patients—white and black—whites from every section, and blacks from 'nigger town,' as the west portion of the place was called.

"The people about me were most of them coarse and rough, but they were simple and generous, and as time passed on I had about abandoned my intention of seeking distinction in wider fields and determined to settle into the place of a modest country doctor. This was rather a strange conclusion for a young man to arrive at, and I will not deny that the presence in the house of my host's beautiful young daughter, Annie, had something to do with my decision. She was a beautiful young girl of seventeen or eighteen, and very far superior to her surroundings. She had a native grace and a pleasing way about her that made everybody that came under her spell her abject slave. White and black who knew her loved her, and none, I thought, more deeply and respectfully than Jube Benson, the black man of all work about the place.

"He was a fellow whom everybody trusted; an apparently steady-going, grinning sort, as we used to call him. Well, he was completely under Miss Annie's thumb, and would fetch and carry for her like a faithful dog. As soon as he saw that I began to care for Annie, and anybody could see that, he transferred some of his allegiance to me and became my faithful servitor also. Never did a man have a more devoted adherent in his wooing than did I, and many a one of Annie's tasks which he volunteered to do gave her an extra hour with me. You can imagine that I liked the boy and you need not wonder any more that as both wooing and my practice waxed apace, I was content to give up my great ambitions and stay just where I was.

"It wasn't a very pleasant thing, then, to have an epidemic of typhoid break out in the town that kept me going so that I hardly had time for the courting that a fellow wants to carry on with his sweetheart while he is still young enough to call her his girl. I fumed, but duty was duty, and I kept to my work night and day. It was now that Jube proved how invaluable he was as a coadjutor. He not only took messages to Annie, but brought sometimes little ones from her to me, and he would tell me little secret things that he had overheard her say that made me throb with joy and swear at him for repeating his mistress' conversation. But best of all, Jube was a perfect Cerberus, and no one on earth could have been more effective in keeping away or deluding the other young fellows who visited the Dalys. He would tell me of it afterwards, chuckling softly to himself. 'An,' Doctah, I say to Mistah Hemp Stevens, "'Scuse us, Mistah Stevens, but Miss Annie, she des gone out," an' den he go outer de gate lookin' moughty lonesome. When Sam Elkins come, I say, "Sh, Mistah Elkins, Miss Annie, she done tuk down," an' he say, "What, Jube, you don' reckon hit de—" Den he stop an' look skeert, an' I say, "I feared hit is, Mistah Elkins," an' sheks my haid ez solemn. He goes outer de gate lookin' lak his bes' frien' done daid, an' all de time Miss Annie behine de cu'tain ovah de po'ch des' a laffin' fit to kill.'

"Jube was a most admirable liar, but what could I do? He knew that I was a young fool of a hypocrite, and when I would rebuke him for these deceptions, he would give way and roll on the floor in an excess of delighted laughter until from very contagion I had to join him—and, well, there was no need of my preaching when there had been no beginning to his repentance and when there must ensue a continuance of his wrong-doing.

"This thing went on for over three months, and then, pouf! I was down like a shot. My patients were nearly all up, but the reaction from overwork made me an easy victim of the lurking germs. Then Jube loomed up as a nurse. He put everyone else aside, and with the doctor, a friend of mine from a neighbouring town, took entire charge of me. Even Annie herself was put aside, and I was cared for as tenderly as a baby. Tom, that was my physician and friend, told me all about it afterward with tears in his eyes. Only he was a big, blunt man and his expressions did not convey all that he meant. He told me how my nigger had nursed me as if I were a sick kitten and he my mother. Of how fiercely he guarded his right to be the sole one to 'do' for me, as he called it, and how, when the crisis came, he hovered, weeping, but hopeful, at my bedside, until it was safely passed, when they drove him, weak and exhausted, from the room. As for me, I knew little about it at the time, and cared less. I was too busy in my fight with death. To my chimerical vision there was only a black but gentle demon that came and went, alternating with a white fairy, who would insist on coming in on her head, growing larger and larger and then dissolving. But the pathos and devotion in the story lost nothing in my blunt friend's telling.

"It was during the period of a long convalescence, however, that I came to know my humble ally as he really was, devoted to the point of abjectness. There were times when for very shame at his goodness to me, I would beg him to go away, to do something else. He would go, but before I had time to realise that I was not being ministered to, he would be back at my side, grinning and pottering just the same. He manufactured duties for the joy of performing them. He pretended to see desires in me that I never had, because he liked to pander to them, and when I became entirely exasperated, and ripped out a good round oath, he chuckled with the remark, 'Dah, now, you sholy is gittin' well. Nevah did hyeah a man anywhaih nigh Jo'dan's sho' cuss lak dat.'

"Why, I grew to love him, love him, oh, yes, I loved him as well—oh, what am I saying? All human love and gratitude are damned poor things; excuse me, gentlemen, this isn't a pleasant story. The truth is usually a nasty thing to stand.

"It was not six months after that that my friendship to Jube, which he had been at such great pains to win, was put to too severe a test.

"It was in the summer time again, and as business was slack, I had ridden over to see my friend, Dr. Tom. I had spent a good part of the day there, and it was past four o'clock when I rode leisurely into Bradford. I was in a particularly joyous mood and no premonition of the impending catastrophe oppressed me. No sense of sorrow, present or to come, forced itself upon me, even when I saw men hurrying through the almost deserted streets. When I got within sight of my home and saw a crowd surrounding it, I was only interested sufficiently to spur my horse into a jog trot, which brought me up to the throng, when something in the sullen, settled horror in the men's faces gave me a sudden, sick thrill. They whispered a word to me, and without a thought, save for Annie, the girl who had been so surely growing into my heart, I leaped from the saddle and tore my way through the people to the house.

"It was Annie, poor girl, bruised and bleeding, her face and dress torn from struggling. They were gathered round her with white faces, and, oh, with what terrible patience they were trying to gain from her fluttering lips the name of her murderer. They made way for me and I knelt at her side. She was beyond my skill, and my will merged with theirs. One thought was in our minds.

"Who?" I asked.

"Her eyes half opened, 'That black –' She fell back into my arms dead.

"We turned and looked at each other. The mother had broken down and was weeping, but the face of the father was like iron.

"'It is enough,' he said; 'Jube has disappeared.' He went to the door and said to the expectant crowd, 'She is dead.'

"I heard the angry roar without swelling up like the noise of a flood, and then I heard the sudden movement of many feet as the men separated into searching parties, and laying the dead girl back upon her couch, I took my rifle and went out to join them.

"As if by intuition the knowledge had passed among the men that Jube Benson had disappeared, and he, by common consent, was to be the object of our search. Fully a dozen of the citizens had seen him hastening toward the woods and noted his skulking air, but as he had grinned in his old good-natured way they had, at the time, thought nothing of it. Now, however, the diabolical reason of his slyness was apparent. He had been shrewd enough to disarm suspicion, and by now was far away. Even Mrs. Daly, who was visiting with a neighbour, had seen him stepping out by a back way, and had said with a laugh, 'I reckon that black rascal's a-running off somewhere.' Oh, if she had only known.

"'To the woods! To the woods!' that was the cry, and away we went, each with the determination not to shoot, but to bring the culprit alive into town, and then to deal with him as his crime deserved.

"I cannot describe the feelings I experienced as I went out that night to beat the woods for this human tiger. My heart smouldered within me like a coal, and I went forward under the impulse of a will that was half my own, half some more malignant power's. My throat throbbed drily, but water nor whiskey would not have quenched my thirst. The thought has come to me since that now I could interpret the panther's desire for blood and sympathise with it, but then I thought nothing. I simply went forward, and watched, watched with burning eyes for a familiar form that I had looked for as often before with such different emotions.

"Luck or ill-luck, which you will, was with our party, and just as dawn was graying the sky, we came upon our quarry crouched in the corner of a fence. It was only half light, and we might have passed, but my eyes had caught sight of him, and I raised the cry. We levelled our guns and he rose and came toward us.

"'I t'ought you wa'n't gwine see me,' he said sullenly, 'I didn't mean no harm.'

"Harm!

"Some of the men took the word up with oaths, others were ominously silent.

"We gathered around him like hungry beasts, and I began to see terror dawning in his eyes. He turned to me, 'I's moughty glad you's hyeah, doc,' he said, 'you ain't gwine let 'em whup me.'

"'Whip you, you hound,' I said, 'I'm going to see you hanged,' and in the excess of my passion I struck him full on the mouth. He made a motion as if to resent the blow against even such great odds, but controlled himself.

"'W'y, doctah,' he exclaimed in the saddest voice I have ever heard, 'w'y, doctah! I ain't stole nuffin' o' yo'n, an' I was comin' back. I only run off to see my gal, Lucy, ovah to de Centah.'

"'You lie!' I said, and my hands were busy helping the others bind him upon a horse. Why did I do it? I don't know. A false education, I reckon, one false from the beginning. I saw his black face glooming there in the half light, and I could only think of him as a monster. It's tradition. At first I was told that the black man would catch me, and when I got over that, they taught me that the devil was black, and when I had recovered from the sickness of that belief, here were Jube and his fellows with faces of menacing blackness. There was only one conclusion: This black man stood for all the powers of evil, the result of whose machinations had been gathering in my mind from childhood up. But this has nothing to do with what happened.

"After firing a few shots to announce our capture, we rode back into town with Jube. The ingathering parties from all directions met us as we made our way up to the house. All was very quiet and orderly. There was no doubt that it was as the papers would have said, a gathering of the best citizens. It was a gathering of stern, determined men, bent on a terrible vengeance.

"We took Jube into the house, into the room where the corpse lay. At sight of it, he gave a scream like an animal's and his face went the colour of storm-blown water. This was enough to condemn him. We divined, rather than heard, his cry of 'Miss Ann, Miss Ann, oh, my God, doc, you don't t'ink I done it?'

"Hungry hands were ready. We hurried him out into the yard. A rope was ready. A tree was at hand. Well, that part was the least of it, save that Hiram Daly stepped aside to let me be the first to pull upon the rope. It was lax at first. Then it tightened, and I felt the quivering soft weight resist my muscles. Other hands joined, and Jube swung off his feet.

"No one was masked. We knew each other. Not even the Culprit's face was covered, and the last I remember of him as he went into the air was a look of sad reproach that will remain with me until I meet him face to face again.

"We were tying the end of the rope to a tree, where the dead man might hang as a warning to his fellows, when a terrible cry chilled us to the marrow.

"'Cut 'im down, cut 'im down, he ain't guilty. We got de one. Cut him down, fu' Gawd's sake. Here's de man, we foun' him hidin' in de barn!'

"Jube's brother, Ben, and another Negro, came rushing toward us, half dragging, half carrying a miserable-looking wretch between them. Someone cut the rope and Jube dropped lifeless to the ground.

"'Oh, my Gawd, he's daid, he's daid!' wailed the brother, but with blazing eyes he brought his captive into the centre of the group, and we saw in the full light the scratched face of Tom Skinner – the worst white ruffian in the town – but the face we saw was not as we were accustomed to see it, merely smeared with dirt. It was blackened to imitate a Negro's.

"God forgive me; I could not wait to try to resuscitate Jube. I knew he was already past help, so I rushed into the house and to the dead girl's side. In the excitement they had not yet washed or laid her out. Carefully, carefully, I searched underneath her broken finger nails. There was skin there. I took it out, the little curled pieces, and went with it to my office.

"There, determinedly, I examined it under a powerful glass, and read my own doom. It was the skin of a white man, and in it were embedded strands of short, brown hair or beard.

"How I went out to tell the waiting crowd I do not know, for something kept crying in my ears, 'Blood guilty! Blood guilty!'

"The men went away stricken into silence and awe. The new prisoner attempted neither denial nor plea. When they were gone I would have helped Ben carry his brother in, but he waved me away fiercely, 'You he'ped murder my brothah, you dat was his frien', go 'way,' go 'way! I'll tek him home myse'f.' I could only respect his wish, and he and his comrade took up the dead man and between them bore him up the street on which the sun was now shining full.

"I saw the few men who had not skulked indoors uncover as they passed, and I – I – stood there between the two murdered ones, while all the while something in my ears kept crying, 'Blood guilty! Blood guilty!'"

The doctor's head dropped into his hands and he sat for some time in silence, which was broken by neither of the men, then he rose, saying, "Gentlemen, that was my last lynching."

Near the end of this story, the physician narrator notes "this isn't a pleasant story. The truth is usually a nasty thing to stand." Such nasty, standing "truths" are the antitheses to the goals and commitments of healthcare, and cognizance of their existence, implicit and explicit, can and should be part of the education of healthcare providers.

Related Poem

The related poem for this chapter is an American Slave Spiritual. Its figure for the lament of loneliness—the expression of feeling like a "motherless child"—should not be taken too figuratively. One of the most terrible crimes of chattel slavery in America was the systematic destruction of families, the selling off of parents and children, husbands and wives. The emancipation of Russian serfs—announced by Tsar Alexander II in 1861—almost directly coincides historically with the war to abolish slavery in the United States. The difference between serfs and American slaves, however, is that serfs belong to the land they lived on: like trees and houses, they were bought and sold as part of an estate. American slaves, however, were treated like articles of personal possessions, completely "alienable" property. Thus, serfdom, while enslaving human beings, does not destroy a sense of family belonging, while American slavery literally created motherless children. The comprehension of this difference, we believe, can help healthcare providers comprehend possibilities of a terrible "foreign" experience. In large part, such "comprehension" is at the base of the powerful empathy provoked by Toni Morrison's novel *Beloved*, discussed in Chap. 4. That is, a "literal" understanding of this poem can help healthcare providers more fully comprehend the many kinds of "differences" that are examined in this chapter focused on the role of the doctor.

Poem: "Sometimes I Feel Like A Motherless Child" (Probably Nineteenth Century), Traditional Spiritual

Author Note: This song/poem, like so many folk songs in every culture, is "authorless," part of the oral tradition that expresses a culture as a whole. In this case, it is the expression of slave culture in the United States, in which Christian "spiritual" music was combined with the horrors of slavery. More specifically, American chattel slavery regularly and consistently broke up families, tearing children from their parents, brothers from sisters, husbands and wives.

Sometimes I Feel Like a Motherless Child

Sometimes I feel like a motherless child,
Sometimes I feel like a motherless child,
Sometimes I feel like a motherless child,
A long way from home,

A long way from home.
(True believer.)
A long way from home,
A long way from home.

Sometimes I feel like I'm almost gone,
Sometimes I feel like I'm almost gone,
Sometimes I feel like I'm almost gone,
Way up in the heav'nly land,
Way up in the heav'nly land.
(True believer.)
Way up in the heav'nly land,
Way up in the heav'nly land.

Lessons for Providers

In every patient encounter there are cultural differences between provider and patient. No matter how small they may be, we must constantly struggle to understand how the patient's culture and background—life experiences both embraced and imposed by others—shade their perception of reality; and how, as healthcare providers, our own culture and background do the same to our perceptions.

Bibliography

LaCombe, Michael. 2010. *Beside: The Art of Medicine*. Orono: University of Maine Press.

Schleifer, Ronald, and Jerry Vannatta. 2013. *The Chief Concern of Medicine: The Integration of the Medical Humanities and Narrative Knowledge into Medical Practices*. Ann Arbor: University of Michigan Press.

Tweedy, Damon. 2015. *Black Man in a White Coat: A Doctor's Reflections on Race and Medicine*. New York: Picador.

Section V
Everyday Ethics of Medical Practices

Everyday Ethics of Medical Practices

The phrase "everyday ethics" in the title of this chapter is designed to distinguish itself from two other ways of understanding judgements of value in relation to human conduct—such judgments of value being the domain of ethics—namely "normative" ("deontological" or "principle-based") ethics, which judges action by means of universal ("normative") truths about what is right or wrong, and "utilitarian" ethics, which judges action by means of cost-benefit calculations to determine what is right or wrong. In the first case, judgments are based upon universal truths or moral imperatives (rules), such as the absolute sanctity of human life (e.g., "thou shalt not kill") or the necessity of recognizing the value of others (e.g., "do unto others as you would have them do unto you"). In the second case, judgments are based upon the outcomes or consequences of choosing one action over another action, and such outcomes are usually measured by means of a calculation, developed by Jeremy Bentham and other eighteenth-century philosophers, of the greatest happiness of the greatest number (e.g., vaccinations are ethically good because they insure the widespread suppression of contagious diseases, even if they entail a very small number of individual negative outcomes). A third school of ethical judgment is described as "virtue ethics," which focuses on "virtues" (or moral character) as they manifest themselves in the behavior of individuals in relation to others. We are calling this focus "everyday ethics" because virtues—as opposed to stand-alone rules or cost-benefit moral calculations—can only be discerned in patterns of behavior, which is the substance of narrative art (including the implicit narratives in poetry). Virtue ethics finds its origin in the ancient work of Aristotle, yet many of the virtues that promote good healthcare (such as the "Pediatric Professionalism Milestones" spelled out in Appendix 5, namely empathy, duty and accountability, professional boundaries, self-awareness, trustworthiness, the ability to cope with uncertainty) are not abstract categories or calculations, but grow out of everyday interactions between and among people. This is why, as we see in Appendix 5, that workshop-training on professionalism benefits from engagements with literary narratives.

© The Author(s) 2019
R. Schleifer, J. B. Vannatta, *Literature and Medicine*,
https://doi.org/10.1007/978-3-030-19128-3_8

The discernment of virtues and "character" is clear in the vignette in this chapter, the negative example of a resident-physician's behavior toward one of his patients that allows us to feel and discern virtues that are markedly absent from its narrative.

The Patient with Diabetic Ketoacidosis: A Vignette (Excerpt *The Chief Concern of Medicine*)

The thirty-two year old woman had been ravaged by type 1 diabetes since age thirteen. She had married at age eighteen, and gave birth to three daughters. This particular week she had spent in the hospital, yet again suffering from diabetic ketoacidosis – caused in large part by several infected boils on her skin. For the first thirty-six hours she was in the intensive care unit, then out on the medicine wards to gain enough strength to return home. Home, however, had a new meaning recently: she had moved out of the family home, accusing her husband of abusing her physically. She smoked two packages of cigarettes a day, and often was outside the hospital smoking so that the intern and resident, as well as the attending physician, commonly had a difficult time locating her.

Caring for this woman proved very difficult. The intern and resident found her angry in general, irrational in her decision making, and commonly attempting to play one caretaker against another. She believed that her stomach was "dead" – she "remembered" that her previous doctor told her so – and since her stomach was dead, she obviously couldn't take the oral antibiotics. Because of this, the resident had to keep her in the hospital.

On one particular day, the resident burst into the patient's room without knocking. With very little introduction, the resident began telling the patient that she was ready to go home, began listing a series of actions the patient must do to take good care of diabetes, such as checking her blood sugar twice a day, exercising daily, dieting, and taking the medications exactly as prescribed. The patient, having had this disease for nearly twenty years, had heard this all before, and yet she rarely performed any of these tasks. Consequently, the patient began to answer with reasons why she could not check her blood sugar – the strips were too expensive. She could not exercise – she must look after three young daughters. She found it difficult to eat the diabetic diet because of lack of funds, and besides she usually vomited most of her food.

In response to the patient's excuses, the resident interrupted her twice, raised the volume of his voice in response, and proceeded to argue with the patient concerning every point of the discussion. This exchange escalated in a very short time into a fight. The patient told the resident to get out of the room and in five minutes, the nursing staff paged the attending physician because the patient was threatening to leave the hospital with an intravenous line in place.

Virtues in Action

The rancor between physician and patient represented in this vignette is not uncommon, even if this might seem an extreme case. Chronic illness is difficult, and it has clearly taken its toll on this patient. Still, for the healthcare worker to competently care for this patient, he needs not only sufficient biomedical knowledge, but also certain interpersonal "virtues," including discernment, conscientiousness, and compassion, which would exhibit themselves in his behavior. A *discerning* physician would recognize in the patient's "unspoken" story, namely that this patient is not behaving rationally, and such irrational behavior requires different strategies from the demand for compliance the resident expresses. That is, a discerning engagement with this patient would be more mindful of the patient's limitations. (This virtue of discernment is clearly absent from the narrator's physician-husband in "The Yellow Wallpaper" in Chap. 6.) Moreover, a *conscientious* physician would modify her actions in dealing with her patient and replace confrontation with negotiation of one form or another. Such conscientious action would begin with asking the patient what immediate and longer-term outcomes she hopes for, and build the interchange around the patient's needs and desires rather than the physician's requirements. Finally, a *compassionate* physician would recognize and acknowledge his recognition of the patient's plight as an abused wife and single parent coping with a terrible chronic illness and an inadequate income. Such recognition and active verbal acknowledgment is the *conscientious action* of verbal empathy.

Literature and Virtues

One of the important things that narrative—and literary narrative more particularly—allows us is the ability to discern human virtues of character in everyday actions; the narrative organization of phenomena fosters this kind of attention. That is, in the same way that literature encourages us to grasp patterns of action (narrative events) and representation (discursive language) as we noted in Chap. 1, it also encourages us to discern within such patterns the qualities of "character" that are at the heart of virtue ethics. When virtue ethics talks of "moral character," it is describing formations of judgements and discriminations of behavior that precipitate out of a series of actions, which now can be understood as a patterned "whole" of actions grasped as the narrative "overall meaning," the literary feature with which we end Chap. 1. When we talk of the virtue of "compassion," for instance, we find it in patterns of empathetic responses, reassuring gestures, engaged listening, and so on, which "add up," so to speak, to the "overall meaning" mentioned in Chap. 1, but here understood not as the overall meaning of a story, but a person's "overall meaning," her "moral character." This is why virtue or everyday ethics is closely aligned with literary narrative, just as normative ethics is closely aligned with universal—and often religious—truths and utilitarian ethics is closely aligned with measure and calculation. Moreover, these alignments should make clear how everyday virtue ethics is at the heart of clinical medicine—just as, perhaps, utilitarian ethics is at the heart of public health and normative ethics is at the heart of working conceptions of medical malpractice.

Ethics and Literature

Training in the discernment of "character" and ethics is particularly useful in healthcare education because it allows healthcare providers to *reflect upon* their own behavior in salutary ways (including reflecting on possibilities of "implicit bias" we discussed in Chap. 7). This, after all, is the purpose of the workshop described in Appendix 5, which trains people to be aware of qualities and levels of virtuous behavior. Such training in the context of healthcare is useful whether or not literary narratives are explicitly focused on healthcare as such. The literary narrative in this chapter, Dr. Anton Chekhov's "Enemies," is superficially "about" healthcare, but more profoundly it is a story about what Chekhov calls "that subtle, almost elusive beauty of human sorrow" and his powerful descriptions of the ways pain and the "egoism of the unhappy"—like the unhappiness of the woman suffering from diabetic ketoacidosis and also her resident-physician—erase many of the virtues we set forth in this chapter.

Literary Narrative: "Enemies" (1887) by Dr. Anton Chekhov

Author Note: Dr. Anton Chekhov (1860–1904), grandson of a Russian serf, was a physician, short-story writer, and playwright. He is considered to be among the greatest writers of short fiction, and his innovations in drama in the early twentieth century are also considered to have transformed the theater. Throughout his writing career he practiced medicine. He once said that "medicine is my lawful wife, and literature is my mistress."

Enemies

Between nine and ten on a dark September evening the only son of the district doctor, Kirilov, a child of six, called Andrey, died of diphtheria. Just as the doctor's wife sank on her knees by the dead child's bedside and was overwhelmed by the first rush of despair there came a sharp ring at the bell in the entry.

All the servants had been sent out of the house that morning on account of the diphtheria. Kirilov went to open the door just as he was, without his coat on, with his waistcoat unbuttoned, without wiping his wet face or his hands which were scalded with carbolic. It was dark in the entry and nothing could be distinguished in the man who came in but medium height, a white scarf, and a large, extremely pale face, so pale that its entrance seemed to make the passage lighter.

"Is the doctor at home?" the newcomer asked quickly.

"I am at home," answered Kirilov. "What do you want?"

"Oh, it's you? I am very glad," said the stranger in a tone of relief, and he began feeling in the dark for the doctor's hand, found it and squeezed it tightly in his own. "I am very … very glad! We are acquainted. My name is Abogin, and I had the honor of meeting you in the summer at Gnutchev's. I am very glad I have found you at home. For God's sake don't refuse to come back with me at once …. My wife has been taken dangerously ill …. And the carriage is waiting…."

From the voice and gestures of the speaker, it could be seen that he was in a state of great excitement. Like a man terrified by a house on fire or a mad dog, he could hardly restrain his rapid breathing and spoke quickly in a shaking voice, and there was a note of unaffected sincerity and childish alarm in his voice. As people always do who are frightened and overwhelmed, he spoke in brief, jerky sentences and uttered a great many unnecessary, irrelevant words.

"I was afraid I might not find you in," he went on. "I was in a perfect agony as I drove here. Put on your things and let us go, for God's sake…. This is how it happened. Alexandr

Semyonovitch Paptchinsky, whom you know, came to see me.... We talked a little and then we sat down to tea; suddenly my wife cried out, clutched at her heart, and fell back on her chair. We carried her to bed and ... and I rubbed her forehead with ammonia and sprinkled her with water ... she lay as though she were dead.... I am afraid it is aneurism .... Come along ... her father died of aneurism."

Kirilov listened and said nothing, as though he did not understand Russian.

When Abogin mentioned again Paptchinsky and his wife's father and once more began feeling in the dark for his hand, the doctor shook his head and said apathetically, dragging out each word:

"Excuse me, I cannot come ... my son died ... five minutes ago!"

"Is it possible!" whispered Abogin, stepping back a pace. "My God, at what an unlucky moment I have come! A wonderfully unhappy day ... wonderfully. What a coincidence.... It's as though it were on purpose!"

Abogin took hold of the door handle and bowed his head. He was evidently hesitating and did not know what to do—whether to go away or to continue entreating the doctor.

"Listen," he said fervently, catching hold of Kirilov's sleeve. "I well understand your position! God is my witness that I am ashamed of attempting at such a moment to intrude on your attention, but what am I to do? Only think, to whom can I go? There is no other doctor here, you know. For God's sake come! I am not asking you for myself.... I am not the patient!"

A silence followed. Kirilov turned his back on Abogin, stood still a moment, and slowly walked into the drawing-room. Judging from his unsteady, mechanical step, from the attention with which he set straight the fluffy shade on the unlighted lamp in the drawing-room and glanced into a thick book lying on the table, at that instant he had no intention, no desire, was thinking of nothing and most likely did not remember that there was a stranger in the entry. The twilight and stillness of the drawing-room seemed to increase his numbness. Going out of the drawing-room into his study he raised his right foot higher than was necessary, and felt for the doorposts with his hands, and as he did so there was an air of perplexity about his whole figure as though he were in somebody else's house, or were drunk for the first time in his life and were now abandoning himself with surprise to the new sensation. A broad streak of light stretched across the bookcase on one wall of the study; this light came together with the close, heavy smell of carbolic and ether from the door into the bedroom, which stood a little way open.... The doctor sank into a low chair in front of the table; for a minute he stared drowsily at his books, which lay with the light on them, then got up and went into the bedroom.

Here in the bedroom reigned a dead silence. Everything to the smallest detail was eloquent of the storm that had been passed through, of exhaustion, and everything was at rest. A candle standing among a crowd of bottles, boxes, and pots on a stool and a big lamp on the chest of drawers threw a brilliant light over all the room. On the bed under the window lay a boy with open eyes and a look of wonder on his face. He did not move, but his open eyes seemed every moment growing darker and sinking further into his head. The mother was kneeling by the bed with her arms on his body and her head hidden in the bedclothes. Like the child, she did not stir; but what throbbing life was suggested in the curves of her body and in her arms! She leaned against the bed with all her being, pressing against it greedily with all her might, as though she were afraid of disturbing the peaceful and comfortable attitude she had found at last for her exhausted body. The bedclothes, the rags and bowls, the splashes of water on the floor, the little paint-brushes and spoons thrown down here and there, the white bottle of lime water, the very air, heavy and stifling – were all hushed and seemed plunged in repose.

The doctor stopped close to his wife, thrust his hands in his trouser pockets, and slanting his head on one side fixed his eyes on his son. His face bore an expression of indifference, and only from the drops that glittered on his beard it could be seen that he had just been crying.

That repellent horror which is thought of when we speak of death was absent from the room. In the numbness of everything, in the mother's attitude, in the indifference on the doctor's face there was something that attracted and touched the heart, that subtle, almost elusive beauty of human sorrow which men will not for a long time learn to understand and describe, and which it seems only music can convey. There was a feeling of beauty, too, in the austere stillness. Kirilov and his wife were silent and not weeping, as though besides the bitterness of their loss they were conscious, too, of all the tragedy of their position; just as once their youth had passed away, so now together with this boy their right to have children had gone for ever to all eternity! The doctor was forty-four, his hair was grey and he looked like an old man; his faded and invalid wife was thirty-five. Andrey was not merely the only child, but also the last child.

In contrast to his wife the doctor belonged to the class of people who at times of spiritual suffering feel a craving for movement. After standing for five minutes by his wife, he walked, raising his right foot high, from the bedroom into a little room which was half filled up by a big sofa; from there he went into the kitchen. After wandering by the stove and the cook's bed he bent down and went by a little door into the passage.

There he saw again the white scarf and the white face.

"At last," sighed Abogin, reaching towards the door-handle. "Let us go, please."

The doctor started, glanced at him, and remembered... .

"Why, I have told you already that I can't go!" he said, growing more animated. "How strange!"

"Doctor, I am not a stone, I fully understand your position ... I feel for you," Abogin said in an imploring voice, laying his hand on his scarf. "But I am not asking you for myself. My wife is dying. If you had heard that cry, if you had seen her face, you would understand my pertinacity. My God, I thought you had gone to get ready! Doctor, time is precious. Let us go, I entreat you."

"I cannot go," said Kirilov emphatically and he took a step into the drawing room.

Abogin followed him and caught hold of his sleeve.

"You are in sorrow, I understand. But I'm not asking you to a case of toothache, or to a consultation, but to save a human life!" he went on entreating like a beggar. "Life comes before any personal sorrow! Come, I ask for courage, for heroism! For the love of humanity!"

"Humanity – that cuts both ways," Kirilov said irritably. "In the name of humanity I beg you not to take me. And how queer it is, really! I can hardly stand and you talk to me about humanity! I am fit for nothing just now.... Nothing will induce me to go, and I can't leave my wife alone. No, no..."

Kirilov waved his hands and staggered back.

"And ... and don't ask me," he went on in a tone of alarm. "Excuse me. By No. XIII of the regulations I am obliged to go and you have the right to drag me by my collar ... drag me if you like, but ... I am not fit ... I can't even speak ... excuse me."

"There is no need to take that tone to me, doctor!" said Abogin, again taking the doctor by his sleeve. "What do I care about No. XIII! To force you against your will I have no right whatever. If you will, come; if you will not – God forgive you; but I am not appealing to your will, but to your feelings. A young woman is dying. You were just speaking of the death of your son. Who should understand my horror if not you?"

Abogin's voice quivered with emotion; that quiver and his tone were far more persuasive than his words. Abogin was sincere, but it was remarkable that whatever he said his words sounded stilted, soulless, and inappropriately flowery, and even seemed an outrage on the atmosphere of the doctor s home and on the woman who was somewhere dying. He felt this himself, and so, afraid of not being understood, did his utmost to put softness and tenderness into his voice so that the sincerity of his tone might prevail if his words did not. As a rule, however fine and deep a phrase may be, it only affects the indifferent, and cannot fully satisfy those who are happy or unhappy; that is why dumbness is most often the highest expression of happiness or unhappiness; lovers understand each other better when they are silent, and a fervent, passionate speech delivered by the grave only touches outsiders, while to the widow and children of the dead man it seems cold and trivial.

Kirilov stood in silence. When Abogin uttered a few more phrases concerning the noble calling of a doctor, self-sacrifice, and so on, the doctor asked sullenly: "Is it far?"

"Something like eight or nine miles. I have capital horses, doctor! I give you my word of honour that I will get you there and back in an hour. Only one hour."

These words had more effect on Kirilov than the appeals to humanity or the noble calling of the doctor. He thought a moment and said with a sigh: "Very well, let us go!"

He went rapidly with a more certain step to his study, and afterwards came back in a long frock-coat. Abogin, greatly relieved, fidgeted round him and scraped with his feet as he helped him on with his overcoat, and went out of the house with him.

It was dark out of doors, though lighter than in the entry. The tall, stooping figure of the doctor, with his long, narrow beard and aquiline nose, stood out distinctly in the darkness. Abogin's big head and the little student's cap that barely covered it could be seen now as well as his pale face. The scarf showed white only in front; behind, it was hidden by his long hair.

"Believe me, I know how to appreciate your generosity," Abogin muttered as he helped the doctor into the carriage. "We shall get there quickly. Drive as fast as you can, Luka, there's a good fellow! Please!"

The coachman drove rapidly. At first there was a row of indistinct buildings that stretched alongside the hospital yard; it was dark everywhere except for a bright light from a window that gleamed through the fence into the furthest part of the yard while three windows of the upper storey of the hospital looked paler than the surrounding air. Then the carriage drove into dense shadow; here there was the smell of dampness and mushrooms and the sound of rustling trees; the crows, awakened by the noise of the wheels, stirred among the foliage and uttered prolonged plaintive cries as though they knew the doctor's son was dead and that Abogin's wife was ill. Then came glimpses of separate trees, of bushes; a pond, on which great black shadows were slumbering, gleamed with a sullen light – and the carriage rolled over a smooth level ground. The clamour of the crows sounded dimly far away and soon ceased altogether.

Kirilov and Abogin were silent almost all the way. Only once Abogin heaved a deep sigh and muttered:

"It's an agonizing state! One never loves those who are near one so much as when one is in danger of losing them."

And when the carriage slowly drove over the river, Kirilov started all at once as though the splash of the water had frightened him, and made a movement.

"Listen – let me go," he said miserably. "I'll come to you later. I must just send my assistant to my wife. She is alone, you know!"

Abogin did not speak. The carriage swaying from side to side and crunching over the stones drove up the sandy bank and rolled on its way. Kirilov moved restlessly and looked about him in misery. Behind them in the dim light of the stars the road could be seen and the riverside willows vanishing into the darkness. On the right lay a plain as uniform and as boundless as the sky; here and there in the distance, probably on the peat marshes, dim lights were glimmering. On the left, parallel with the road, ran a hill tufted with small bushes, and above the hill stood motionless a big, red half-moon, slightly veiled with mist and encircled by tiny clouds, which seemed to be looking round at it from all sides and watching that it did not go away.

In all nature there seemed to be a feeling of hopelessness and pain. The earth, like a ruined woman sitting alone in a dark room and trying not to think of the past, was brooding over memories of spring and summer and apathetically waiting for the inevitable winter. Wherever one looked, on all sides, nature seemed like a dark, infinitely deep, cold pit from which neither Kirilov nor Abogin nor the red half-moon could escape....

The nearer the carriage got to its goal the more impatient Abogin became. He kept moving, leaping up, looking over the coachman's shoulder. And when at last the carriage stopped before the entrance, which was elegantly curtained with striped linen, and when he looked at the lighted windows of the second storey, there was an audible catch in his breath.

"If anything happens ... I shall not survive it," he said, going into the hall with the doctor, and rubbing his hands in agitation. "But there is no commotion, so everything must be going well so far," he added, listening in the stillness.

There was no sound in the hall of steps or voices and all the house seemed asleep in spite of the lighted windows. Now the doctor and Abogin, who till then had been in darkness, could see each other clearly. The doctor was tall and stooped, was untidily dressed, and not good-looking. There was an unpleasantly harsh, morose, and unfriendly look about his lips, thick as a negro's, his aquiline nose, and listless, apathetic eyes. His unkempt head and sunken temples, the premature greyness of his long, narrow beard through which his chin was visible, the pale grey hue of his skin and his careless, uncouth manners – the harshness of all this was suggestive of years of poverty, of ill fortune, of weariness with life and with men. Looking at his frigid figure one could hardly believe that this man had a wife, that he was capable of weeping over his child. Abogin presented a very different appearance. He was a thick-set, sturdy-looking, fair man with a big head and large, soft features; he was elegantly dressed in the very latest fashion. In his carriage, his closely buttoned coat, his long hair, and his face there was a suggestion of something generous, leonine; he walked with his head erect and his chest squared, he spoke in an agreeable baritone, and there was a shade of refined almost feminine elegance in the manner in which he took off his scarf and smoothed his hair. Even his paleness and the childlike terror with which he looked up at the stairs as he took off his coat did not detract from his dignity nor diminish the air of sleekness, health, and aplomb which characterized his whole figure.

"There is nobody and no sound," he said going up the stairs. "There is no commotion. God grant all is well."

He led the doctor through the hall into a big drawing-room where there was a black piano and a chandelier in a white cover; from there they both went into a very snug, pretty little drawing-room full of an agreeable, rosy twilight.

"Well, sit down here, doctor, and I ... will be back directly. I will go and have a look and prepare them."

Kirilov was left alone. The luxury of the drawing-room, the agreeably subdued light and his own presence in the stranger's unfamiliar house, which had something of the character of an adventure, did not apparently affect him. He sat in a low chair and scrutinized his hands, which were burnt with carbolic. He only caught a passing glimpse of the bright red lamp-shade and the violoncello case, and glancing in the direction where the clock was ticking, he noticed a stuffed wolf as substantial and sleek-looking as Abogin himself.

It was quiet.... Somewhere far away in the adjoining rooms someone uttered a loud exclamation:

"Ah!" There was a clang of a glass door, probably of a cupboard, and again all was still. After waiting five minutes Kirilov left off scrutinizing his hands and raised his eyes to the door by which Abogin had vanished.

In the doorway stood Abogin, but he was not the same as when he had gone out. The look of sleekness and refined elegance had disappeared – his face, his hands, his attitude were contorted by a revolting expression of something between horror and agonizing physical pain. His nose, his lips, his moustache, all his features were moving and seemed trying to tear themselves from his face, his eyes looked as though they were laughing with agony....

Abogin took a heavy stride into the drawing room, bent forward, moaned, and shook his fists.

"She has deceived me," he cried, with a strong emphasis on the second syllable of the verb. "Deceived me, gone away. She fell ill and sent me for the doctor only to run away with that clown Paptchinsky! My God!"

Abogin took a heavy step towards the doctor, held out his soft white fists in his face, and shaking them went on yelling:

"Gone away! Deceived me! But why this deception? My God! My God! What need of this dirty, scoundrelly trick, this diabolical, snakish farce? What have I done to her? Gone away!"

Tears gushed from his eyes. He turned on one foot and began pacing up and down the drawing room. Now in his short coat, his fashionable narrow trousers which made his legs look disproportionately slim, with his big head and long mane he was extremely like a lion. A gleam of curiosity came into the apathetic face of the doctor. He got up and looked at Abogin.

"Excuse me, where is the patient?" he said.

"The patient! The patient!" cried Abogin, laughing, crying, and still brandishing his fists. "She is not ill, but accursed! The baseness! The vileness! The devil himself could not have imagined anything more loathsome! She sent me off that she might run away with a buffoon, a dull-witted clown, an Alphonse! Oh God, better she had died! I cannot bear it! I cannot bear it!"

The doctor drew himself up. His eyes blinked and filled with tears, his narrow beard began moving to right and to left together with his jaw.

"Allow me to ask what's the meaning of this?" he asked, looking round him with curiosity. "My child is dead, my wife is in grief alone in the whole house.... I myself can scarcely stand up, I have not slept for three nights.... And here I am forced to play a part in some vulgar farce, to play the part of a stage property! I don't ... don't understand it!"

Abogin unclenched one fist, flung a crumpled note on the floor, and stamped on it as though it were an insect he wanted to crush.

"And I didn't see, didn't understand," he said through his clenched teeth, brandishing one fist before his face with an expression as though someone had trodden on his corns. "I did not notice that he came every day! I did not notice that he came today in a closed carriage! What did he come in a closed carriage for? And I did not see it! Noodle!"

"I don't understand ..." muttered the doctor. "Why, what's the meaning of it? Why, it's an outrage on personal dignity, a mockery of human suffering! It's incredible.... It's the first time in my life I have had such an experience!"

With the dull surprise of a man who has only just realized that he has been bitterly insulted the doctor shrugged his shoulders, flung wide his arms, and not knowing what to do or to say sank helplessly into a chair.

"If you have ceased to love me and love another – so be it; but why this deceit, why this vulgar, treacherous trick?" Abogin said in a tearful voice. "What is the object of it? And what is there to justify it? And what have I done to you? Listen, doctor," he said hotly, going up to Kirilov. "You have been the involuntary witness of my misfortune and I am not going to conceal the truth from you. I swear that I loved the woman, loved her devotedly, like a slave! I have sacrificed everything for her; I have quarreled with my own people, I have given up the service and music, I have forgiven her what I could not have forgiven my own mother or sister... I have never looked askance at her.... I have never gainsaid her in anything. Why this deception? I do not demand love, but why this loathsome duplicity? If she did not love me, why did she not say so openly, honestly, especially as she knows my views on the subject? ..."

With tears in his eyes, trembling all over, Abogin opened his heart to the doctor with perfect sincerity. He spoke warmly, pressing both hands on his heart, exposing the secrets of his private life without the faintest hesitation, and even seemed to be glad that at last these secrets were no longer pent up in his breast. If he had talked in this way for an hour or two, and opened his heart, he would undoubtedly have felt better. Who knows, if the doctor had listened to him and had sympathized with him like a friend, he might perhaps, as often happens, have reconciled himself to his trouble without protest, without doing anything needless and absurd.... But what happened was quite different. While Abogin was speaking, the outraged doctor perceptibly changed. The indifference and wonder on his face gradually gave way to an expression of bitter resentment, indignation, and anger. The features of his face became even harsher, coarser, and more unpleasant. When Abogin held out before his eyes the photograph of a young woman with a handsome face as cold and expressionless as a nun's and asked him whether, looking at that face, one could conceive that it was capable of duplicity, the doctor suddenly flew out, and with flashing eyes said, rudely rapping out each word:

"What are you telling me all this for? I have no desire to hear it! I have no desire to!" he shouted and brought his fist down on the table. "I don't want your vulgar secrets! Damnation take them! Don't dare to tell me of such vulgar doings! Do you consider that I have not been insulted enough already? That I am a flunkey whom you can insult without restraint? Is that it?"

Abogin staggered back from Kirilov and stared at him in amazement.

"Why did you bring me here?" the doctor went on, his beard quivering. "If you are so puffed up with good living that you go and get married and then act a farce like this, how do I come in? What have I to do with your love affairs? Leave me in peace! Go on squeezing money out of the poor in your gentlemanly way. Make a display of humane ideas, play (the doctor looked sideways at the violoncello case) play the bassoon and the trombone, grow as fat as capons, but don't dare to insult personal dignity! If you cannot respect it, you might at least spare it your attention!"

"Excuse me, what does all this mean?" Abogin asked, flushing red.

"It means that it's base and low to play with people like this! I am a doctor; you look upon doctors and people generally who work and don't stink of perfume and prostitution as your menials and mauvais ton; well, you may look upon them so, but no one has given you the right to treat a man who is suffering as a stage property!"

"How dare you say that to me!" Abogin said quietly, and his face began working again, and this time unmistakably from anger.

"No, how dared you, knowing of my sorrow, bring me here to listen to these vulgarities!" shouted the doctor, and he again banged on the table with his fist. "Who has given you the right to make a mockery of another man's sorrow?"

"You have taken leave of your senses," shouted Abogin. "It is ungenerous. I am intensely unhappy myself and … and …"

"Unhappy!" said the doctor, with a smile of contempt. "Don't utter that word, it does not concern you. The spendthrift who cannot raise a loan calls himself unhappy, too. The capon, sluggish from over-feeding, is unhappy, too. Worthless people!"

"Sir, you forget yourself," shrieked Abogin. "For saying things like that … people are thrashed! Do you understand?"

Abogin hurriedly felt in his side pocket, pulled out a pocket-book, and extracting two notes flung them on the table.

"Here is the fee for your visit," he said, his nostrils dilating. "You are paid."

"How dare you offer me money?" shouted the doctor and he brushed the notes off the table on to the floor. "An insult cannot be paid for in money!"

Abogin and the doctor stood face to face, and in their wrath continued flinging undeserved insults at each other. I believe that never in their lives, even in delirium, had they uttered so much that was unjust, cruel, and absurd. The egoism of the unhappy was conspicuous in both. The unhappy are egoistic, spiteful, unjust, cruel, and less capable of understanding each other than fools. Unhappiness does not bring people together but draws them apart, and even where one would fancy people should be united by the similarity of their sorrow, far more injustice and cruelty is generated than in comparatively placid surroundings.

"Kindly let me go home!" shouted the doctor, breathing hard.

Abogin rang the bell sharply. When no one came to answer the bell he rang again and angrily flung the bell on the floor; it fell on the carpet with a muffled sound, and uttered a plaintive note as though at the point of death. A footman came in.

"Where have you been hiding yourself, the devil take you?" His master flew at him, clenching his fists. "Where were you just now? Go and tell them to bring the victoria round for this gentleman, and order the closed carriage to be got ready for me. Stay," he cried as the footman turned to go out. "I won't have a single traitor in the house by to-morrow! Away with you all! I will engage fresh servants! Reptiles!"

Abogin and the doctor remained in silence waiting for the carriage. The first regained his expression of sleekness and his refined elegance. He paced up and down the room, tossed his head elegantly, and was evidently meditating on something. His anger had not cooled, but he tried to appear not to notice his enemy…. The doctor stood, leaning with one hand on the edge of the table, and looked at Abogin with that profound and somewhat cynical, ugly contempt only to be found in the eyes of sorrow and indigence when they are confronted with well-nourished comfort and elegance.

When a little later the doctor got into the victoria and drove off there was still a look of contempt in his eyes. It was dark, much darker than it had been an hour before. The red half-moon had sunk behind the hill and the clouds that had been guarding it lay in dark patches near the stars. The carriage with red lamps rattled along the road and soon overtook the doctor. It was Abogin driving off to protest, to do absurd things….

All the way home the doctor thought not of his wife, nor of his Andrey, but of Abogin and the people in the house he had just left. His thoughts were unjust and inhumanly cruel. He condemned Abogin and his wife and Paptchinsky and all who lived in rosy, subdued light among sweet perfumes, and all the way home he hated and despised them till his head ached. And a firm conviction concerning those people took shape in his mind.

Time will pass and Kirilov's sorrow will pass, but that conviction, unjust and unworthy of the human heart, will not pass, but will remain in the doctor's mind to the grave.

—translated by Constance Garnett

Character, Ethics, and Mystery

In Chekhov's writing—we've seen it already in Chap. 4—his characters confront particular *crises* in life, which is precisely where a career in healthcare frequently positions those who, as Anatole Broyard notes, routinely face "the crisis of [their patient's] life" (1992: online). The power of Chekhov's story here is that it is "twice told": two characters, physician and aristocrat, face the terrible loss of child and spouse so that Chekhov is able to describe the *workings* of the absence of virtues—including what we take to be an overriding virtue of "decency"—that are encountered, but not fully understood, in the vignette in this chapter.

In *The Chief Concern of Medicine*, we note six virtues that are particularly useful in healthcare practices. They are Decency, Discernment, Conscientiousness, Trustworthiness, Compassion, and Competence (see *The Chief Concern*: 294–295). (It is instructive to note their relation to the Professional Milestones set forth in Appendix 5 since the latter were developed in order to recognize and instill professional healthcare comportment while the former grew out of engagements with literature.) In *The Chief Concern* we even offer an acronym to help people remember these enumerated virtues: "*Doctor Dogood Comforts The Crying Child.*" One possible exercise in reading "Enemies" might be to consider the degree to which Dr. Kirilov exhibits these virtues, even in the face of his devastating loss, much as the Professional Workshop considers the degree to which Dr. Franciscus in Richard Selzer's story exhibits professional milestones. Do the "undeserved insults" Kirilov and Abogin fling at one another—"unjust, cruel, and absurd"—help us discern "virtuous" human interaction? Would virtues, then, be "just, kind, sensible"? Is "the egoism of the unhappy" a form of arrogance? What is the relationship of arrogance to virtue?

Related Poem

The poem we have chosen for this chapter touches on the mysterious nature of "character" and everyday ethics that Chekhov represents in relation to "the feeling of beauty … in the austere stillness" of grief. William Blake's poem "A Poison Tree" is another depiction of enemies, and it describes the seeming mysterious growth of enmity and evil, replete with biblical imagery from the Garden of Eden in a manner that emphasizes the absence of virtues.

Poem: "A Poison Tree" (1794) by William Blake

> *Author Note*: William Blake (1757–1827) was an English poet, painter, and printmaker. Blake, writing at the time of the French and American Revolutions, was a central figure in English Romantic poetry. His poetry offers a curious combination of politics—he was a great supporter of the democratic revolutions of his age—and a strong commitment to a sense of religion as personal revelation. Some commentators suggest that Blake's true God was the Human Imagination.

A Poison Tree

> I was angry with my friend;
> I told my wrath, my wrath did end.
> I was angry with my foe:
> I told it not, my wrath did grow.
>
> And I waterd it in fears,
> Night & morning with my tears:
> And I sunned it with smiles,
> And with soft deceitful wiles.
>
> And it grew both day and night.
> Till it bore an apple bright.
> And my foe beheld it shine,
> And he knew that it was mine.
>
> And into my garden stole,
> When the night had veild the pole;
> In the morning glad I see;
> My foe outstretched beneath the tree.

Blake included this poem in his *Song of Experience* in 1794, which answered his earlier book *Songs of Innocence*. Note how in the poem what can be imagined, anger conceived of as a growing tree, becomes real. How does Blake's poem accomplish this? Can the same be said of a person's "moral character" or of the "virtues" that manifest that character?

Lessons for Providers

Arrogance is a common poison in our profession. Understanding its origins and constant self-auditing are its only antidote.

Bibliography

Broyard, Anatole. 1992. Doctor Talk to Me. *New York Times*. https://www.nytimes.com/1990/08/26/magazine/doctor-talk-to-me.html

Schleifer, Ronald, and Jerry Vannatta. 2013. *The Chief Concern of Medicine: The Integration of the Medical Humanities and Narrative Knowledge into Medical Practices*. Ann Arbor: University of Michigan Press.

Section VI
Vicarious Experiences

Culture

9

The four chapters of this section offer literary narratives that provoke vicarious experience that can aid in the ability of healthcare providers to understand and comprehend the life experiences of people different from themselves. We begin here with the notion of "culture," by which we mean the *horizon* of possible experiences we discover growing up in an environment of shared assumptions and values that people hold, which, for this reason, do not seem to be "assumptions" and "values" but simply the nature of things, matters of fact. People live and experience assumptions and values almost unconsciously, so that some thinkers have described these beliefs as "habits of thought" that feel, to members of a culture, to be simply how the world is. In the short story of this chapter, Demetria Martinez's "The Annunciation: Lupe," the main character and narrator notes that "the gringos believe in cholesterol the way Mexicanos believe in the existence of God." The ability to grasp and respect another person's (and another culture's) belief system, as the vignette of this chapter suggests, is a skill that is important in healthcare. Moreover, the ability to grasp and understand that "self-evident" truths one holds—such as the truth of chemical analysis implicit in "believing" in cholesterol—do not necessarily govern the understanding of others is also important. (In *The Spirit Catches You and You Fall Down* [1998], a case history of American physicians treating a Hmong child in California, the author Anne Fadiman demonstrates this in her extended narrative.) Religious and political beliefs fall under the category of cultural values, but so do self-evident social roles, such as those of "patient" and "doctor" that we encountered in Section IV of this book. This present chapter focuses on these kinds of cultural experiences, but subsequent chapters in this section focus on particular life experiences that might be outside the experience of healthcare providers: sexual abuse, excruciating—and sometimes chronic—pain, ageing. This chapter also brings up two important topics in medicine not fully suggested by other chapters of *Literature and Medicine*, namely birthing and the ways that healthcare "medicalizes" this natural process; and, more subtly, the ways that sexuality shapes and inflects experience and value.

The sense of "culture" we are describing is recoverable in literary text, which often records what we are too enmeshed in to see distinctly and thus too enmeshed in to imagine could be otherwise. Many years ago, the literary critic Lionel Trilling described this under the category of "manners" in fiction, the "manners" or everyday behaviors by which a culture manifests itself. He describes this as

> all the buzz of implication which always surrounds us in the present, coming to us from what never gets fully stated, coming in the tone of greetings and the tone of quarrels, in slang and humor and popular songs, in the way children play, in the gesture the waiter makes when he puts down the plate, in the nature of the very food we prefer.
>   [...] What I understand as manners, then, is a culture's hum and buzz of implication[, ...] that part of a culture [...] hinted at by small actions [... .] They are things that for good or bad draw the people of a culture together and separate them from the people of another culture. It is the part of a culture which is not art, nor religion, nor morals, nor politics, and yet it relates to all these highly formulated departments of culture. It is modified by them; it modifies them; it is generated by them; it generates them. (1950: 200)

Trilling's description is close to the "body memories" Alicia Gaspar de Alba presents in the poem of this chapter. He is suggesting that the ordinary behaviors he describes are recorded in literary narrative in such a way that "people from another culture"—or, as in Alba's poem, from another sexuality—can see and even experience what would otherwise be separate and foreign, or simply not part of their experience at all. Moreover, we are suggesting that such understanding and experience can help healthcare providers to more fully engage with patients by apprehending the *concerns* that patients bring to their healthcare providers along with their ailments.

The Patient's Chief Concern: A Vignette (Excerpt from *The Chief Concern of Medicine*)

> Accompanied by her daughter in the hospital, Mrs. Jones, an elderly woman with serious bedsores, was faced with the necessity of surgery. But hearing the plan from the attending physician, she refused surgery even after the doctor informed her that without surgery, these sores would not heal and she would die of infection. Both mother and daughter listened carefully and the mother stated, "I can't have surgery today, because the moon is over my chest and I will die of a heart attack in surgery." Her daughter agreed and explained that she had her mother's durable power of attorney and that they both make all major decisions based on the Farmer's Almanac and the major zodiac signs. Frustrated, the physician asked them to think it over and let him know when they could schedule surgery. Every day Mrs. Jones and her daughter rejected surgery because of some problem with the moon and its phase. In the face of this seeming stubbornness, the physician confronted the patient, and the following morning the daughter informed him that they had hired a different doctor, one whom they trusted.
>   The attending physician contacted the new physician by telephone, and he laughed and said yes he would see Mrs. Jones, but only after her illness was taken care of and she left the hospital. Two days later, Mrs. Jones refused surgery again, the moon was over her abdomen, and surgery would ruin her bowels. The new doctor was at the nursing station, and the attending physician explained the situation and asked him if he would go into the room with him and talk to them. He reluctantly agreed. Upon entering the room the patient and her daughter smiled, the room immediately warmed.

"Hello, Mrs. Jones," he said and shook her hand. He then turned to the daughter and introduced himself. They were all aglow.

"I hear you need some surgery," he said enthusiastically.

"Not so sure," Mrs. Jones replied.

"Let me look," showing concern. "Yep, you sure will need surgery on this. When do you suppose we can do this?"

The patient looked nervously at the daughter. The daughter shifted in her chair, looked briefly at both doctors.

"Well, my daughter makes all my decisions, and the moon is over my abdomen now, so…"

"Oh!" he responded, "you use the almanac?"

"Yes," the patient said.

"Do you plant your garden by it too? How was your garden this year?" This was followed by a three-minute colloquy on tomatoes, corn, and turnip greens.

"So, the moon is over your abdomen?" The new doctor moved closer to the bed and touched her belly, "and this would mean…"

"Bowel trouble," Mrs. Jones replied.

"And if it's over your head?"

"A stroke."

"Oh my!"

"And your chest?"

"Heart attack."

"Boy, then we can't do that!" the doctor replied.

He turned to the daughter and asked, "Do you have an almanac?"

"Yes," she said slowly.

Then back to the patient, "Where would the moon need to be to do surgery?"

The patient looked shocked, glancing quickly around the room. "Well, I guess, uh, oh, well…"

He moved back to the bed, kindly touched her lower leg.

"How about your lower leg here?"

"I guess so."

Looking at the daughter, the doctor inquired, "When is the moon over the lower legs?" Immediately he moved to the chair where she sat and helped her look it up. They studied and discussed, changed their minds a few times. It was a negotiation to behold.

"The seventeenth that's it," the daughter said emphatically.

"Yes, that would be a safe day. Three days from now. We will get it scheduled, and Mrs. Jones, you are going to do so well." With the patient looking surprised at her daughter, the new doctor left the room looking enthusiastically over his right shoulder and said, "I'll drop by every day and make sure of it."

What is striking about this vignette is the manner in which the seasoned physician enters into—without disparaging or dismissing—the values his patient assumes about health and about the world.

In this chapter, we turn to literary texts that are enmeshed in Mexican American culture (or what has recently been describe as *Mestizo* culture [see Davis-Undiano 2017]) in order to participate, vicariously, in a culture that might be different from our own. (The whole of Damon Tweedy's book, *Black Man in a White Coat*, offers insights into African American culture, and many other vignettes and stories throughout this book function similarly to present and represent the manners and behaviors that follow from cultural assumptions.) The story itself remarkably brings three cultures into context: Mexican, Mexican American, and White American, and all "the buzz of implication" that creates the *quality* of cultural experience in each of these cultural contexts. Moreover, it is a notable story written in the second person—something rare in literary works—and its discussion of American medicine, in its description of the physician "infomercial," also raises the question of healthcare work for life-enhancing rather than other purposes. Its almost unreflected-upon feelings of prospective birthing and motherhood, like Gaspar de Alba's feelings mixing together love-making and cooking, offer a sense of the felt-experiences of everyday life—a sense of the qualities of experience we call "culture"—which is captured and shared in literary works.

Literary Narrative: "The Annunciation: Lupe" (2012) by Demetria Martinez

Author Note: Demetria Martinez (b. 1960) is a Mexican American poet, fiction writer, and activist. In 1988 she was charged with conspiracy for allegedly transporting two Salvadoran women into the United States, but she was acquitted of the charge. Her works include *Three Times a Woman: Chacana Poetry* (with Alicia Gaspar de Alba and Maria Herrera-Sobek, 1989), *The Devil's Workshop* (2002), and *The Block Captain's Daughter* (2012), from which this story is taken. In 1994 her novel *Mother Tongue* won Western States Book Award for fiction.

The Annunciation: Lupe

You can't believe the ninth month will ever arrive. But it will, and you know you'd better break the news without further delay.

Stretched out on the couch, watching a spider skittering across the ceiling, you say, "Precious one, the doctors took another picture yesterday. And it turns out … well, it turns out that you don't have a pee-pee after all. You, my love, are a girl."

Placing your hands on your belly, you wait for baby to stir. Nothing.

You go on. "Little one, all the time I took coming up with a name for you – Jesús Paul – was in vain. So I set about finding a replacement, no easy thing."

You look across the living room at the TV set and bite your lip. Every afternoon – after long days of waiting on tables at La Tropical – you watch infomercials to unwind. The one you enjoy the most features a doctor in a white coat advertising plastic surgery procedures. Face and butt, abs and boobs. Only in America, you think. No need to be embalmed at death when you can be embalmed throughout life. The doctor carries on for half an hour. Surgery can improve a woman's self-esteem, he crows. It can even change the course of her destiny.

"Now listen up, *mi preciosa*," you say, stroking your belly. "After much prayer I've decided that your name will be Destiny. Destiny Jane Anaya."

The baby kicks not once, not twice, but three times. You have no idea if the baby understands a word of what you've said. Still, you worry. Thinking back to the names of your family in Mexico, you wonder if you've made a terrible mistake.

Adelina, Maudi, Encarnación, Consuelo, Lucinda, and Belén. There's even a Telesfora in there – a great-aunt who joined the Sisters of Loretto, where her name was changed to Crucita. The old-time names make you think of a cast-iron pot, unbreakable, with a lifetime guarantee. Destiny? For an instant it sounds light as cotton candy, too lightweight to pin the child to earth when she lands – a spirit no more, but a human being.

You feel around beside you in the folds of the couch and pull out your cell phone. You point it at the TV to turn it off – then catch yourself and reach for the remote on the coffee table. It has been this way for months – hormones scrambled, moods seesawing – leaving you unable to think clearly, especially at work where the gringos' orders have grown increasingly complex.

"Beans and cheese burrito, hold the cheese." "Huevos rancheros, egg whites only." "Tortillas, the kind without lard." "That's whole beans, please, not fried." Everyone's on one kind of diet or another. When you take orders you feel like a doctor scribbling out a prescription, life and death in your hands. What is the world coming to? The gringos believe in cholesterol the way Mexicanos believe in the existence of God. It's enough to make you ravenous.

You pull yourself up, go to the freezer, and take out two burritos, one for you and one for the child. Your mouth waters. You can just taste the trans-fatty acids.

"Hey, Lupe, have a good one!" the mailman shouts through the screen door. "*Y tu también, Juan*," you answer.

For three days you've let mail pile up: phone, electricity, and gas bills addressed to Guadalupe Gabriela Anaya. Some days you wish you could take a blade to those bills, cutting your name, so heavy with history, into confetti. Five hundred years ago Our Lady of Guadalupe appeared to a Nahuatl-speaking Indian. Two thousand years ago Gabriel appeared to Mary. Visitations, annunciations. You understand such things all too well. Like Juan Diego and Mary, you had no choice but to say yes.

You crossed the Mexican border into Arizona on foot, the phone number of a cousin's cousin hidden in your bra, the sun a broken compass pointing you for days in all the wrong directions – forcing you, finally, to curl up beneath a Palo Verde tree to wait for death.

"Hey Lupe. It's Juan again. Somehow your *Time* magazine got in the wrong bundle. I'll add it to the rest of the stuff. Better take your mail in. Someone will think you're not home and break in."

"*Gracias*," you say, opening the screen door. "It would be embarrassing, no? I was just elected Block Captain. I'm in charge of raising awareness about safety. My campaign platform was 'God helps those who help themselves.'"

You take the mail from the box as Juan moves on to the next house. Forgetting to lock your screen door, you return to your kitchen and set the burritos in the microwave. A few minutes later you take them, steaming, on a plate to a small round table covered with a lace tablecloth topped with a sheet of clear plastic. At the center of the table: a glazed, lime-green pitcher you spent a week's worth of tips on is filled to the brim with cold water.

After you curled up under the Palo Verde tree, you gripped your stomach to try to stop the cramping, which you feared was caused by drinking water out of a cattle trough. You fell asleep and dreamed of the things you'd seen on your journey: plastic water bottles scattered like headstones, empty sardine cans, a perfume bottle, toothbrush, toothpaste, a packet Spanish-English dictionary, and a booklet of prayers to St. Anthony, finder of lost things.

When you woke up the stars shone like coins. They shone like the stars over China where the factory you had worked for relocated, leaving you and hundreds of women with no way to earn a living. The one star fell so close you smelled it, then touched it. You put your finger in your mouth and savored: The star was made of lard, which you once spread on tortillas like it was butter, the main meal for you and your mother during the hard times. You pointed to the sky again and waited for another star to fall, but it did not. You thought of your mother. What will she do, you wondered, if I can't work and send money home? Even lard will be out of reach for her.

Phoenix is just around the bend, you said to the Palo Verde tree, only to realize no words had come out of your mouth. I will freshen up and apply for a job, you said, but again no words emerged. You closed your eyes and thought, I must be dead, and the words came out, sung sweetly in Chinese – your voice and those of hundreds of other women.

You made the sign of the cross and again fell asleep. You dreamed that your bones had turned to dust. In your dreams you heard the Palo Verde tree say, "Potential renal failure." And another tree answered, "Let's get her to the hospital in Tucson. Call the doctor from the church and have him meet us there." You dreamed you opened your eyes and saw a man and woman – your arms over their shoulders – walking you to a van.

"I'm Daniel," said the man, but you heard, Michael the Archangel. "I'm Shanti," said the woman, putting a wet rag on your forehead, as you rested your head on her lap. Like the man, she had fiery winds so large they hung out of the van's open windows. "We're from Southside Church," they said in unison, but you heard, Upon this rock you will build my church. "We're not going to turn you in to *la migra*," they said. But you heard, We were once strangers in the land of Egypt, therefore we must welcome the stranger.

* * *

"Lupe, it's Cory."

"Come in, come in. I'm sitting here daydreaming while my burritos are getting cold. Let me thaw one out for you."

"Sounds good."

"They're bean, cheese, and red chile – nothing too hot. But I want this baby to get used to the red stuff now. Otherwise she'll grow up to be a ketchup Mexican. It happens to the best of us."

"Good news, Lupe. Virginia doesn't need her stroller any more. Don't worry about buying one."

You pull Cory's burrito out of the microwave and touch it to see if it is warm enough. Perfect. "*Gracias, chica*, but I don't need it. The neighbor gave me hers. One of those ones the gringos use to run around the golf course with."

You open the refrigerator and reach in the back for the bottle of Taco Hell salsa, in case Cory wants to spice up her burrito.

Someday, you think, you might tell Cory the truth. That you dipped into your savings and bought the stroller brand-new from Kmart. That one of the things you saw in the desert was a stroller, abandoned by a mother and her child whose fate you can only imagine. Your baby will have a different destiny.

"Okay, Cory. Don't forget our vow. We're going to speak only Spanish for an hour every week. You're coming along so well."

"Ay, Lupe, how would I make it without you?"

"You'd make it just fine."

You take two glasses and the green pitcher, imagining that it is filled with wine that some miracle worker turned into water, clean and cool for you and Cory to drink as you lead her, word by word, into the Spanish language.

Related Poem

For this chapter, the related poem, written by Alicia Gaspar de Alba, continues and enriches the sense of manners and values that Martinez represents, and it nicely complements the flashback of Martinez's narrator's memories of coming to the US with its description of its speaker confusing the everyday activity of cooking with the warmth and memory of love making. Throughout her story, Martinez describes, in passing, the experience of the working-class job of her narrator—she is a waitress in a Mexican restaurant—and Alba's poem develops the *quality* of that experience in depicting household chores. Thus, just as "The Annunciation: Lupe" offers a second-person narrative of a Mexican American woman who is teaching her neighbor how to speak Spanish—and how to become part of the community—so the related poem in this chapter captures, in its first-person narrative of quiet everyday activity after love-making, a sense of quotidian community for a Mexican American woman.

Poem: "Making Tortillas" (1989) by Alicia Gaspar de Alba

Author Note: Alicia Gaspar de Alba (b. 1958) is a prize-winning poet and novelist as well as an interdisciplinary scholar at UCLA, where she was a founding faculty member of the César E. Chávez Department of Chicana/o Studies. Her novels range from historical to noir, and her academic books explore Chicana/o art, sexuality, cultural studies, and gender studies. She has served as Chair of the Lesbian, Gay, Bisexual, Transgender, and Queer Studies Program at UCLA. Her work includes *Desert Blood: The Juarez Murders*, which won the Lambda Literary Award for Best Lesbian Mystery. This poem appeared in *Three Times a Woman: Chicana Poetry*, by herself, Maria Herrera-Sobek, and Demetria Martinez.

Making Tortillas

My body remembers
what it means to love slowly,
what it means to start from scratch:
to soak the maíz,
scatter bonedust in the limewater,
and let the seeds soften
overnight.

Sunrise is the best time
for grinding masa,
cornmeal rolling out
on the metate like a flannel sheet.
Smell of wet corn, lard, fresh
morning love and the light
sound of clapping.

                Pressed between the palms,
  clap-clap
                thin yellow moons—
  clap-clap
                still moist, heavy still
                from last night's soaking
  clap-clap
                slowly start finding their shape
  clap-clap.

My body remembers
the feel of the griddle,
beads of grease sizzling
under the skin, a cry gathering
like an air bubble in the belly
of the unleavened cake. Smell
of baked tortillas all over the house,
all over the hands still
hot from clapping, cooking.

Tortilleras, we are called,
grinders of maíz, makers, bakers,
slow lovers of women.
The secret is starting from scratch.

Lessons for Providers
1. When the healthcare provider isn't connecting because the patient has a different belief system, she must use her imagination. The imagination asks the question: How can I use the patient's perspective to solve the problem? Or under what circumstances could this patient with such and odd belief system solve the problem?
2. The birthing process, while sometimes dangerous for humans and properly considered within a healthcare system, nevertheless is notably different from most conditions that bring patients to healthcare providers. An awareness of this special case is important.
3. Sexuality, like "culture," shapes the experience and shapes the *values* of people. Like culture—like human beings themselves—it commands respect. Such respectfulness is literally part of healthcare; it is the basis of the successful *working together* of patient and healthcare provider.

Bibliography

Con Davis-Undiano, Robert. 2017. *Mestizos Come Home!: Making and Claiming Mexican American Identity*. Norman: University of Oklahoma Press.

Fadiman, Anne. 1998. *The Spirit Catches You and You Fall Down: A Hmong Child, Her American Doctors, and the Collision of Two Cultures*. New York: Noonday.

Trilling, Lionel. 1950. *The Liberal Imagination*. New York: Anchor.

Sexual and Domestic Abuse

10

One issue that confronts healthcare providers too often, we are sorry to say, is sexual and domestic abuse. Although these phenomena are widespread in our culture, very often healthcare professionals have little personal experience of abuse. Moreover, insofar as they encompass psycho-social aspects of a patient's life, and insofar as in our culture we generally share a respect for the privacy of people's personal lives, many physicians and healthcare professionals are particularly shy about confronting the possibility of domestic or sexual abuse as the cause of a patient's condition. In addition, patients themselves are often deeply ashamed of being the victim of abuse perpetrated by people close to them—members of their family, relatives, family friends—and so, often, there is a context in which both patients and healthcare providers are reluctant to directly make possibilities of violent abuse an explicit part of the patient-caretaker interview. In this chapter, we analyze a vignette of a physician encountering wounds that are the result of domestic abuse, which includes the physician's self-conscious deliberation about reasons why it might be best not to ask about domestic abuse. As part of that vignette, the physician-narrator mentions how his experience of reading Roddy Doyle's novel *The Woman Who Walked into Doors* helped shape his interaction with his patient. This novel traces the life of a poor Irish working-class woman, Paula Spencer, from her childhood, when she is subject to verbal abuse and physical harassment from her fellow students and teachers, through 17 years of marriage to Charlo Spencer, during which time he repeatedly beat her and raped her. Her story—written by a man and narrated in the first-person voice of Paula herself—powerfully provokes in readers, both men and women, the terrible vicarious experience of events and thoughts in Paula's life. (For a detailed analysis of the way that Doyle's literary narrative represents the terrible violence of sexual abuse, see Schleifer 2018b.)

You Don't Deserve This: A Vignette by Dr. Jerry Vannatta

Here is a vignette of a physician encountering a patient with serious facial bruises. Note that the patient initially misrepresents what happened to her and tries to misrepresent the emotions she brings to this encounter.

> As I entered the exam room it was obvious that my long-time patient, Sharon, was upset. The room was permeated with anxiety and stress, void of her usual effusive delight that she usually displayed when we met. Sharon sat on the exam table. As I approached, she turned to greet me and showed the ugly, black, swollen left eye.
> "My God, Sharon, what happened to your eye?" I exclaimed. I had not waited for her to give me her chief complaint. It seemed to me it was obvious.
> "Oh, I had a car wreck and hit my head on the steering wheel," she replied in an uplifted voice that belied the demeanor of sadness and stress she presented.
> "A car wreck?" I asked incredulously. "That doesn't look like a car wreck to me." I hesitated, but risking a lot, I quickly added, "It looks to me more like a fist."
> Sharon broke into tears. I remained silent. The risk of guessing that a fist had created the havoc on her face seemed huge: my observation hung like an accusation, not only of whoever had struck Sharon, but an accusation of Sharon herself. On some level it also felt like it was none of my business. Still, explicitly stating my judgment also felt appropriate, probably because of the long term and strong relationship we had built as patient and physician over many years. Even so, the waiting after I spoke seemed interminable (though probably it was only a few seconds).
> I reached out and touched her shoulder. She leaned against my arm. Her hand came up – grasped it. She sobbed.
> I spoke, but only a few words. "Sharon I do not know who did this to you, but I do know that you don't deserve this." Sharon continued to cry for a few moments, and then began to relay a sad story about domestic abuse. She said that the abuse had begun as emotional abuse, then became physical and now also sexual. She had made the appointment because she didn't know what else to do. Even when she came in, she didn't know what to do.
> A few years later, after Sharon had left the dangerous environment, divorced, and moved to another state, she reported in a public venue – while working for a shelter for abused women – that those few words from her physician were what it took to motivate her to act: to leave her husband, file for divorce, and leave the state.
> Reflecting on the risk of making that statement as a physician brought several thoughts to my mind. When making a diagnosis, we are always "guessing" or abducing what might be the cause. When faced with highly charged emotional situations the risk of acting seems high – insulting and hurting a patient, taking the responsibility for changing someone's life, provoking an angry denial or simply, as with Sharon, tears and an emotional confrontation that shouldn't be part of the job. But the possible pay-off is likewise high. Also, such fears are overblown – and perhaps simply a way of avoiding a highly charged emotional situation: after all, usually the worst that can happen is you can be wrong, for which the patient will always forgive you with rapid correction.
> On further reflection of that day with Sharon I understand that the courage to take the risk was brought about by having recently read Roddy Doyle's novel, *The Woman Who Walked into Doors*. As I thought about the novel – its emotional effect on me – I realized that the experience of reading Paula's story had changed me as a physician. The change was in my "automatic" behavior. I could not ignore my educated guess of what had happened to her. I had no choice but to take the risk.

This vignette describes a middle-class victim of abuse, just as *The Woman Who Walked into Doors* describes a lower-class victim. This is important because, in fact, sexual and domestic abuse occurs in our society equally in all social classes.

This vignette presents a powerful narrative that describes the patient-physician encounter from the vantage of the physician, just as Doyle's novel is written from the vantage of the patient-survivor of domestic abuse and just as the literary example from Edgar Allan Poe is written from the vantage of the abuser himself. The physician vantage is important: it spells out some of the reasons why healthcare professionals are reluctant to bring up questions about domestic abuse in the interchange with patients who themselves are reluctant to bring up the topic. As this narrative suggests, the issue raises all kinds of emotional—and even *personal*—issues that reasonably do not seem to be part of a strictly biomedical conception of clinical work. Moreover, it suggests that the very fact that a patient does not accurately describe what happened is a call for a physician to respect his privacy, and that pointing out such concealment is itself some kind of violation of ordinary decency. Finally, as is sometimes the case—though it is not in the situation described in this vignette—it is easy simply to wonder why in the world a woman (or a man) would put up with abuse rather than simply walk away from "the dangerous environment" this vignette describes. That is, it is easy for some healthcare providers to imagine that a victim of abuse is somehow "responsible" for their ailments, just as a generation ago large numbers of healthcare professionals (and ordinary people as well) blamed people with AIDS for their health condition. (A fine dramatic narrative of this situation—again from the point of view of the physician—in Abraham Verghese's *My Own Country*.)

A second important aspect of this vignette is what the physician says to his patient: "you don't deserve this." As Doyle demonstrates in his full-length novel, many victims of abuse feel terribly alone, ashamed, and, as Paula says repeatedly in the novel, that they somehow are responsible for the violence they experience. We learn that *all her life* Paula was the object of abuse: even her father calls her a "slut" when she puts on makeup as a young teenager. The doctor's few words—"you don't deserve this"—is a powerful statement coming from a person in authority. Moreover, words such as these can "percolate"; that is, they don't have to have an immediate effect, but can contribute to someone's developing a different way of seeing things, thinking about things, figuring out what to do some time after she hears these words. In a contrary fashion, the physicians whom Paula encounters in Doyle's novel reinforce what the world had already told her: their neglect and indifference suggests that she *does* deserve the abuse she suffers. Time and again when healthcare providers treat her, she silently says to herself "Ask me. Ask me. Ask me" (1996: pp. 164 [see also p. 23], 175, 187, 202). She silently pleas for healthcare workers to see her as a person rather than an alcoholic poor person who walks into doors.

Finally, an important part of this vignette is how it represents the effects of vicarious experience itself. The *experience* of Paula's world by the physician in this vignette, who read Doyle's novel—his vicarious experience of the abuse she suffers, of her family's and physicians' willful ignoring of its obvious signs, of Paula's sense of being trapped in a situation which allows no reasonable alternatives—changed the way he encounters obvious abuse presented by a patient. That sense of the imaginative grasping of experience—as in Dr. Charon's imaginative grasping of the practical experience of a 98-year-old demented woman we saw in Chap. 1—allows physicians and healthcare professionals to comprehend the *plight* of a patient and to focus on that and *to act* on that understanding as part of caring for those in distress. As this might suggest, if time allows we recommend reading a novel like *The Woman Who Walked into Doors* in the study of literature and medicine, or even simply Chapters 25 and 26, which graphically depict the pain and bewilderment Paula feels when she is first attacked by her husband.

The literary narrative presented here is written neither from the vantage of the healthcare provider nor that of one of the victims of abuse. (We should add that such victims are not exclusively women. In 2017 the US Center for Disease Control reported that "1 in 4 women and 1 in 7 men will experience severe physical violence by an intimate partner in their lifetime" [Safehorizon 2018].) The literary example we are offering is a true "horror" story of a man who is so obsessed with his fiancée—and particularly with conceiving of her as "body parts"—that upon seeing that she was mistakenly pronounced dead, attacks and mutilates her. In Poe's story, Berenice, the victim of this crime, like Paula, never speaks to those in authority—often in her interactions with the narrator, her cousin and fiancé Egaeus, "she spoke, however, no word" and never says a word in the story itself—and, as the narrator notes, he "revelled in the less important but more startling changes wrought in the *physical* frame of Berenice" rather than in her "*moral*" existence. Poe aims to horrify his readers in the narrative presented by a self-proclaimed "monomaniac," but the horror of this story, like that of Paula and that Dr. Vannatta's patient, creates an important context in which to understand the responsibilities and caretaking of healthcare.

Literary Narrative: "Berenice – A Tale" (1835) by Edgar Allan Poe

Author Note: Edgar Allan Poe (1809–1849) was an American writer in the mid-nineteenth century, the time of the "American Renaissance," best known for his poetry and short stories. Those stories fall into the categories of "tales of ratiocination" in the first detective stories written in English, such as "The Murders in the Rue Morgue" discussed in Chap. 2; "tales of the grotesque and arabesque," that is horror stories, such as "Berenice"; and "hoaxes," designed to fool his readers (newspaper readers, where many of his stories first appeared). He is widely regarded as central to American Romanticism and American literature more generally, an early American practitioner of the short story form, which flowered in the nineteenth century.

Berenice – A Tale

Misery is manifold. The wretchedness of earth is multiform. Overreaching the wide horizon as the rainbow, its hues are as various as the hues of that arch—as distinct too, yet as intimately blended. Overreaching the wide horizon as the rainbow! How is it that from beauty I have derived a type of unloveliness?—from the covenant of peace, a simile of sorrow? But as, in ethics, evil is a consequence of good, so, in fact, out of joy is sorrow born. Either the memory of past bliss is the anguish of to-day, or the agonies which are, have their origin in the ecstasies which *might have been*. I have a tale to tell in its own essence rife with horror—I would suppress it were it not a record more of feelings than of facts.

My baptismal name is Egaeus—that of my family I will not mention. Yet there are no towers in the land more time-honored than my gloomy, gray, hereditary halls. Our line has been called a race of visionaries; and in many striking particulars—in the character of the family mansion—in the frescos of the chief saloon—in the tapestries of the dormitories—in the chiselling of some buttresses in the armory—but more especially in the gallery of antique paintings—in the fashion of the library chamber—and, lastly, in the very peculiar nature of the library's contents—there is more than sufficient evidence to warrant the belief.

The recollections of my earliest years are connected with that chamber, and with its volumes—of which latter I will say no more. Here died my mother. Herein was I born. But it is mere idleness to say that I had not lived before—that the soul has no previous existence. You deny it. Let us not argue the matter. Convinced myself, I seek not to convince. There is, however, a remembrance of aerial forms—of spiritual and meaning eyes—of sounds, musical yet sad—a remembrance which will not be excluded; a memory like a shadow—vague, variable, indefinite, unsteady; and like a shadow, too, in the impossibility of my getting rid of it while the sunlight of my reason shall exist.

In that chamber was I born. Thus awaking from the long night of what seemed, but was not, nonentity, at once into the very regions of fairy land—into a palace of imagination—into the wild dominions of monastic thought and erudition—it is not singular that I gazed around me with a startled and ardent eye—that I loitered away my boyhood in books, and dissipated my youth in reverie; but it is singular that as years rolled away, and the noon of manhood found me still in the mansion of my fathers—it is wonderful what stagnation there fell upon the springs of my life—wonderful how total an inversion took place in the character of my commonest thought. The realities of the world affected me as visions, and as visions only, while the wild ideas of the land of dreams became, in turn, not the material of my every-day existence, but in very deed that existence utterly and solely in itself.

* * *

Berenice and I were cousins, and we grew up together in my paternal halls. Yet differently we grew—I, ill of health, and buried in gloom—she, agile, graceful, and overflowing with energy; hers, the ramble on the hill-side—mine the studies of the cloister; I, living within my own heart, and addicted, body and soul, to the most intense and painful meditation—she, roaming carelessly through life, with no thought of the shadows in her path, or the silent flight of the raven-winged hours. Berenice!—I call upon her name—Berenice!—and from the gray ruins of memory a thousand tumultuous recollections are startled at the sound! Ah, vividly is her image before me now, as in the early days of her light-heartedness and joy! Oh, gorgeous yet fantastic beauty! Oh, sylph amid the shrubberies of Arnheim! Oh, Naiad among its fountains! And then—then all is mystery and terror, and a tale which should not be told. Disease—a fatal disease, fell like the Simoon upon her frame; and, even while I gazed upon her, the spirit of change swept over her, pervading her mind, her habits, and her character, and, in a manner the most subtle and terrible, disturbing even the identity of her person! Alas! the destroyer came and went!—and the victim—where is she? I knew her not—or knew her no longer as Berenice.

Among the numerous train of maladies superinduced by that fatal and primary one which effected a revolution of so horrible a kind in the moral and physical being of my cousin, may be mentioned as the most distressing and obstinate in its nature, a species of epilepsy not unfrequently terminating in *trance* itself—trance very nearly resembling positive dissolution, and from which her manner of recovery was in most instances, startlingly abrupt. In the mean time my own disease—for I have been told that I should call it by no other appellation—my own disease, then, grew rapidly upon me, and assumed finally a monomaniac character of a novel and extraordinary form—hourly and momently gaining vigor—and at length obtaining over me the most incomprehensible ascendancy. This monomania, if I must so term it, consisted in a morbid irritability of those properties of the mind in metaphysical science termed the *attentive*. It is more than probable that I am not understood—but I fear, indeed, that it is in no manner possible to convey to the mind of the merely general reader, an adequate idea of that nervous *intensity of interest* with which, in my case, the powers of meditation (not to speak technically) busied and buried themselves, in the contemplation of even the most ordinary objects of the universe.

To muse for long unwearied hours, with my attention riveted to some frivolous device on the margin, or in the typography of a book—to become absorbed, for the better part of a summer's day, in a quaint shadow falling aslant upon the tapestry or upon the floor—to lose myself, for an entire night, in watching the steady flame of a lamp, or the embers of a fire—to dream away whole days over the perfume of a flower—to repeat, monotonously, some common word, until the sound, by dint of frequent repetition, ceased to convey any idea whatever to the mind—to lose all sense of motion or physical existence, by means of absolute bodily quiescence long and obstinately persevered in—Such were a few of the most common and least pernicious vagaries induced by a condition of the mental faculties, not, indeed, altogether unparalleled, but certainly bidding defiance to anything like analysis or explanation.

Yet let me not be misapprehended. The undue, earnest, and morbid attention thus excited by objects in their own nature frivolous, must not be confounded in character with that ruminating propensity common to all mankind, and more especially indulged in by persons of ardent imagination. It was not even, as might be at first supposed, an extreme condition, or exaggeration of such propensity, but primarily and essentially distinct and different. In the one instance, the dreamer, or enthusiast, being interested by an object usually not frivolous, imperceptibly loses sight of this object in a wilderness of deductions and suggestions issuing therefrom, until, at the conclusion of a day dream *often replete with luxury*, he finds the *incitamentum*, or first cause of his musings, entirely vanished and forgotten. In my case, the primary object was *invariably frivolous*, although assuming, through the medium of my distempered vision, a refracted and unreal importance. Few deductions—if any—were made; and those few pertinaciously returning in upon the original object as a centre. The meditations were *never* pleasurable; and, at the termination of the reverie, the first cause, so far from being out of sight, had attained that supernaturally exaggerated interest which was the prevailing feature of the disease. In a word, the powers of mind more particularly exercised were, with me, as I have said before, the *attentive*, and are, with the day-dreamer, the *speculative*.

My books, at this epoch, if they did not actually serve to irritate the disorder, partook, it will be perceived, largely, in their imaginative and inconsequential nature, of the characteristic qualities of the disorder itself. I well remember, among others, the treatise of the noble Italian, Coelius Secundus Curio, "*De amplitudine beati regni Dei*"—St. Austin's great work, the "City of God"—and Tertullian's "*De Carne Christi*," in which the paradoxical sentence "*Mortuus est Dei filius; credible est quia ineptum est: et sepultus resurrexit; certum est quia impossibile est* [That the Son of God died is entirely believable simply because it seems so absurd; that He should do so; and that he rose from the dead is certain simply because it is impossible to do so]," occupied my undivided time, for many weeks of laborious and fruitless investigation.

Thus it will appear that, shaken from its balance only by trivial things, my reason bore resemblance to that ocean-crag spoken of by Ptolemy Hephestion, which steadily resisting the attacks of human violence, and the fiercer fury of the waters and the winds, trembled only to the touch of the flower called Asphodel. And although, to a careless thinker, it might appear a matter beyond doubt, that the alteration produced by her unhappy malady, in the *moral* condition of Berenice, would afford me many objects for the exercise of that intense and abnormal meditation whose nature I have been at some trouble in explaining, yet such was not in any degree the case. In the lucid intervals of my infirmity, her calamity, indeed, gave me pain, and, taking deeply to heart that total wreck of her fair and gentle life, I did not fall to ponder, frequently and bitterly, upon the wonder-working means by which so strange a revolution had been so suddenly brought to pass. But these reflections partook not of the idiosyncrasy of my disease, and were such as would have occurred, under similar circumstances, to the ordinary mass of mankind. True to its own character, my disorder revelled in the less important but more startling changes wrought in the *physical* frame of Berenice, and in the singular and most appalling distortion of her personal identity.

During the brightest days of her unparalleled beauty, most surely I had never loved her. In the strange anomaly of my existence, feelings with me, *had never been* of the heart, and my passions always were of the mind. Through the gray of the early morning—among the trellised shadows of the forest at noonday—and in the silence of my library at night—she had flitted by my eyes, and I had seen her—not as the living and breathing Berenice, but as the Berenice of a dream—not as a being of the earth—earthy—but as the abstraction of such a being—not as a thing to admire, but to analyze—not as an object of love, but as the theme of the most abstruse although desultory speculation. And *now*—now I shuddered in her presence, and grew pale at her approach; yet, bitterly lamenting her fallen and desolate condition, I called to mind that she had loved me long, and, in an evil moment, I spoke to her of marriage.

And at length the period of our nuptials was approaching, when, upon an afternoon in the winter of the year—one of those unseasonably warm, calm, and misty days which are the nurse of the beautiful Halcyon—I sat, (and sat, as I thought, alone,) in the inner apartment of the library. But, uplifting my eyes, I saw that Berenice stood before me.

Was it my own excited imagination—or the misty influence of the atmosphere—or the uncertain twilight of the chamber—or the gray draperies which fell around her figure—that caused in it so vacillating and indistinct an outline? I could not tell. She spoke, however, no word; and I—not for worlds could I have uttered a syllable. An icy chill ran through my frame; a sense of insufferable anxiety oppressed me; a consuming curiosity pervaded my soul; and sinking back upon the chair, I remained for some time breathless and motionless, with my eyes riveted upon her person. Alas! its emaciation was excessive, and not one vestige of the former being lurked in any single line of the contour. My burning glances at length fell upon the face.

The forehead was high, and very pale, and singularly placid; and the once jetty hair fell partially over it, and overshadowed the hollow temples with innumerable ringlets, now of a vivid yellow, and jarring discordantly, in their fantastic character, with the reigning melancholy of the countenance. The eyes were lifeless, and lustreless, and seemingly pupilless, and I shrank involuntarily from their glassy stare to the contemplation of the thin and shrunken lips. They parted; and in a smile of peculiar meaning, the teeth of the changed Berenice disclosed themselves slowly to my view. Would to God that I had never beheld them, or that, having done so, I had died!

* * *

The shutting of a door disturbed me, and, looking up, I found that my cousin had departed from the chamber. But from the disordered chamber of my brain, had not, alas! departed, and would not be driven away, the white and ghastly *spectrum* of the teeth. Not a speck on their surface—not a shade on their enamel—not an indenture in their edges—but what that period of her smile had sufficed to brand in upon my memory. I saw them now even more unequivocally than I beheld them then. The teeth!—the teeth!—they were here, and there, and everywhere, and visibly and palpably before me; long, narrow, and excessively white, with the pale lips writhing about them, as in the very moment of their first terrible development. Then came the full fury of my *monomania*, and I struggled in vain against its strange and irresistible influence. In the multiplied objects of the external world I had no thoughts but for the teeth. For these I longed with a phrenzied desire. All other matters and all different interests became absorbed in their single contemplation. They—they alone were present to the mental eye, and they, in their sole individuality, became the essence of my mental life. I held them in every light—I turned them in every attitude. I surveyed their characteristics—I dwelt upon their peculiarities—I pondered upon their conformation—I mused upon the alteration in their nature—and shuddered as I assigned to them in imagination a sensitive and sentient power, and even when unassisted by the lips, a capability of moral expression. Of Mademoiselle Salle it has been well said, "*Que tous ses pas etaient des sentiments* [All her (ballet) steps were sentiments]," and of Berenice I more seriously believed *que toutes ses dents etaient des idees. Des idees!* [All of her teeth were ideas. Ideas!]—ah here was the idiotic thought that destroyed me! *Des idees!*—ah therefore it was that I coveted them so madly! I felt that their possession could alone ever restore me to peace, in giving me back to reason.

And the evening closed in upon me thus—and then the darkness came, and tarried, and went—and the day again dawned—and the mists of a second night were now gathering around—and still I sat motionless in that solitary room, and still I sat buried in meditation, and still the *phantasma* of the teeth maintained its terrible ascendancy, as, with the most vivid hideous distinctness, it floated about amid the changing lights and shadows of the chamber. At length there broke in upon my dreams a cry as of horror and dismay; and thereunto, after a pause, succeeded the sound of troubled voices, intermingled with many low moanings of sorrow or of pain. I arose from my seat, and throwing open one of the doors of the library, saw standing out in the ante-chamber a servant maiden, all in tears, who told me that Berenice was—no more. She had been seized with epilepsy in the early morning, and now, at the closing in of the night, the grave was ready for its tenant, and all the preparations for the burial were completed.

With a heart full of grief, yet reluctantly, and oppressed with awe, I made my way to the bed-chamber of the departed. The room was large, and very dark, and at every step within its gloomy precincts I encountered the paraphernalia of the grave. The coffin, so a menial told me, lay surrounded by the curtains of yonder bed, and in that coffin, he whisperingly assured me, was all that remained of Berenice. Who was it asked me would I not look upon the corpse? I had seen the lips of no one move, yet the question had been demanded, and the echo of the syllables still lingered in the room. It was impossible to refuse; and with a sense of suffocation I dragged myself to the side of the bed. Gently I uplifted the sable draperies of the curtains.

As I let them fall they descended upon my shoulders, and shutting me thus out from the living, enclosed me in the strictest communion with the deceased.

The very atmosphere was redolent of death. The peculiar smell of the coffin sickened me; and I fancied a deleterious odor was already exhaling from the body. I would have given worlds to escape—to fly from the pernicious influence of mortality—to breathe once again the pure air of the eternal heavens. But I had no longer the power to move—my knees tottered beneath me—and I remained rooted to the spot, and gazing upon the frightful length of the rigid body as it lay outstretched in the dark coffin without a lid.

God of heaven!—is it possible? Is it my brain that reels—or was it indeed the finger of the enshrouded dead that stirred in the white cerement that bound it? Frozen with unutterable awe I slowly raised my eyes to the countenance of the corpse. There had been a band around the jaws, but, I know hot how, it was broken asunder. The livid lips were wreathed into a species of smile, and, through the enveloping gloom, once again there glared upon me in too palpable reality, the white and glistening, and ghastly teeth of Berenice. I sprang convulsively from the bed, and, uttering no word, rushed forth a maniac from that apartment of triple horror, and mystery, and death.

* * *

I found myself sitting in the library, and again sitting there alone. It seemed that I had newly awakened from a confused and exciting dream. I knew that it was now midnight, and I was well aware, that since the setting of the sun, Berenice had been interred. But of that dreary period which intervened I had no positive, at least no definite comprehension. Yet its memory was replete with horror—horror more horrible from being vague, and terror more terrible from ambiguity. It was a fearful page in the record my existence, written all over with dim, and hideous, and unintelligible recollections. I strived to decypher them, but in vain; while ever and anon, like the spirit of a departed sound, the shrill and piercing shriek of a female voice seemed to be ringing in my ears. I had done a deed—what was it? I asked myself the question aloud, and the whispering echoes of the chamber answered me—"what was it?"

On the table beside me burned a lamp, and near it lay a little box. It was of no remarkable character, and I had seen it frequently before, for it was the property of the family physician; but how came it there, upon my table, and why did I shudder in regarding it? These things were in no manner to be accounted for, and my eyes at length dropped to the open pages of a book, and to a sentence underscored therein. The words were the singular but simple ones of the poet Ebn Zaiat:—"*Dicebant mihi sodales si sepulchrum amicae visitarem, curas meas aliquantulum fore levatas* [My companions told me I might find some little alleviation of my misery, in visiting the grave of my beloved]." Why then, as I perused them, did the hairs of my head erect themselves on end, and the blood of my body become congealed within my veins?

There came a light tap at the library door, and, pale as the tenant of a tomb, a menial entered upon tiptoe. His looks were wild with terror, and he spoke to me in a voice tremulous, husky, and very low. What said he?—some broken sentences I heard. He told of a wild cry disturbing the silence of the night—of the gathering together of the household—of a search in the direction of the sound; and then his tones grew thrillingly distinct as he whispered me of a violated grave—of a disfigured body enshrouded, yet still breathing—still palpitating—still alive!

He pointed to my garments—they were muddy and clotted with gore. I spoke not, and he took me gently by the hand—but it was indented with the impress of human nails. He directed my attention to some object against the wall—I looked at it for some minutes—it was a spade. With a shriek I bounded to the table, and grasped the box that lay upon it. But I could not force it open; and in my tremor it slipped from my hands, and fell heavily, and burst into pieces; and from it, with a rattling sound, there rolled out some instruments of dental surgery, intermingled with many white and ivory-looking substances that were scattered to and fro about the floor.

Vicarious Experience

As we have suggested, Poe's story lends itself to analyses in relation to the "features" of narrative. In Chap. 1, we offered the story as an example of "patterned repetition of narrative events," but in fact Poe's *goal* of creating a horror story—his goal of provoking emotional responses from his readers—makes his work a powerful example of the ways in which literary narrative provokes vicarious experience. ("The Murders in the Rue Morgue" is a powerful example of the ways in which literary narrative allows us to grasp intellectual structures.) Thus, "Berenice" displays many of the features of narrative we describe in Chap. 1: the scene specifically displays the dynamic of form and content with the narrator's rambling discourse (represented by Poe's many dashes); in his obsession, he tells the same story over and over, in a twice-told story; he displays patterned repetition of sound (it opens with "Misery is manifold"); of events (as we note in Chap. 1), of theme (of violent obsession); and the very aim at horror requires the strategy of "defamiliarization," which makes the obsessed narrator's experience *real* to Poe's readers, just as in the vignette in this chapter the quality and characteristics of his patient's wound makes his patient's experience real to the physician—and just as, we might add, Doyle's *The Woman Who Walked into Doors*, makes Paula's experience *real* to readers such as Dr. Vannatta. The whole becomes a "moral education"—perhaps an unintentional effect of the story—as Poe presents Egaeus's distorted violent thinking in an extraordinary act of Theory of Mind in which readers are "transported" into a life that many could hardly imagine. We are not sure that Poe hopes to "transport" his readers into the "mind" of Berenice, who is hardly presented as a person Egaeus's distorted thinking, yet our ability as readers to do so—to reimagine this story in order to empathize with Berenice, as Dr. Vannatta empathizes with his patient and readers empathize with Paula—offers a strong example of the ways that readers can imaginatively grasp an "overall meaning" that lies dormant in the *situation* of a narrative.

Related Poem

A poem somewhat related to the situation of abuse and violence is Yeats's famous poem, "Leda and the Swan," published in 1924 but written earlier after the poet had lived through the terrible violence of World War I (1914–1918), the Irish war for independence from British rule after the war (1918–1922), and the ongoing Irish Civil War (1922–1923) that accompanied the establishment of the Irish Free State based upon the Anglo-Irish Treaty in 1922. This poem and Yeats himself are trying to make sense out of what can only seem to be senseless violence. It asks us to imagine—again, as a physician might try to imagine the state of mind of an aged demented woman or a young woman with a bruised face—what unimaginable violence from the sky might feel like to someone in a world without airplane bombers. (World War I was the first time in history air warfare took place, and Yeats wrote a moving elegy for the son of a close friend who had died as an air pilot in the war.)

"Leda and the Swan" (1924) by W. B. Yeats

Author Note: W. B. Yeats (1865–1939), an Irish poet and playwright, is considered to be one of the great twentieth-century poets in English. In 1923 he was awarded the Nobel Prize in Literature, and in helping to create the Irish Literary Revival in the early twentieth century, he is often considered an architect of Cultural Nationalism in Ireland. He is noted for powerful poetry focused on love, spirituality, and Irish cultural traditions. He composed this poem during the Irish Civil War, after the British partition of Ireland in 1922.

Leda and the Swan

A sudden blow: the great wings beating still
Above the staggering girl, her thighs caressed
By the dark webs, her shape caught in his bill,
He holds her helpless breast upon his breast.

"Leda and the Swan" (1924) by W. B. Yeats

> How can those terrified vague fingers push
> The feathered glory from her loosening thighs?
> And how can body, laid in that white rush,
> But feel the strange heart beating where it lies?
>
> A shudder in the loins engenders there
> The broken wall, the burning roof and tower
> And Agamemnon dead.
>   Being so caught up,
> So mastered by the brute blood of the air,
> Did she put on his knowledge with his power
> Before the indifferent beak could let her drop?

In order to situate the violence he experienced—and to make sense of it—Yeats turns to ancient Greek mythology, where the god Zeus would often take the form of animals in order to rape human women. In this instance, he takes the form of a swan and descends upon Leda, who subsequently gave birth to Helen of Troy. Thus, in this poem, Yeats juxtaposes the violence of the first two stanzas to the images of the Trojan War in the subsequent three lines. The last three lines are an extended rhetorical question, asking whether any sense can be made of violence: did Leda comprehend the god's knowledge—namely, that this seeming senseless act of violence would lead to the world-changing event of the establishment of Greek and, later, Western culture in the Trojan War—while she experiences his brutal power?

Again, many of the features of Chap. 1 can help us comprehend the implicit narrative in this poem and how it deploys language to provoke feelings in its readers. The dynamic of form and content is played out in the traditional form of the poem, which is a 14-line sonnet in the Italian (or "Petrarchan") form, where, using a strict rhyme scheme, the first eight lines ("octave") sets a situation and the last six lines ("sestet") comments upon what it might mean. Yeats modifies this form somewhat by dividing the sestet into two paragraphs that allows his rhetorical question to seem a more general observation than simply commenting on the mythical event. Similarly, the act of violence is "twice-told": in the immediate violence ("the sudden blow") and the reflection on that violence ("the brute blood of the air") in the rhetorical questions. Systematic repetition—in rhymes, events, semantics (such as the odd use of "caressed")—is the hallmark of poetry; and the unsaid manifests itself in the *power* of a rhetorical question, which seems to make a statement, but being in the form of a question, it does not quite fully "say" what it suggests. The "relational facts" of this poem are *historical*: the presentation of "mythical" violence in a time of real violence. That rhetorical question also helps determine who is the witness who learns. The great question of the poem, however, is that of its "overall meaning": what is Yeats striving for? Most commentators relate the poem to Yeats's cosmological thinking—his sense that there is a great cultural crisis in the twentieth century that will lead to a new cosmic "dispensation" of human life. But we are suggesting that it is much more closely related to the *experience* of Yeats's (and our) time, which is a time of seeming senseless violence wherever we turn. In any case, Yeats's poetic impulse can shed light on the impulse often felt in victims of violence—and, as in Paula's doctors and Berenice's grotesque narrator, the witnesses of violence—to explain it away as the better alternative than to confront it and figure out what kind of practical actions that confrontation demands. In other words, to try to answer Yeats's question by saying, "there is no knowledge that can justify brutal violence – no excuse (such as Berenice's 'gorgeous yet fantastic beauty!'), no rationalization (such as 'she never resisted'), no convenience (such as 'it's not my business')" becomes itself an occasion for moral education, for self-consciously articulating value in the face of experience.

Lessons for Providers

While it is easy as a medical provider to ignore the obvious abuse we see in these stories (just as it is easy to ignore the actual violence Berenice endures), vicarious experience through narratives such as we see in this chapter can arm the provider with courage to engage in a helpful manner.

Bibliography

Doyle, Roddy. 1996. *The Woman Who Walked into Doors*. New York: Viking.
Safehorizon, 2018. Domestic Violence Statistics and Fact. https://www.safehorizon.org/get-informed/domestic-violence-statistics-facts/#description/
Schleifer, Ronald. 2018b. The Aesthetics of Pain: Semiotics and Affective Comprehension in Music, Literature, and Sensate Experience. *Configurations* 26 (2018): 471–491.

Pain 11

We began by noting in Chap. 10 that one issue that confronts physicians too often is one form of abuse or another. Another issue that confronts healthcare providers even more regularly is pain and suffering. Illness, more often than not, gives rise to pain and suffering. From time immemorial, caretaking, first and foremost, has been attempts to relieve, care for, or simply console people wracked with pain and suffering. One of the great successes of scientific medicine in the last 150 years has been the development of treatments and, frequently, cures for acute ailments. One result of this success is the ability of the suffering patient to put her pain and the suffering associated with acute somatic conditions into a context of possible and sometimes probable relief. Such reasonable hope for acute suffering goes a long way in alleviating pain and suffering. Another result, however, is that—with lives not so frequently cut short by acute illnesses and accidents—there has been a marked increase in chronic pain associated with long-term illnesses and conditions. Biomedicine in the last century has focused on acute illnesses, with great success. But the problems of chronic illnesses—and especially chronic pain—has not regularly responded to the same strategies that characterize the success with acute ailments.

One reason for this is the nature of pain itself. In relation to acute ailments, pain functions as a signal, a warning sign that something is wrong and needs to be attended to. But once that attention has been gained, pain seems to no longer be the focus on healthcare treatment. In large part, that is why the physician in Ernest Hemingway's famous story "Indian Camp" tells his son his patient's screams of pain are not important:

> "Oh, Daddy, can't you give her something to make her stop screaming?" asked Nick.
> "No. I haven't any anaesthetic," his father said. "But her screams are not important. I don't hear them because they are not important."
> The husband in the upper bunk rolled over against the wall. (1972: 19)

As the last line of this passage suggests insofar as we remember the husband soon after commits suicide as he helplessly faces his wife's screams, such expressions of pain *are* important: to the human suffering, to those connected to her, to those committed to caring for her. All this is multiplied in the face of chronic pain. Chronic pain, one physician, who has committed his career to the treatment of chronic pain, notes "does not alert a person to possible damage, but rather 'is just ugly noise'; he further remarks that 'acute pain is a symptom of disease; chronic pain itself is a disease'" (Dr. Scott Fishman 2000, cited in Schleifer 2014: 39). Chronic pain is "defined as pain that extends beyond normal recovery time, is cyclical, like a migraine headache, or last longer than six months" (Schleifer 2014: 8).

In this chapter, we offer a vignette written by a person who has suffered chronic pain over many, many years: "approximately 8700 hours of severe pain, 7200 hours of considerable pain, and 11,500 hours of light pain. In all, 27,000 hours of pain over an eleven-year period" (Heshusius 2009: 7). We also offer a chapter from a novel that represents the moment of intense acute pain resulting from a gunshot wound and a subsequent operation before the use of anesthetic. And finally, we offer a short lyric poem that meditates on the nature of pain itself, without distinguishing between acute and chronic pain. This chapter is important because in our time, as Dr. Fishman notes, chronic pain is not a symptom of disease but a disease itself, and it is one which, for a number of reasons, has not leant itself to regular successful treatment. These reasons include an understanding, held by many (and, perhaps, implicit in training in biomedicine), that pain is simply a symptom of an underlying disease, and can be ignored, as Nick's father ignores it in "Indian Camp," once one can focus on the underlying disease; the complicated psycho-physiological nature of pain, as both a physiological "event" and an individual and/or social "experience" (see Schleifer 2014); and the "invisibility" of pain, as Lous Heshusius notes in the vignette, to those encountering it in others. Thus, the engagements with the *phenomenology* of pain—that is, attempts in narrative and poetry to present the very "feel" of pain and suffering—is and should be an important element in the education of people preparing themselves for careers in healthcare. Finally, a focus on pain can allow healthcare providers—and, indeed, all of us—to meditate on the relationship between pain and suffering. To this end, we recommend an important and widely available article written by Dr. Eric Cassel and published in *The New England Journal of Medicine* in 1982 entitled "The Nature of Suffering and the Goals of Medicine" (see Bibliography: Cassel 1982).

"That Which Has No Words, That Which Cannot Be Seen": A Vignette (Excerpt from Lous Heshusius, *Inside Chronic Pain: An Intimate and Critical Account* [2009])

> *Author Note*: Lous Heshusius (b. 1940), Dutch-born now living in Canada, is Professor Emeritus of Education at York University, where she focused on Disability Studies in Education. She is the author of *Inside Chronic Pain*, a chronical of her terrible experience of chronic pain extended for more than a decade. In his "Clinical Commentary" to this book, Dr. Scott Fishman notes that "while no two patients' experiences [of pain] are the same, I have heard innumerable stories that highlight the central dilemma of this book: that, for many patients like Lous Heshusius, the health-care system itself is a major source of pain and suffering."

The most pervasive problem of the chronic pain experience, apart from the torment of the pain itself, is its inexpressibility and its invisibility. Much of my story must be understood in the light of this difficulty.

Researcher David Morris tells of his reaction when first talking to chronic pain patients: "What surprised me most … was the apparent normal faces of the patients. I had steeled myself to expect agonized expressions and frightful cries." Indeed, we appear normal. That is our liability. […]

Palliative care physician David Kuhl, author of *What Dying People Want*, says he came to understand that when physical pain is present it is virtually impossible to address psychological and spiritual concerns. When I read Dr. Kuhl's words I felt relief because I had always felt that, somehow, by the sheer force of will, I should be able to address other matters while in great pain. But it is not possible. Serious pain itself *is* an interfering pattern: it floods the brain, which is then no longer available for thought. […]

Elaine Scarry's analysis of the complications of pain's invisibility [in *The Body in Pain*] mirrors my own experience:

> For the person whose pain it is, it is "effortlessly" grasped (that is, even with the most heroic effort it cannot *not* be grasped); while for the person outside the sufferer's body, what is "effortless" is *not* grasping it (it is easy to remain wholly unaware of its existence […]).

This difficulty with expression also holds for my medical appointments. I try to speak to doctors about the severity of my pain. My words float strangely in the air. As I pronounce them, I myself become a spectator. As soon as I begin to speak, I am no longer there. Someone else is speaking these words. Someone who has not suffered the pain, for it is so much worse than she says. […] In the meantime, I am watching the doctor. Trying to see how he reacts. Did he get it? Should she be more dramatic? More detailed? But how? How can she, how can I, express this prelanguage of torment?

Of my many doctors, only three, perhaps four, seemed to grasp what chronic pain does to a life. Perhaps others also did, but they never gave any sign of it. […]

Someone else's pain, then, can never be confirmed and is, therefore, often denied and always underestimated. These truths echo in the stories told by those in chronic pain who speak of doctors, employers, friends, and even family members who think the sufferer is exaggerating, who can't believe it can be all that bad.

Can Pain Be Represented?

What is most striking about this passage—Elaine Scarry repeats this in her groundbreaking study of pain in 1987, *The Body in Pain*—is its claim that pain is somehow necessarily private, un-transmissible, and for that reason a further source of pain and suffering. Thus Hushusius notes that the inability of healthcare providers to *acknowledge* her pain increased it. As we noted in Chap. 1, there is a physiological basis to empathy under the category of "mirror neurons," which respond not only to sensational events that affect a human subject but also to their seeing or hearing about such events affecting a cohort. This, neurologists who study mirror neurons suggest, accounts for the ways we might cringe when we see a child fall off a bike or try to balance ourselves as we watch tight-rope walkers. Such responses, we noted earlier, can also be occasioned by the experiences of reading or watching a movie. These facts seem to suggest that the absolute privacy of pain, which Heshusius has experienced—reinforced in her narrative by the non-acknowledging responses of healthcare providers she encountered—can be modified by the *shared* experiences that literary texts provoke. We imagine that many readers will cringe empathetically when reading Herman Melville's representation of a pre-anesthetic amputation in *White Jacket* published in 1850 in this chapter. (Anesthesia was first demonstrated—at the instigation of a dentist—in 1846 at Harvard University. Melville based his novel on his 14-month service in the United States Navy from 1843 to 1844.) Moreover, the pursuit and achievement of such shared experience can, we believe, contribute to the preparation for a career in healthcare in a signal fashion.[1] In this chapter, Melville's "The Operation," in its satirical account of a "routine" amputation, suggests the patient's overwhelming pain and anxiety that eventually leads to his death.

Literary Narrative: "The Operation" from *White Jacket* (1850) by Herman Melville

> *Author Note*: Herman Melville (1819–1891) was an American novelist, short-story writer and poet in the mid-nineteenth century, the time of the "American Renaissance." He is best known for *Moby-Dick* (1851), often understood to be one of the great American novels. Many of his novels, including *White Jacket*, explores the life of sailing in the nineteenth century. Much of the command of Melville's writing comes from his intense focus on the widest meanings of ordinary occurrences.

The Operation

> Next morning, at the appointed hour, the surgeons arrived in a body. They were accompanied by their juniors, young men ranging in age from nineteen years to thirty. Like the senior surgeons, these young gentlemen were arrayed in their blue navy uniforms, displaying a profusion of bright buttons, and several broad bars of gold lace about the wristbands. As in honour of the occasion, they had put on their best coats; they looked exceedingly brilliant.
>
> The whole party immediately descended to the half-deck, where preparations had been made for the operation. A large garrison-ensign was stretched across the ship by the mainmast, so as completely to screen the space behind. This space included the whole extent aft to the bulk-head of the Commodore's cabin, at the door of which the marine-orderly paced, in plain sight, cutlass in hand.
>
> Upon two gun-carriages, dragged amidships, the Death-board (used for burials at sea) was horizontally placed, covered with an old royal-stun'-sail. Upon this occasion, to do duty as an amputation-table, it was widened by an additional plank. Two match-tubs, near by, placed one upon another, at either end supported another plank, distinct from the table, whereon was exhibited an array of saws and knives of various and peculiar shapes and sizes; also, a sort of steel, something like the dinner-table implement, together with long needles, crooked at the end for taking up the arteries, and large darning-needles, thread and bee's-wax, for sewing up a wound.
>
> At the end nearest the larger table was a tin basin of water, surrounded by small sponges, placed at mathematical intervals. From the long horizontal pole of a great-gun rammer—

---

[1] As mentioned in Chap. 10, Roddy Doyle's disturbing novel *The Woman Who Walked into Doors* powerfully represents the pain Paula Spencer experienced as a result of her suffering domestic abuse. For a detailed literary analysis of such pain—touching on many of the features of literary narrative descripted in Chap. 1—see Schleifer (2018b).

fixed in its usual place overhead—hung a number of towels, with "U.S." marked in the corners.

All these arrangements had been made by the "Surgeon's steward," a person whose important functions in a man-of-war will, in a future chapter, be entered upon at large. Upon the present occasion, he was bustling about, adjusting and readjusting the knives, needles, and carver, like an over-conscientious butler fidgeting over a dinner-table just before the convivialists enter.

But by far the most striking object to be seen behind the ensign was a human skeleton, whose every joint articulated with wires. By a rivet at the apex of the skull, it hung dangling from a hammock-hook fixed in a beam above. Why this object was here, will presently be seen; but why it was placed immediately at the foot of the amputation-table, only Surgeon Cuticle can tell.

While the final preparations were being made, Cuticle stood conversing with the assembled Surgeons and Assistant Surgeons, his invited guests.

"Gentlemen," said he, taking up one of the glittering knives and artistically drawing the steel across it; "Gentlemen, though these scenes are very unpleasant, and in some moods, I may say, repulsive to me—yet how much better for our patient to have the contusions and lacerations of his present wound—with all its dangerous symptoms—converted into a clean incision, free from these objections, and occasioning so much less subsequent anxiety to himself and the Surgeon. Yes," he added, tenderly feeling the edge of his knife, "amputation is our only resource. Is it not so, Surgeon Patella?" turning toward that gentleman, as if relying upon some sort of an assent, however clogged with conditions.

"Certainly," said Patella, "amputation is your only resource, Mr. Surgeon of the Fleet; that is, I mean, if you are fully persuaded of its necessity."

The other surgeons said nothing, maintaining a somewhat reserved air, as if conscious that they had no positive authority in the case, whatever might be their own private opinions; but they seemed willing to behold, and, if called upon, to assist at the operation, since it could not now be averted.

The young men, their Assistants, looked very eager, and cast frequent glances of awe upon so distinguished a practitioner as the venerable Cuticle.

"They say he can drop a leg in one minute and ten seconds from the moment the knife touches it," whispered one of them to another.

"We shall see," was the reply, and the speaker clapped his hand to his fob, to see if his watch would be forthcoming when wanted.

"Are you all ready here?" demanded Cuticle, now advancing to his steward; "have not those fellows got through yet?" pointing to three men of the carpenter's gang, who were placing bits of wood under the gun-carriages supporting the central table.

"They are just through, sir," respectfully answered the steward, touching his hand to his forehead, as if there were a cap-front there.

"Bring up the patient, then," said Cuticle.

"Young gentlemen," he added, turning to the row of Assistant Surgeons, "seeing you here reminds me of the classes of students once under my instruction at the Philadelphia College of Physicians and Surgeons. Ah, those were happy days!" he sighed, applying the extreme corner of his handkerchief to his glass-eye. "Excuse an old man's emotions, young gentlemen; but when I think of the numerous rare cases that then came under my treatment, I cannot but give way to my feelings. The town, the city, the metropolis, young gentlemen, is the place for you students; at least in these dull times of peace, when the army and navy furnish no inducements for a youth ambitious of rising in our honourable profession. Take an old man's advice, and if the war now threatening between the States and Mexico should break out, exchange your navy commissions for commissions in the army. From having no military marine herself, Mexico has always been backward in furnishing subjects for the amputation-tables of foreign navies. The cause of science has languished in her hands. The army, young gentlemen, is your best school; depend upon it. You will hardly believe it, Surgeon Bandage," turning to that gentleman, "but this is my first important case of surgery in a nearly three years' cruise. I have been almost wholly confined in this ship to doctor's practice prescribing for fevers and fluxes. True, the other day a man fell from the mizzen-top-sail-yard; but that was merely an aggravated case of dislocations and bones splintered and broken. No one, sir, could have made an amputation of it, without severely contusing his conscience. And mine—I may say it, gentlemen, without ostentation is—peculiarly susceptible."

And so saying, the knife and carver touchingly dropped to his sides, and he stood for a moment fixed in a tender reverie but a commotion being heard beyond the curtain, he started, and, briskly crossing and recrossing the knife and carver, exclaimed, "Ali, here comes our patient; surgeons, this side of the table, if you please; young gentlemen, a little further off, I beg. Steward, take off my coat—so; my neckerchief now; I must be perfectly unencumbered, Surgeon Patella, or I can do nothing whatever."

These articles being removed, he snatched off his wig, placing it on the gun-deck capstan; then took out his set of false teeth, and placed it by the side of the wig; and, lastly, putting his forefinger to the inner angle of his blind eye, spirited out the glass optic with professional dexterity, and deposited that, also, next to the wig and false teeth.

Thus, divested of nearly all inorganic appurtenances, what was left of the Surgeon slightly shook itself, to see whether anything more could be spared to advantage.

"Carpenter's mates," he now cried, "will you never get through with that job?"

"Almost through, sir—just through," they replied, staring round in search of the strange, unearthly voice that addressed them; for the absence of his teeth had not at all improved the conversational tones of the Surgeon of the Fleet.

With natural curiosity, these men had purposely been lingering, to see all they could; but now, having no further excuse, they snatched up their hammers and chisels, and—like the stage-builders decamping from a public meeting at the eleventh hour, after just completing the rostrum in time for the first speaker—the Carpenter's gang withdrew.

The broad ensign now lifted, revealing a glimpse of the crowd of man-of-war's-men outside, and the patient, borne in the arms of two of his mess-mates, entered the place. He was much emaciated, weak as an infant, and every limb visibly trembled, or rather jarred, like the head of a man with the palsy. As if an organic and involuntary apprehension of death had seized the wounded leg, its nervous motions were so violent that one of the mess-mates was obliged to keep his hand upon it.

The top-man was immediately stretched upon the table, the attendants steadying his limbs, when, slowly opening his eyes, he glanced about at the glittering knives and saws, the towels and sponges, the armed sentry at the Commodore's cabin-door, the row of eager-eyed students, the meagre death's-head of a Cuticle, now with his shirt sleeves rolled up upon his withered arms, and knife in hand, and, finally, his eyes settled in horror upon the skeleton, slowly vibrating and jingling before him, with the slow, slight roll of the frigate in the water.

"I would advise perfect repose of your every limb, my man," said Cuticle, addressing him; "the precision of an operation is often impaired by the inconsiderate restlessness of the patient. But if you consider, my good fellow," he added, in a patronising and almost sympathetic tone, and slightly pressing his hand on the limb, "if you consider how much better it is to live with three limbs than to die with four, and especially if you but knew to what torments both sailors and soldiers were subjected before the time of Celsus, owing to the lamentable ignorance of surgery then prevailing, you would certainly thank God from the bottom of your heart that *your* operation has been postponed to the period of this enlightened age, blessed with a Bell, a Brodie, and a Lally. My man, before Celsus's time, such was the general ignorance of our noble science, that, in order to prevent the excessive effusion of blood, it was deemed indispensable to operate with a red-hot knife"—making a professional movement toward the thigh—"and pour scalding oil upon the parts"—elevating his elbow, as if with a tea-pot in his hand—"still further to sear them, after amputation had been performed."

Literary Narrative: "The Operation" from *White Jacket* (1850) by Herman Melville

"He is fainting!" said one of his mess-mates; "quick! some water!" The steward immediately hurried to the top-man with the basin.

Cuticle took the top-man by the wrist, and feeling it a while, observed, "Don't be alarmed, men," addressing the two mess-mates; "he'll recover presently; this fainting very generally takes place." And he stood for a moment, tranquilly eyeing the patient.

Now the Surgeon of the Fleet and the top-man presented a spectacle which, to a reflecting mind, was better than a church-yard sermon on the mortality of man.

Here was a sailor, who four days previous, had stood erect—a pillar of life—with an arm like a royal-mast and a thigh like a windlass. But the slightest conceivable finger-touch of a bit of crooked trigger had eventuated in stretching him out, more helpless than an hour-old babe, with a blasted thigh, utterly drained of its brawn. And who was it that now stood over him like a superior being, and, as if clothed himself with the attributes of immortality, indifferently discoursed of carving up his broken flesh, and thus piecing out his abbreviated days. Who was it, that in capacity of Surgeon, seemed enacting the part of a Regenerator of life? The withered, shrunken, one-eyed, toothless, hairless Cuticle; with a trunk half dead—a memento mori to behold!

And while, in those soul-sinking and panic-striking premonitions of speedy death which almost invariably accompany a severe gun-shot wound, even with the most intrepid spirits; while thus drooping and dying, this once robust top-man's eye was now waning in his head like a Lapland moon being eclipsed in clouds—Cuticle, who for years had still lived in his withered tabernacle of a body—Cuticle, no doubt sharing in the common self-delusion of old age—Cuticle must have felt his hold of life as secure as the grim hug of a grizzly bear. Verily, Life is more awful than Death; and let no man, though his live heart beat in him like a cannon—let him not hug his life to himself; for, in the predestinated necessities of things, that bounding life of his is not a whit more secure than the life of a man on his death-bed. To-day we inhale the air with expanding lungs, and life runs through us like a thousand Niles; but to-morrow we may collapse in death, and all our veins be dry as the Brook Kedron in a drought.

"And now, young gentlemen," said Cuticle, turning to the Assistant Surgeons, "while the patient is coming to, permit me to describe to you the highly-interesting operation I am about to perform."

"Mr. Surgeon of the Fleet," said Surgeon Bandage, "if you are about to lecture, permit me to present you with your teeth; they will make your discourse more readily understood." And so saying, Bandage, with a bow, placed the two semicircles of ivory into Cuticle's hands.

"Thank you, Surgeon Bandage," said Cuticle, and slipped the ivory into its place.

"In the first place, now, young gentlemen, let me direct your attention to the excellent preparation before you. I have had it unpacked from its case, and set up here from my stateroom, where it occupies the spare berth; and all this for your express benefit, young gentlemen. This skeleton I procured in person from the Hunterian department of the Royal College of Surgeons in London. It is a masterpiece of art. But we have no time to examine it now. Delicacy forbids that I should amplify at a juncture like this"—casting an almost benignant glance toward the patient, now beginning to open his eyes; "but let me point out to you upon this thigh-bone"—disengaging it from the skeleton, with a gentle twist—"the precise place where I propose to perform the operation. *Here*, young gentlemen, *here* is the place. You perceive it is very near the point of articulation with the trunk."

"Yes," interposed Surgeon Wedge, rising on his toes, "yes, young gentlemen, the point of articulation with the *acetabulum* of the *os innominatum*."

"Where's your Bell on Bones, Dick?" whispered one of the assistants to the student next to him. "Wedge has been spending the whole morning over it, getting out the hard names."

"Surgeon Wedge," said Cuticle, looking round severely, "we will dispense with your commentaries, if you please, at present. Now, young gentlemen, you cannot but perceive, that the point of operation being so near the trunk and the vitals, it becomes an unusually beautiful one, demanding a steady hand and a true eye; and, after all, the patient may die under my hands."

"Quick, Steward! water, water; he's fainting again!" cried the two mess-mates.

"Don't be alarmed for your comrade; men," said Cuticle, turning round. "I tell you it is not an uncommon thing for the patient to betray some emotion upon these occasions—most usually manifested by swooning; it is quite natural it should be so. But we must not delay the operation. Steward, that knife—no, the next one—there, that's it. He is coming to, I think"—feeling the top-man's wrist. "Are you all ready, sir?"

This last observation was addressed to one of the Never-sink's assistant surgeons, a tall, lank, cadaverous young man, arrayed in a sort of shroud of white canvas, pinned about his throat, and completely enveloping his person. He was seated on a match-tub—the skeleton swinging near his head—at the foot of the table, in readiness to grasp the limb, as when a plank is being severed by a carpenter and his apprentice.

"The sponges, Steward," said Cuticle, for the last time taking out his teeth, and drawing up his shirt sleeves still further. Then, taking the patient by the wrist, "Stand by, now, you mess-mates; keep hold of his arms; pin him down. Steward, put your hand on the artery; I shall commence as soon as his pulse begins to—*now, now!*" Letting fall the wrist, feeling the thigh carefully, and bowing over it an instant, he drew the fatal knife unerringly across the flesh. As it first touched the part, the row of surgeons simultaneously dropped their eyes to the watches in their hands while the patient lay, with eyes horribly distended, in a kind of waking trance. Not a breath was heard; but as the quivering flesh parted in a long, lingering gash, a spring of blood welled up between the living walls of the wounds, and two thick streams, in opposite directions, coursed down the thigh. The sponges were instantly dipped in the purple pool; every face present was pinched to a point with suspense; the limb writhed; the man shrieked; his mess-mates pinioned him; while round and round the leg went the unpitying cut.

"The saw!" said Cuticle.

Instantly it was in his hand.

Full of the operation, he was about to apply it, when, looking up, and turning to the assistant surgeons, he said, "Would any of you young gentlemen like to apply the saw? A splendid subject!"

Several volunteered; when, selecting one, Cuticle surrendered the instrument to him, saying, "Don't be hurried, now; be steady."

While the rest of the assistants looked upon their comrade with glances of envy, he went rather timidly to work; and Cuticle, who was earnestly regarding him, suddenly snatched the saw from his hand. "Away, butcher! you disgrace the profession. Look at *me!*"

For a few moments the thrilling, rasping sound was heard; and then the top-man seemed parted in twain at the hip, as the leg slowly slid into the arms of the pale, gaunt man in the shroud, who at once made away with it, and tucked it out of sight under one of the guns.

"Surgeon Sawyer," now said Cuticle, courteously turning to the surgeon of the Mohawk, "would you like to take up the arteries? They are quite at your service, sir."

"Do, Sawyer; be prevailed upon," said Surgeon Bandage.

Sawyer complied; and while, with some modesty he was conducting the operation, Cuticle, turning to the row of assistants said, "Young gentlemen, we will now proceed with our Illustration. Hand me that bone, Steward." And taking the thigh-bone in his still bloody hands, and holding it conspicuously before his auditors, the Surgeon of the Fleet began:

"Young gentlemen, you will perceive that precisely at this spot—*here*—to which I previously directed your attention—at the corresponding spot precisely—the operation has been performed. About here, young gentlemen, here"—lifting his hand some inches from the bone—"about *here* the great artery was. But you noticed that I did not use the tourniquet; I never do. The forefinger of my steward is far better than a tourniquet, being so much more manageable, and leaving the smaller veins uncompressed. But I have been told, young gentlemen, that a certain Seignior Seignioroni, a surgeon of Seville, has recently invented an admirable substitute for the clumsy, old-fashioned tourniquet. As I understand it, it is something like a pair of *calipers*, working with a small Archimedes screw—a very clever invention, according to all accounts. For the padded points at the end of the arches"—arching his forefinger and thumb—"can be so worked as to approximate in such a way, as to—but you don't attend to me, young gentlemen," he added, all at once starting.

Being more interested in the active proceedings of Surgeon Sawyer, who was now threading a needle to sew up the overlapping of the stump, the young gentlemen had not scrupled to turn away their attention altogether from the lecturer.

A few moments more, and the top-man, in a swoon, was removed below into the sick-bay. As the curtain settled again after the patient had disappeared, Cuticle, still holding the thigh-bone of the skeleton in his ensanguined hands, proceeded with his remarks upon it; and having concluded them, added, "Now, young gentlemen, not the least interesting consequence of this operation will be the finding of the ball, which, in case of non-amputation, might have long eluded the most careful search. That ball, young gentlemen, must have taken a most circuitous route. Nor, in cases where the direction is oblique, is this at all unusual. Indeed, the learned Henner gives us a most remarkable—I had almost said an incredible—case of a soldier's neck, where the bullet, entering at the part called Adam's Apple—".

"Yes," said Surgeon Wedge, elevating himself, "the *pomum Adami*."

"Entering the point called *Adam's Apple*," continued Cuticle, severely emphasising the last two words, "ran completely round the neck, and, emerging at the same hole it had entered, shot the next man in the ranks. It was afterward extracted, says Renner, from the second man, and pieces of the other's skin were found adhering to it. But examples of foreign substances being received into the body with a ball, young gentlemen, are frequently observed. Being attached to a United States ship at the time, I happened to be near the spot of the battle of Ayacucho, in Peru. The day after the action, I saw in the barracks of the wounded a trooper, who, having been severely injured in the brain, went crazy, and, with his own holster-pistol, committed suicide in the hospital. The ball drove inward a portion of his woollen night-cap——".

"In the form of a *cul-de-sac*, doubtless," said the undaunted Wedge.

"For once, Surgeon Wedge, you use the only term that can be employed; and let me avail myself of this opportunity to say to you, young gentlemen, that a man of true science"—expanding his shallow chest a little—"uses but few hard words, and those only when none other will answer his purpose; whereas the smatterer in science"—slightly glancing toward Wedge—"thinks, that by mouthing hard words, he proves that he understands hard things. Let this sink deep in your minds, young gentlemen; and, Surgeon Wedge" —with a stiff bow—"permit me to submit the reflection to yourself. Well, young gentlemen, the bullet was afterward extracted by pulling upon the external parts of the *cul-de-sac*—a simple, but exceedingly beautiful operation. There is a fine example, somewhat similar, related in Guthrie; but, of course, you must have met with it, in so well-known a work as his Treatise upon Gun-shot Wounds. When, upward of twenty years ago, I was with Lord Cochrane, then Admiral of the fleets of this very country"—pointing shoreward, out of a port-hole—"a sailor of the vessel to which I was attached, during the blockade of Bahia, had his leg——" But by this time the fidgets had completely taken possession of his auditors, especially of the senior surgeons; and turning upon them abruptly, he added, "But I will not detain you longer, gentlemen"—turning round upon all the surgeons—"your dinners must be waiting you on board your respective ships. But, Surgeon Sawyer, perhaps you may desire to wash your hands before you go. There is the basin, sir; you will find a clean towel on the rammer. For myself, I seldom use them"—taking out his handkerchief. "I must leave you now, gentlemen"—bowing. "To-morrow, at ten, the limb will be upon the table, and I shall be happy to see you all upon the occasion. Who's there?" turning to the curtain, which then rustled.

"Please, sir," said the Steward, entering, "the patient is dead."

"The body also, gentlemen, at ten precisely," said Cuticle, once more turning round upon his guests. "I predicted that the operation might prove fatal; he was very much run down. Good-morning;" and Cuticle departed.

"He does not, surely, mean to touch the body?" exclaimed Surgeon Sawyer, with much excitement.

"Oh, no!" said Patella, "that's only his way; he means, doubtless, that it may be inspected previous to being taken ashore for burial."

The assemblage of gold-laced surgeons now ascended to the quarter-deck; the second cutter was called away by the bugler, and, one by one, they were dropped aboard of their respective ships.

The following evening the mess-mates of the top-man rowed his remains ashore, and buried them in the ever-vernal Protestant cemetery, hard by the Beach of the Flamingoes, in plain sight from the bay.

Pain and Value

In this satirical account of pre-anesthetic surgery—note the satirical names Melville attributes to healthcare workers—emphasizes the manner in which the physiological condition of pain is somehow related to meaning and value. Thus, in this narrative Melville strives to represent and make meaningful—by means of allusion, metaphorical language, and charged language—the situation of pain, represented by terrifying upcoming surgery, that he represents.

We should mention one other aspect of this narrative. In Chap. 8, we discussed everyday virtue ethics in relation to other approaches to judging moral behavior. In this story, the surgeon, Dr. Cuticle, repeatedly dismisses the terror of his patient, the "top-man" who isn't even named. Instead, he asserts his own importance, even as Melville describes "the withered, shrunken, one-eyed, toothless, hairless Cuticle" in stark contrast to the able-bodied sailor before he was shot in the leg. In this, Melville contrasts risible caricature and satire with the memento mori—the remembrance of death—that give rise to "soul-sinking and panic-striking premonitions" that constitute much of the debilitating power of the pain and suffering patients feel.

Related Poem

For this chapter, the related poem, written by Emily Dickinson, does not distinguish between chronic and acute pain but simply meditates on the experience of pain altogether. In this poem, Dickinson describes how pain "has an element of blank," which blocks out the consideration of anything else. Heshusius also records this aspect of the pain experience—one it is wise that healthcare providers be cognizant of, if only to understand how patients' narratives can be inflected and distorted in the presence of great pain.

Poem: "Pain has an Element of Blank" (1890; poem #650) by Emily Dickinson

Author Note: Emily Dickinson (1830–1886) was an American poet who was intensely private: less than a dozen of her 1800 poems were published during her lifetime. Family members published her poems after her death, and these poems established her as a major poet in American culture. Throughout her life she suffered from various ailments—she did suffer from a prolonged and painful condition of her eye—but source of the pain in this poem is not altogether clear.

Pain has an Element of Blank

> Pain — has an Element of Blank —
> It cannot recollect
> When it begun — or if there were
> A time when it was not —
>
> It has no Future — but itself —
> Its Infinite realms contain
> Its Past — enlightened to perceive
> New Periods — of Pain.

What makes this poem powerful even as it creates a sense of uneasiness might be the way it puts together the ordinariness of pain and its stark and overwhelming power. Dickinson, like the other narratives in this chapter, is attempting to represent and acknowledge the power of pain.

Lessons for Providers

Expressing empathy for someone suffering is a skill which requires much practice. Engaging with narratives of illness and suffering is good practice.

Bibliography

Cassel, Eric. 1982. The Nature of Suffering and the Goals of Medicine. www.ericcassell.com/download/NatureOfSuffering.pdf

Fishman, Scott (with Lisa Berger). 2000. *The War on Pain*. New York: Quill.

Hemingway, Ernest. 1972. *The Nick Adam Stories*. Ed. Philip Young. New York: Scribner's.

Heshusius, Lous. 2009. *Inside Chronic Pain: An Intimate and Critical Account*. Ithaca: Cornell University Press.

Schleifer, Ronald. 2014. *Pain and Suffering*. New York: Routledge.

———. 2018b. The Aesthetics of Pain: Semiotics and Affective Comprehension in Music, Literature, and Sensate Experience. *Configurations* 26 (2018): 471–491.

Ageing   12

A significant aspect of medical treatment—now and in the foreseeable future—is the constant ageing of the patient population. Patients are getting older and older, and the age difference between the age of healthcare providers and of their patients continues to grow. It is necessary for healthcare providers to understand the experience and limitations of their patients to develop understanding and patience when dealing with older, slower people. The vignette for this chapter gives a good sense of the problems that arise in treating older patients and, perhaps, a strategy for addressing those problems.

Treating a Very Old Woman: A Vignette by Dr. Jerry Vannatta

The patient, Mrs. Gilbert, was a 92-year-old woman who had been brought to the 32 year-old internist by her daughter, who was a primary care patient and thought her mother needed a new doctor. The internist, treating her like he did all new patients, performed a complete history and physical examination. Following this procedure he recommended a few blood tests and some x-rays and a follow-up appointment in two weeks. On return to clinic he went over the test results and recommended several things that this older woman should do. In reality, she was in pretty good health for someone over the age of 90. She had osteoarthritis of knees, hips and shoulders – to be expected at this age. She had mild high blood pressure currently being handled by a once a day diuretic. She was being treated for hypothyroidism and that was under good control. The internist felt confident of his process, assessment, and plan.

Four months later she returned for a follow-up regarding her high blood pressure.

"Hello Mrs. Gilbert, how are you doing?"

"I am doing OK, but these hips are killing me."

"Yes, I know your hips have arthritis, anything else?"

"I am having trouble getting to the bathroom in the middle of the night."

"You could empty your bladder before going to bed, Mrs. Gilbert."

Becoming visibly upset, the patient looks into the eyes of the young internist with a note of disdain and says. "Will you just be quiet and listen to me?"

The internist was flabbergasted. Never had he been addressed like this by a patient. "You have no idea what it is like to be 92 years old. How old are you anyway? 27?" she asked in

© The Author(s) 2019
R. Schleifer, J. B. Vannatta, *Literature and Medicine*,
https://doi.org/10.1007/978-3-030-19128-3_12

a huff. "Ever since I met you, you ask me questions but you don't really listen to the answers. You do not seem to wonder how my complaints affect my life. Did you know that I have two masters' degrees and taught school for 42 years? I am smart, probably as smart as you are, and I demand to be understood."

The internist squirmed in his chair. "Oh my, Mrs. Gilbert, I am so sorry. I had no idea I was coming across this way."

"Yes, well just be quiet and listen to me, young man. You might just learn something about being old."

The internist pushed his chair back and put down his pen. "Please Mrs. Gilbert, tell me what I need to hear."

"For the past 20 years it seems to me that I have lost something of myself each and every week. I have difficulty getting up out of a chair. I have difficulty reading – an activity that used to bring me great joy. The ophthalmologist tells me I have macular degeneration and there is nothing that can be done. I am cold all the time, and when my daughter visits she burns up in my house I keep it so hot. My hips are killing me and what I really want to do is go out to eat with my daughter and granddaughter at least once a month. I have difficulty walking to the restaurant and sitting through a meal. I do empty my bladder before I go to bed, I am not an idiot, but I must get up to use it and I cannot see, I am unsteady and am afraid of falling. I realize I am not young and I realize that my body is old and is wearing out. I do not expect you and modern medicine to make me young again or make all my aches and pains to go away. But I do expect you to listen to what my concerns are and address them in a way that attempts to accomplish what I expect out of life. I want to reduce the pain in my hips and get around well enough to go out to lunch. I want to feel safe in the middle of the night when I get up to urinate. I want to be comfortable but be able to have younger people come by to visit. And I want to be able to enjoy the great literature of the world, which I spent so many years teaching."

The patient was rescheduled for an hour in one week. The internist promised to spend the hour addressing all her concerns she mentioned today and to listen more carefully in the future to what exactly she wanted addressed.

The Disparities of Ageing

This intelligent patient nicely catalogues the *situation* of ageing, the sense of loss and, implicit here, the sense of disparity between self-conception and physical limitations. Thus, an 84 year-old man told Ronald Blythe in his collection of interviews with aging people, *The View in Winter*, that

> old age doesn't necessarily mean that one is entirely old – *all* old, if you follow me. It doesn't mean that for many people, which is why it is so very difficult. It is complicated by the retention of a lot of one's youth in an old body. I tend to look upon other old men as *old men* – and not include myself. It is not vanity; it is just that it is still natural for me to be young in some respects. What is generally assumed to have happened to a man in his eighties has not happened to me.... Yet I resent it all in some ways, this being very old, yes, I resent it. (1979: 185)

This man is talking about the ambiguity of ageing. Moreover, this ambiguity often leads to inappropriate assertions of power, taking the form sometimes of very much talk. But this talk is also a function of the loneliness of old age, a matter touched upon in the chapter from Nathaniel Hawthorne and Thomas Hardy's poem included in this chapter. (A black-humor novel, *The House of God* by Samuel Shem [1978], portrays and mocks the loneliness of old people, who have few other people to talk besides the healthcare providers they seek out.) And finally, there is the despair of very old age, its combination of hopelessness and a younger sensibility, which Hardy chillingly captures in his poem. That feeling is less notable, but still present in the literary narrative of this chapter.

Literary Narrative: "The Little Shop Window," from *The House of Seven Gables* (1851) by Nathaniel Hawthorne

Author Note: Nathaniel Hawthorne (1804–1864) was an American novelist and short-story writer in the mid-nineteenth century, the time of the "American Renaissance." He is best known for *The Scarlet Letter* (1851), often understood to be one of the great American novels. In 1837 he published a collection of stories entitled *Twice Told Tales*. He was a close college friend of Franklin Pierce, the 14th President of the United States, for whom he wrote a campaign biography. *The House of Seven Gables*, like many of his novels, focuses on New England life.

The Little Shop Window

It still lacked half an hour of sunrise, when Miss Hepzibah Pyncheon – we will not say awoke, it being doubtful whether the poor lady had so much as closed her eyes during the brief night of midsummer – but, at all events, arose from her solitary pillow, and began what it would be mockery to term the adornment of her person. Far from us be the indecorum of assisting, even in imagination, at a maiden lady's toilet! Our story must therefore await Miss Hepzibah at the threshold of her chamber; only presuming, meanwhile, to note some of the heavy sighs that labored from her bosom, with little restraint as to their lugubrious depth and volume of sound, inasmuch as they could be audible to nobody save a disembodied listener like ourself. The Old Maid was alone in the old house. Alone, except for a certain respectable and orderly young man, an artist in the daguerreotype line, who, for about three months back, had been a lodger in a remote gable, – quite a house by itself, indeed, – with locks, bolts, and oaken bars on all the intervening doors. Inaudible, consequently, were poor Miss Hepzibah's gusty sighs. Inaudible the creaking joints of her stiffened knees, as she knelt down by the bedside. And inaudible, too, by mortal ear, but heard with all-comprehending love and pity in the farthest heaven, that almost agony of prayer – now whispered, now a groan, now a struggling silence – wherewith she besought the Divine assistance through the day! Evidently, this is to be a day of more than ordinary trial to Miss Hepzibah, who, for above a quarter of a century gone by, has dwelt in strict seclusion, taking no part in the business of life, and just as little in its intercourse and pleasures. Not with such fervor prays the torpid recluse, looking forward to the cold, sunless, stagnant calm of a day that is to be like innumerable yesterdays.

The maiden lady's devotions are concluded. Will she now issue forth over the threshold of our story? Not yet, by many moments. First, every drawer in the tall, old-fashioned bureau is to be opened, with difficulty, and with a succession of spasmodic jerks then, all must close again, with the same fidgety reluctance. There is a rustling of stiff silks; a tread of backward and forward footsteps to and fro across the chamber. We suspect Miss Hepzibah, moreover, of taking a step upward into a chair, in order to give heedful regard to her appearance on all sides, and at full length, in the oval, dingy-framed toilet-glass, that hangs above her table. Truly! well, indeed! who would have thought it! Is all this precious time to be lavished on the matutinal repair and beautifying of an elderly person, who never goes abroad, whom nobody ever visits, and from whom, when she shall have done her utmost, it were the best charity to turn one's eyes another way?

Now she is almost ready. Let us pardon her one other pause; for it is given to the sole sentiment, or, we might better say, – heightened and rendered intense, as it has been, by sorrow and seclusion, – to the strong passion of her life. We heard the turning of a key in a small lock; she has opened a secret drawer of an escritoire, and is probably looking at a certain miniature, done in Malbone's most perfect style, and representing a face worthy of no less delicate a pencil. It was once our good fortune to see this picture. It is a likeness of a young man, in a silken dressing-gown of an old fashion, the soft richness of which is well adapted to the countenance of reverie, with its full, tender lips, and beautiful eyes, that seem to indicate not so much capacity of thought, as gentle and voluptuous emotion. Of the possessor of such features we shall have a right to ask nothing, except that he would take the rude world easily, and make himself happy in it. Can it have been an early lover of Miss Hepzibah? No; she never had a lover – poor thing, how could she? – nor ever knew, by her own experience, what love technically means. And yet, her undying faith and trust, her fresh remembrance, and continual devotedness towards the original of that miniature, have been the only substance for her heart to feed upon.

She seems to have put aside the miniature, and is standing again before the toilet-glass. There are tears to be wiped off. A few more footsteps to and fro; and here, at last, – with another pitiful sigh, like a gust of chill, damp wind out of a long-closed vault, the door of which has accidentally been set, ajar – here comes Miss Hepzibah Pyncheon! Forth she steps into the dusky, time-darkened passage; a tall figure, clad in black silk, with a long and shrunken waist, feeling her way towards the stairs like a near-sighted person, as in truth she is.

The sun, meanwhile, if not already above the horizon, was ascending nearer and nearer to its verge. A few clouds, floating high upward, caught some of the earliest light, and threw down its golden gleam on the windows of all the houses in the street, not forgetting the House of the Seven Gables, which – many such sunrises as it had witnessed – looked cheerfully at the present one. The reflected radiance served to show, pretty distinctly, the aspect and arrangement of the room which Hepzibah entered, after descending the stairs. It was a low-studded room, with a beam across the ceiling, panelled with dark wood, and having a large chimney-piece, set round with pictured tiles, but now closed by an iron fire-board, through which ran the funnel of a modern stove. There was a carpet on the floor, originally of rich texture, but so worn and faded in these latter years that its once brilliant figure had quite vanished into one indistinguishable hue. In the way of furniture, there were two tables: one, constructed with perplexing intricacy and exhibiting as many feet as a centipede; the other, most delicately wrought, with four long and slender legs, so apparently frail that it was almost incredible what a length of time the ancient tea-table had stood upon them. Half a dozen chairs stood about the room, straight and stiff, and so ingeniously contrived for the discomfort of the human person that they were irksome even to sight, and conveyed the ugliest possible idea of the state of society to which they could have been adapted. One exception there was, however, in a very antique elbow-chair, with a high back, carved elaborately in oak, and a roomy depth within its arms, that made up, by its spacious comprehensiveness, for the lack of any of those artistic curves which abound in a modern chair.

As for ornamental articles of furniture, we recollect but two, if such they may be called. One was a map of the Pyncheon territory at the eastward, not engraved, but the handiwork of some skilful old draughtsman, and grotesquely illuminated with pictures of Indians and wild beasts, among which was seen a lion; the natural history of the region being as little known as its geography, which was put down most fantastically awry. The other adornment was the portrait of old Colonel Pyncheon, at two thirds length, representing the stern features of a Puritanic-looking personage, in a skull-cap, with a laced band and a grizzly beard; holding a Bible with one hand, and in the other uplifting an iron sword-hilt. The latter object, being more successfully depicted by the artist, stood out in far greater prominence than the sacred volume. Face to face with this picture, on entering the apartment, Miss Hepzibah Pyncheon came to a pause; regarding it with a singular scowl, a strange contortion of the brow, which, by people who did not know her, would probably have been interpreted as an expression of bitter anger and ill-will. But it was no such thing. She, in fact, felt a reverence for the pictured visage, of which only a far-descended and time-stricken virgin could be susceptible; and this forbidding scowl was the innocent result of her near-sightedness, and an effort so to concentrate her powers of vision as to substitute a firm outline of the object instead of a vague one.

We must linger a moment on this unfortunate expression of poor Hepzibah's brow. Her scowl, – as the world, or such part of it as sometimes caught a transitory glimpse of her at the window, wickedly persisted in calling it, – her scowl had done Miss Hepzibah a very ill office, in establishing her character as an ill-tempered old maid; nor does it appear improbable that, by often gazing at herself in a dim looking-glass, and perpetually encountering her own frown with its ghostly sphere, she had been led to interpret the expression almost as unjustly as the world did. "How miserably cross I look!" she must often have whispered to herself; and ultimately have fancied herself so, by a sense of inevitable doom. But her heart never frowned. It was naturally tender, sensitive, and full of little tremors and palpitations; all of which weaknesses it retained, while her visage was growing so perversely stern, and even fierce. Nor had Hepzibah ever any hardihood, except what came from the very warmest nook in her affections.

All this time, however, we are loitering faintheartedly on the threshold of our story. In very truth, we have an invincible reluctance to disclose what Miss Hepzibah Pyncheon was about to do.

It has already been observed, that, in the basement story of the gable fronting on the street, an unworthy ancestor, nearly a century ago, had fitted up a shop. Ever since the old gentleman retired from trade, and fell asleep under his coffin-lid, not only the shop-door, but the inner arrangements, had been suffered to remain unchanged; while the dust of ages gathered inch-deep over the shelves and counter, and partly filled an old pair of scales, as if it were of value enough to be weighed. It treasured itself up, too, in the half-open till, where there still lingered a base sixpence, worth neither more nor less than the hereditary pride which had here been put to shame. Such had been the state and condition of the little shop in old Hepzibah's childhood, when she and her brother used to play at hide-and-seek in its forsaken precincts. So it had remained, until within a few days past.

But now, though the shop-window was still closely curtained from the public gaze, a remarkable change had taken place in its interior. The rich and heavy festoons of cobweb, which it had cost a long ancestral succession of spiders their life's labor to spin and weave, had been carefully brushed away from the ceiling. The counter, shelves, and floor had all been scoured, and the latter was overstrewn with fresh blue sand. The brown scales, too, had evidently undergone rigid discipline, in an unavailing effort to rub off the rust, which, alas! had eaten through and through their substance. Neither was the little old shop any longer empty of merchantable goods. A curious eye, privileged to take an account of stock and investigate behind the counter, would have discovered a barrel, yea, two or three barrels and half ditto, – one containing flour, another apples, and a third, perhaps, Indian meal. There was likewise a square box of pine-wood, full of soap in bars; also, another of the same size, in which were tallow candles, ten to the pound. A small stock of brown sugar, some white beans and split peas, and a few other commodities of low price, and such as are constantly in demand, made up the bulkier portion of the merchandise. It might have been taken for a ghostly or phantasmagoric reflection of the old shop-keeper Pyncheon's shabbily provided shelves, save that some of the articles were of a description and outward form which could hardly have been known in his day. For instance, there was a glass pickle-jar, filled with fragments of Gibraltar rock; not, indeed, splinters of the veritable stone foundation of the famous fortress, but bits of delectable candy, neatly done up in white paper. Jim Crow, moreover, was seen executing his world-renowned dance, in gingerbread. A party of leaden dragoons were galloping along one of the shelves, in equipments and uniform of modern cut; and there were some sugar figures, with no strong resemblance to the humanity of any epoch, but less unsatisfactorily representing our own fashions than those of a hundred years ago. Another phenomenon, still more strikingly modern, was a package of lucifer matches, which, in old times, would have been thought actually to borrow their instantaneous flame from the nether fires of Tophet.

In short, to bring the matter at once to a point, it was incontrovertibly evident that somebody had taken the shop and fixtures of the long-retired and forgotten Mr. Pyncheon, and was about to renew the enterprise of that departed worthy, with a different set of customers. Who could this bold adventurer be? And, of all places in the world, why had he chosen the House of the Seven Gables as the scene of his commercial speculations?

We return to the elderly maiden. She at length withdrew her eyes from the dark countenance of the Colonel's portrait, heaved a sigh, – indeed, her breast was a very cave of Aolus that morning, – and stept across the room on tiptoe, as is the customary gait of elderly women. Passing through an intervening passage, she opened a door that communicated with the shop, just now so elaborately described. Owing to the projection of the upper story – and still more to the thick shadow of the Pyncheon Elm, which stood almost directly in front of the gable – the twilight, here, was still as much akin to night as morning. Another heavy sigh from Miss Hepzibah! After a moment's pause on the threshold, peering towards the window with her near-sighted scowl, as if frowning down some bitter enemy, she suddenly projected herself into the shop. The haste, and, as it were, the galvanic impulse of the movement, were really quite startling.

Nervously – in a sort of frenzy, we might almost say – she began to busy herself in arranging some children's playthings, and other little wares, on the shelves and at the shop-window. In the aspect of this dark-arrayed, pale-faced, ladylike old figure there was a deeply tragic character that contrasted irreconcilably with the ludicrous pettiness of her employment. It seemed a queer anomaly, that so gaunt and dismal a personage should take a toy in hand; a miracle, that the toy did not vanish in her grasp; a miserably absurd idea, that she should go on perplexing her stiff and sombre intellect with the question how to tempt little boys into her premises! Yet such is undoubtedly her object. Now she places a gingerbread elephant against the window, but with so tremulous a touch that it tumbles upon the floor, with the dismemberment of three legs and its trunk; it has ceased to be an elephant, and has become a few bits of musty gingerbread. There, again, she has upset a tumbler of marbles, all of which roll different ways, and each individual marble, devil-directed, into the most difficult obscurity that it can find. Heaven help our poor old Hepzibah, and forgive us for taking a ludicrous view of her position! As her rigid and rusty frame goes down upon its hands and knees, in quest of the absconding marbles, we positively feel so much the more inclined to shed tears of sympathy, from the very fact that we must needs turn aside and laugh at her. For here, – and if we fail to impress it suitably upon the reader, it is our own fault, not that of the theme, here is one of the truest points of melancholy interest that occur in ordinary life. It was the final throe of what called itself old gentility. A lady – who had fed herself from childhood with the shadowy food of aristocratic reminiscences, and whose religion it was that a lady's hand soils itself irremediably by doing aught for bread, – this born lady, after sixty years of narrowing means, is fain to step down from her pedestal of imaginary rank. Poverty, treading closely at her heels for a lifetime, has come up with her at last. She must earn her own food, or starve! And we have stolen upon Miss Hepzibah Pyncheon, too irreverently, at the instant of time when the patrician lady is to be transformed into the plebeian woman.

In this republican country, amid the fluctuating waves of our social life, somebody is always at the drowning-point. The tragedy is enacted with as continual a repetition as that of a popular drama on a holiday, and, nevertheless, is felt as deeply, perhaps, as when an hereditary noble sinks below his order. More deeply; since, with us, rank is the grosser substance of wealth and a splendid establishment, and has no spiritual existence after the death of these, but dies hopelessly along with them. And, therefore, since we have been unfortunate enough to introduce our heroine at so inauspicious a juncture, we would entreat for a mood of due solemnity in the spectators of her fate. Let us behold, in poor Hepzibah, the immemorial, lady – two hundred years old, on this side of the water, and thrice as many on the other, – with her antique portraits, pedigrees, coats of arms, records and traditions, and her claim, as joint heiress, to that princely territory at the eastward, no longer a wilderness, but a populous fertility, – born, too, in Pyncheon Street, under the Pyncheon Elm, and in the Pyncheon House, where she has spent all her days, – reduced. Now, in that very house, to be the hucksteress of a cent-shop.

This business of setting up a petty shop is almost the only resource of women, in circumstances at all similar to those of our unfortunate recluse. With her near-sightedness, and those tremulous fingers of hers, at once inflexible and delicate, she could not be a seamstress; although her sampler, of fifty years gone by, exhibited some of the most recondite specimens of ornamental needlework. A school for little children had been often in her thoughts; and, at one time, she had begun a review of her early studies in the New England Primer, with a view to prepare herself for the office of instructress. But the love of children had never been quickened in Hepzibah's heart, and was now torpid, if not extinct; she watched the little people of the neighborhood from her chamber-window, and doubted whether she could tolerate a more intimate acquaintance with them. Besides, in our day, the very ABC has become a science greatly too abstruse to be any longer taught by pointing a pin from letter to letter. A modern child could teach old Hepzibah more than old Hepzibah could teach the child. So – with many a cold, deep heart-quake at the idea of at last coming into sordid contact with the world, from which she had so long kept aloof, while every added day of seclusion had rolled another stone against the cavern door of her hermitage – the poor thing bethought herself of the ancient shop-window, the rusty scales, and dusty till. She might have held back a little longer; but another circumstance, not yet hinted at, had somewhat hastened her decision. Her humble preparations, therefore, were duly made, and the enterprise was now to be commenced. Nor was she entitled to complain of any remarkable singularity in her fate; for, in the town of her nativity, we might point to several little shops of a similar description, some of them in houses as ancient as that of the Seven Gables; and one or two, it may be, where a decayed gentlewoman stands behind the counter, as grim an image of family pride as Miss Hepzibah Pyncheon herself.

It was overpoweringly ridiculous, – we must honestly confess it, – the deportment of the maiden lady while setting her shop in order for the public eye. She stole on tiptoe to the window, as cautiously as if she conceived some bloody-minded villain to be watching behind the elm-tree, with intent to take her life. Stretching out her long, lank arm, she put a paper of pearl-buttons, a jew's-harp, or whatever the small article might be, in its destined place, and straightway vanished back into the dusk, as if the world need never hope for another glimpse of her. It might have been fancied, indeed, that she expected to minister to the wants of the community unseen, like a disembodied divinity or enchantress, holding forth her bargains to the reverential and awe-stricken purchaser in an invisible hand. But Hepzibah had no such flattering dream. She was well aware that she must ultimately come forward, and stand revealed in her proper individuality; but, like other sensitive persons, she could not bear to be observed in the gradual process, and chose rather to flash forth on the world's astonished gaze at once.

The inevitable moment was not much longer to be delayed. The sunshine might now be seen stealing down the front of the opposite house, from the windows of which came a reflected gleam, struggling through the boughs of the elm-tree, and enlightening the interior of the shop more distinctly than heretofore. The town appeared to be waking up. A baker's cart had already rattled through the street, chasing away the latest vestige of night's sanctity with the jingle-jangle of its dissonant bells. A milkman was distributing the contents of his cans from door to door; and the harsh peal of a fisherman's conch shell was heard far off, around the corner. None of these tokens escaped Hepzibah's notice. The moment had arrived. To delay longer would be only to lengthen out her misery. Nothing remained, except to take down the bar from the shop-door, leaving the entrance free – more than free – welcome, as if all were household friends – to every passer-by, whose eyes might be attracted by the commodities at the window. This last act Hepzibah now performed, letting the bar fall with what smote upon her excited nerves as a most astounding clatter. Then – as if the only barrier betwixt herself and the world had been thrown down, and a flood of evil consequences would come tumbling through the gap – she fled into the inner parlor, threw herself into the ancestral elbow-chair, and wept.

Our miserable old Hepzibah! It is a heavy annoyance to a writer, who endeavors to represent nature, its various attitudes and circumstances, in a reasonably correct outline and true coloring, that so much of the mean and ludicrous should be hopelessly mixed up with the purest pathos which life anywhere supplies to him. What tragic dignity, for example, can be wrought into a scene like this! How can we elevate our history of retribution for the sin of long ago, when, as one of our most prominent figures, we are compelled to introduce – not a young and lovely woman, nor even the stately remains of beauty, storm-shattered by affliction – but a gaunt, sallow, rusty-jointed maiden, in a long-waisted silk gown, and with the strange horror of a turban on her head! Her visage is not even ugly. It is redeemed from insignificance only by the contraction of her eyebrows into a near-sighted scowl. And, finally, her great life-trial seems to be, that, after sixty years of idleness, she finds it convenient to earn comfortable bread by setting up a shop in a small way. Nevertheless, if we look through all the heroic fortunes of mankind, we shall find this same entanglement of something mean and trivial with whatever is noblest in joy or sorrow. Life is made up of marble and mud. And, without all the deeper trust in a comprehensive sympathy above us, we might hence be led to suspect the insult of a sneer, as well as an immitigable frown, on the iron countenance of fate. What is called poetic insight is the gift of discerning, in this sphere of strangely mingled elements, the beauty and the majesty which are compelled to assume a garb so sordid.

The Grief of Ageing

This chapter from *The House of Seven Gables*, like the poem to follow, is about the grief of ageing, something that people, in the midst of committed lives and careers, find difficult to comprehend. Throughout the narrator treats Hepzibah Pyncheon with soft irony, although he does catalogue her "stiffened knees," "the agony of prayer," "sorrow and seclusion," her appearance of scowling that seems, but is not, "an expression of bitter anger and ill-will," and how there is something "overpoweringly ridiculous" in her comportment. Still, the final paragraph offers a wide overview of the condition of ageing. The vicarious experiences of this literary narrative, like the other narratives of Section VI in their different ways, aim at granting healthcare providers a fuller comprehension of the lives and apprehensions of their older patients.

Related Poem

For this chapter, the related poem, written by Thomas Hardy, entitled "I Look into My Glass," captures the loneliness and grief—the loneliness that is part of the grief—of ageing, with all kinds of living things "gone all slowly."

Poem: "I Look into My Glass" (1898) by Thomas Hardy

Author Note: Thomas Hardy (1840–1928) was one of the most renowned novelists and poets in English. His novels—he wrote 14—are considered among the most important Victorian novels, even while his poetry—he wrote almost 1000 poems—is considered an important part of Modern English poetry. Both are focused on southern England. His poetry, like his novels, describes a fatalism, even though the style of his poems, its wonderfully innovative stanza forms and language (see "undistrest" below), seems to create an alternative to his pervasive philosophical fatalism. The combination of fatalism and a glimmer of hope imbues the poem below.

I Look into My Glass

I look into my glass,
And view my wasting skin,
And say, "Would God it came to pass
My heart had shrunk as thin!"

For then, I, undistrest
By hearts grown cold to me,
Could lonely wait my endless rest
With equanimity.

But Time, to make me grieve,
Part steals, lets part abide;
And shakes this fragile frame at eve
With throbbings of noontide.

Lessons for Providers
It takes more time to care for a person of age. We must be self-conscious of this and slow down so we resonate with the patient's bradykinesia.

Bibliography

Blythe, Ronald. 1979. *The View in Winter: Reflections on Old Age*. New York: Harcourt Brace Jovanovich.

Section VII

Mistakes in Medicine

Mistakes in Medicine 13

The preceding section offered narrative and literary texts that provoke vicarious experience that could be integrated into the training of healthcare providers. And earlier, in Section V, we examined the manner in which literary texts can help us understand and be mindful of everyday ethics in medicine. This chapter examines the practical issue of mistakes in medicine, which also touches upon the issue of ethics, negatively understood as regular *kinds* of practical mistakes that mar healthcare practices and institutions. In this chapter we offer a vignette of a terrible mistake in medicine and analyze it in relation to a ground-breaking (and widely available: see Bibliography) discussion of medical mistakes by Dr. David Hilfiker in 1984 in *The New England Journal of Medicine* entitled "Facing Our Mistakes." But many other vignettes in *Literature and Medicine* also describe mistakes: see "The Woman with Hyponatremia" in Chap. 2; "Playing God" in Chap. 3; excerpt from Dr. Damon Tweedy's *Black Man in a White Coat*, which focuses on the terrible "mistake" of racism in relation to healthcare in Chap. 7; "The Patient with Diabetic Ketoacidosis" in Chap. 8; and "Frenzy Facing Death" in the final chapter, which portrays a patient's and her family's "mistake." All of these "mistakes" can be re-examined in relation to the analyses of mistakes in this chapter.

In his article, Hilfiker catalogs five causes of mistakes in medicine that are mentioned in the vignette: (1) healthcare providers do not know enough, (2) they do not have the necessary skills, (3) they are sometimes careless, (4) they suffer from a failure to make the right judgment in a particular situation, and (5) even though they know the right thing to do, because of distraction, tiredness, time pressures, or other factors, they suffer from a failure of will to do what they know is right. Although he does not note this, a lack of trustworthiness and compassion—which we noted among the "virtues" of healthcare and which are implicit in standards of professionalism in medicine examined in Chaps. 3 and 8—can also be considered mistakes. In addition, Hilfiker also does not catalog mistakes that follow from the failure to follow normal ("normative") behavior, such as (perhaps) the issues arising from "Playing God" or the problem of engaging in direct arguments with patients in "The

© The Author(s) 2019
R. Schleifer, J. B. Vannatta, *Literature and Medicine*,
https://doi.org/10.1007/978-3-030-19128-3_13

Patient with Diabetic Ketoacidosis." Finally, Hilfiker's early essay does not address mistakes arising from *systematic institutions* of professional healthcare noted in the following vignette.

Mistakes: Enough to Spread Around: A Vignette by Dr. Jerry Vannatta

Mr. Johnson was admitted for a blood clot in his right leg. It involved veins above the knee: a problem considered serious because of the likelihood that a small piece might break off and land in the lung, a life threatening condition. My intern, resident, and medical students admitted him. I saw him the next morning. The primary reason for his clotting abnormality was that he had widespread, untreatable lung cancer. He was pleasant, had an attentive family, and seemed to have come to grips with his terminal illness. He was anti-coagulated appropriately by the intern and resident with an intravenous medication, Heparin. This medication was used for approximately three days, at which time an oral medicine was started. The new medicine – Coumadin – is much more difficult to use and regulate, but can be overlapped with the Heparin while the dose is slowly titrated upwards into a "therapeutic" range. There are guidelines utilized to get the job done, and the intern used the guidelines. On the third day of Coumadin the blood test used to guide the dosage was observed and the dosage was increased.

Two days later the patient called the nurse in the middle of the night complaining that he had developed a penile erection. He complained that it would not go away and the pain was debilitating. The nurse discovered that the erection had been present for three hours. The intern on call asked for consultation from the urology department. The urology resident examined the patient and a blood test was ordered to explore the state of his anti-coagulation. Meanwhile the operating room was readied in an attempt to surgically decompress the blood engorged painful penis. The blood test previously ordered showed "Coumadin poisoning," a term used at the time to mean Coumadin overdose. His blood was too thin and he had bled into the base of his penis occluding the normal draining of venous blood from the penis. Therefore, because the blood could not get out but the arteries kept pumping blood in, the penis became painfully engorged – the clinical term of which is priapism.

The attempt to decompress the penis failed and the urology team informed the patient and the family that he would require "penectomy," a penile amputation. The patient agreed to the operation and it was performed soon.

As attending physician, I was notified in the middle of the night of this problem. By the time I arrived he was in the operating room receiving the amputation. My medical team had arrived for morning rounds and we immediately began reviewing the patient's record and the process of caring for him. The resident who had received a call in the middle of the night about this patient came to the hospital immediately and began an investigation.

She found on the day mentioned above an order was indicated to increase the dose of Coumadin, a hand-written order on the medical record – this was before the use of electronic medical records – for an increase of 1.0 mg of Coumadin. She then went to the hospital pharmacy and asked for this patient's records. She found that on that morning (now three days ago) an order was received from the ward to increase the dose of Coumadin by 10.0 mg (instead of the 1.0 which had been written). That order was filled and the new dose of Coumadin was sent to the floor and administered immediately. The patient received two days of this increased dose, which had caused his blood to become completely uncoagulable. A side effect of this condition is bleeding. In his case the bleeding was into the base of the penis. Further investigation found that there was a nursing policy that when a dosage increase for Coumadin was entered into the chart, the order was sent to the pharmacy to be dispensed and given the next day – not the day the order was written. We also found a hospital pharmacy policy that stated that an order written by the doctor, which the pharmacist found out of the ordinary, should cause the pharmacist to contact the ordering physician. A conversation was then to take place to clarify the order.

It appeared to me that an error at every step following the writing of the order by the intern had occurred. These errors had led to a terrible clinical outcome, penile amputation. This was an error in the system. The system failed to have enough safeguards to ensure protocols are followed in order to prevent an error leading to a horrible patient outcome.

Errors in medicine are common. David Hilfiker published examples from his own practice in *The New England Journal of Medicine* in 1984. In his article, "Facing Our Mistakes," he categorizes five medical errors, which are important but do not include an additional category to the list of errors in medical practice, namely errors of the medical system. In our case, in addition to the systemic error—that is, the absence of *systematic* double-checking and review—it is also possible to categorize the error in Hilfiker's category of "carelessness" in understanding the mistake of an order for 1.0 mg as one for 10.0 mg, no matter the condition of the handwriting. There is also an error of carelessness in the pharmacy in filling an order of an outrageous increase in the daily dose of medication. One can also wonder about the carelessness of the nursing staff administering such an unusual dose of a common medication 24 hours too early.

Errors of the system are errors that occur in the system/routines of health care delivery. That system might be the doctor's office, the outpatient surgical center, the hospital, or any other environment where healthcare is delivered. In November of 1999, the Institute of Medicine published a report entitled "To Err Is Human: Building a Safer Health Care System." In this report the Institute reported on two studies that found that there were between 44,000 and 98,000 deaths in American hospitals per year due to medical errors. They found that errors in the system of healthcare delivery were the most common cause of these preventable deaths. The Institute recommended four processes to reduce these deaths, one of which was "Implementing safety systems in health care organizations to ensure safe practices at the delivery level."

This vignette describes a terrible, life-destroying mistake in medicine; it is a powerful and disturbing narrative. Yet as it suggests itself, it could *instructively* be juxtaposed to Dr. Hilfiker's ground-breaking essay "Facing Our Mistakes." In doing so, one can see that some of Hilfiker's categories could be more fully detailed to offer *systematic* responses to errors in medicine. (In his important book *A Checklist Manifesto* [2010], Dr. Atul Gawande offers strategies to build mistake-responses into the *routine practices* of medicine. In the 1857 literary text in this chapter, one can see the complete absence of "systematic" medicine before the institution of professional medicine in the early twentieth century.) In Dr. Vannatta's vignette in this chapter, multiple instances of "carelessness" describe the terrible, seeming "simple" mistake that Mr. Johnson suffered. Moreover, in relation to the vignette, we can see that Hilfiker's categorization both recognizes mistakes and, at the same time, it *personalizes and individualizes* the error: it asks us to recognize (1) individual errors of the individual physician's writing legibility; (2) errors on the part of the individual pharmacist and (3) of the individual nurse, both of whom do not recognizing the dangerous and unusual dose; and (4) error of the individual nurse who does not abide by the usual 24-hour delay in the administration of the medication. Doing so, "carelessness" obscures the *systematic* solution, which would prevent such a catastrophic error in the future, namely to create *impersonal* procedures in the administration of medications. Such procedures would enjoin practitioners—pharmacists, nurses, physicians—to stop and question questionable doses at all stages of a medical procedure.

An explicit one-item checklist—"is this dose appropriate for this condition?"—would have prevented this mistake at various stages of Mr. Johnson's medical care by calling on healthcare providers to double-check their behavior.

Hilfiker's essay was ground-breaking because it called attention to the ways in which the medical profession systematically ignores mistakes (unlike, in Gawande's example, the airline industry, which makes focus on catastrophic mistakes in air transportation part of industry standards). Moreover, Hilfiker includes within his discussion of mistakes in medicine a powerful meditation on the "culture of perfection" in healthcare. Thus, he notes that "because of its technological wonders and near-miraculous drugs, modern medicine has created for the physician an expectation of perfection. The technology seems so exact that error becomes almost unthinkable. We are not prepared for our mistakes, and we don't know how to cope with them when they occur" (1984: 5). Such a culture of perfection manifests itself in the smugness of Dr. Canivet in *Madame Bovary* below (as well as Dr. Cuticle in Melville in Chap. 11, but it is also observable in twentieth-century physicians in Selzer and others). Hilfiker concludes his argument by noting that medicine's cultural of "perfectionism" creates a number of *institutional* inabilities to systematically confront and learn from mistakes. These "inabilities" stem from legal institutions, in terms of malpractice litigation; professional (collegial) institutions, in terms of colleagues' reluctance to chastise other professionals; and personal institutions, in terms of individuals' reluctance to admit mistakes.

Hilfiker begins his analysis with an instructive narrative about his patient, Barb Dailey. Barb, whom Hilfiker had treated during the successful birth of her first child, comes to him and says she "feels" pregnant. Even though he notices her enlarged uterus, all the tests indicate she is not pregnant. Dr. Hilfiker notes: "I could find out [absolutely whether or not Barb was pregnant] by ordering an ultrasound examination. This procedure would give me a 'picture' of the uterus and of the embryo. But Barb would have to go to Duluth for the examination. The procedure is also expensive. I know the Dailys well enough to know they have a modest income. Besides, by waiting a few weeks, I should be able to find out for sure without the ultrasound: either the urine test will be positive or Barb will have a miscarriage" (1984: 1). After negative results of the subsequent test, Hilfiker performs a "Dilation and curettage, or D & C, is a relatively simple surgical procedure performed thousands of times each day in this country" (1984: 3) and discovers he has aborted her living fetus. In this initial narrative, Hilfiker fails to note two other "mistakes" in his experience that he does not include in his catalog of mistakes. (1) By considering, but not discussing the financial aspects of an ultrasound test with his patient, he begins to conceive of his position with his patient as a "financial advisor" rather than as a "medical provider"; that is, he "mistakes" his professional work in relation to his patient. (Dr. Bovary similarly confuses the nature of his job when he puts the prospect of his medical reputation before the well-being of his patient.) (2) He fails to acknowledge the importance of his patient's story, in this case the fact that when Barb, having been pregnant before, states she "feels" pregnant, her patient testimony should be considered to be at least as important as the pregnancy test. (No test is 100% sure, and Hilfiker does not consider in his narrative why these tests were wrong [e.g., a bad "batch" of urine tests?]). As we have already suggested, it might be instructive for readers of *Literature and Medicine* to reconsider many of its literary narratives and vignettes in relation to Hilfiker's catalog of mistakes, those added here, those that might be suggested by the book's discussion of ethics, and by these narratives and vignettes themselves.

The literary narrative of this chapter presents the narrative of a doctor who is swayed by the "agendas" of people other than his patient—including his own—in his medical engagements and decisions early in his career, when he faces a serious problem. Dr. Bovary's mistakes falls under Hilfiker's categories of "lack of knowledge," "lack of skill," "carelessness," and "failure of judgment." Hilfiker's final category—"lack of will to do the right thing" is obviated by Dr. Bovary's lack of knowledge, and the whole episode demonstrates the "mistake" of a "non-systematic" medical organization.

Literary Narrative: Chapter Eleven from *Madame Bovary* (1857) by Gustav Flaubert

> *Author Note*: Gustav Flaubert (1821–1880) was a highly influential French novelist. He is regarded as a prime mover of the "realist" school of French literature and best known for what readers take to be his masterpiece, *Madame Bovary*, a realistic portrayal of bourgeois life, which led to a trial on charges of the novel's immorality (because of its seeming sympathy for its adulterous heroine). His careful attention to the responses of his characters—rather than to the "events" of narrative—led him to be admired and written about by almost every major literary personality of the twentieth century.

Chapter Eleven (from *Madame Bovary*)

> [Monsieur Homais, the town pharmacist] had recently read a eulogy on a new method for curing club-foot, and as he was a partisan of progress, he conceived the patriotic idea that Yonville, in order to keep to the fore, ought to have some operations for strephopody or club-foot.
> "For," said he to Emma [Bovary], "what risk is there? See—" (and he enumerated on his fingers the advantages of the attempt), "success, almost certain relief and beautifying of the patient, celebrity acquired by the operator. Why, for example, should not your husband relieve poor Hippolyte of the 'Lion d'Or'? Note that he would not fail to tell about his cure to all the travellers, and then" (Homais lowered his voice and looked round him) "who is to prevent me from sending a short paragraph on the subject to the paper? Eh! goodness me! an article gets about; it is talked of; it ends by making a snowball! And who knows? who knows?"
> In fact, [her husband, Dr. Charles] Bovary might succeed. Nothing proved to Emma that he was not clever; and what a satisfaction for her to have urged him to a step by which his reputation and fortune would be increased! She only wished to lean on something more solid than love.
> Charles, urged by the druggist and by her, allowed himself to be persuaded. He sent to Rouen for Dr. Duval's volume, and every evening, holding his head between both hands, plunged into the reading of it.
> While he was studying equinus, varus, and valgus, that is to say, katastrephopody, endostrephopody, and exostrephopody (or better, the various turnings of the foot downwards, inwards, and outwards, with the hypostrephopody and anastrephopody), otherwise torsion

downwards and upwards, Monsier Homais, with all sorts of arguments, was exhorting the lad ["poor Hippolyte"] at the inn to submit to the operation.

"You will scarcely feel, probably, a slight pain; it is a simple prick, like a little bloodletting, less than the extraction of certain corns."

Hippolyte, reflecting, rolled his stupid eyes.

"However," continued the chemist, "it doesn't concern me. It's for your sake, for pure humanity! I should like to see you, my friend, rid of your hideous claudication, together with that waddling of the lumbar regions which, whatever you say, must considerably interfere with you in the exercise of your calling."

Then Homais represented to him how much jollier and brisker he would feel afterwards, and even gave him to understand that he would be more likely to please the women; and the stable-boy began to smile heavily. Then he attacked him through his vanity:

"Aren't you a man? Hang it! what would you have done if you had had to go into the army, to go and fight beneath the standard? Ah! Hippolyte!"

And Homais retired, declaring that he could not understand this obstinacy, this blindness in refusing the benefactions of science.

The poor fellow gave way, for it was like a conspiracy. Binet, who never interfered with other people's business, Madame Lefrancois, Artemise, the neighbours, even the mayor, Monsieur Tuvache—everyone persuaded him, lectured him, shamed him; but what finally decided him was that it would cost him nothing. Bovary even undertook to provide the machine for the operation. This generosity was an idea of Emma's, and Charles consented to it, thinking in his heart of hearts that his wife was an angel.

So by the advice of the chemist, and after three fresh starts, he had a kind of box made by the carpenter, with the aid of the locksmith, that weighed about eight pounds, and in which iron, wood, sheer-iron, leather, screws, and nuts had not been spared.

But to know which of Hippolyte's tendons to cut, it was necessary first of all to find out what kind of club-foot he had.

He had a foot forming almost a straight line with the leg, which, however, did not prevent it from being turned in, so that it was an equinus together with something of a varus, or else a slight varus with a strong tendency to equinus. But with this equinus, wide in foot like a horse's hoof, with rugose skin, dry tendons, and large toes, on which the black nails looked as if made of iron, the clubfoot ran about like a deer from morn till night. He was constantly to be seen on the Place, jumping round the carts, thrusting his limping foot forwards. He seemed even stronger on that leg than the other. By dint of hard service it had acquired, as it were, moral qualities of patience and energy; and when he was given some heavy work, he stood on it in preference to its fellow.

Now, as it was an equinus, it was necessary to cut the tendon of Achilles, and, if need were, the anterior tibial muscle could be seen to afterwards for getting rid of the varus; for the doctor did not dare to risk both operations at once; he was even trembling already for fear of injuring some important region that he did not know.

Neither Ambrose Pare, applying for the first time since Celsus, after an interval of fifteen centuries, a ligature to an artery, nor Dupuytren, about to open an abscess in the brain, nor Gensoul when he first took away the superior maxilla, had hearts that trembled, hands that shook, minds so strained as Monsieur Bovary when he approached Hippolyte, his tenotome between his fingers. And as at hospitals, near by on a table lay a heap of lint, with waxed thread, many bandages—a pyramid of bandages—every bandage to be found at the druggist's. It was Monsieur Homais who since morning had been organising all these preparations, as much to dazzle the multitude as to keep up his illusions. Charles pierced the skin; a dry crackling was heard. The tendon was cut, the operation over. Hippolyte could not get over his surprise, but bent over Bovary's hands to cover them with kisses.

"Come, be calm," said the druggist; "later on you will show your gratitude to your benefactor."

And he went down to tell the result to five or six inquirers who were waiting in the yard, and who fancied that Hippolyte would reappear walking properly. Then Charles, having buckled his patient into the machine, went home, where Emma, all anxiety, awaited him at the door. She threw herself on his neck; they sat down to table; he ate much, and at dessert he even wanted to take a cup of coffee, a luxury he only permitted himself on Sundays when there was company.

The evening was charming, full of prattle, of dreams together. They talked about their future fortune, of the improvements to be made in their house; he saw people's estimation of him growing, his comforts increasing, his wife always loving him; and she was happy to refresh herself with a new sentiment, healthier, better, to feel at last some tenderness for this poor fellow who adored her. The thought of Rodolphe for one moment passed through her mind, but her eyes turned again to Charles; she even noticed with surprise that he had not bad teeth.

They were in bed when Monsieur Homais, in spite of the servant, suddenly entered the room, holding in his hand a sheet of paper just written. It was the paragraph he intended for the "Fanal de Rouen." He brought it for them to read.

"Read it yourself," said Bovary.

He read—

"Despite the prejudices that still invest a part of the face of Europe like a net, the light nevertheless begins to penetrate our country places. Thus on Tuesday our little town of Yonville found itself the scene of a surgical operation which is at the same time an act of loftiest philanthropy. Monsieur Bovary, one of our most distinguished practitioners—"

"Oh, that is too much! too much!" said Charles, choking with emotion.

"No, no! not at all! What next!"

"'—Performed an operation on a club-footed man.' I have not used the scientific term, because you know in a newspaper everyone would not perhaps understand. The masses must—'".

"No doubt," said Bovary; "go on!"

"I proceed," said the chemist. "Monsieur Bovary, one of our most distinguished practitioners, performed an operation on a club-footed man called Hippolyte Tautain, stableman for the last twenty-five years at the hotel of the 'Lion d'Or', kept by Widow Lefrancois, at the Place d'Armes. The novelty of the attempt, and the interest incident to the subject, had attracted such a concourse of persons that there was a veritable obstruction on the threshold of the establishment. The operation, moreover, was performed as if by magic, and barely a few drops of blood appeared on the skin, as though to say that the rebellious tendon had at last given way beneath the efforts of art. The patient, strangely enough—we affirm it as an eye-witness—complained of no pain. His condition up to the present time leaves nothing to be desired. Everything tends to show that his convelescence will be brief; and who knows even if at our next village festivity we shall not see our good Hippolyte figuring in the bacchic dance in the midst of a chorus of joyous boon-companions, and thus proving to all eyes by his verve and his capers his complete cure? Honour, then, to the generous savants! Honour to those indefatigable spirits who consecrate their vigils to the amelioration or to the alleviation of their kind! Honour, thrice honour! Is it not time to cry that the blind shall see, the deaf hear, the lame walk? But that which fanaticism formerly promised to its elect, science now accomplishes for all men. We shall keep our readers informed as to the successive phases of this remarkable cure."

This did not prevent Mere Lefrancois, from coming five days after, scared, and crying out—.

"Help! he is dying! I am going crazy!"

Charles rushed to the "Lion d'Or," and the chemist, who caught sight of him passing along the Place hatless, abandoned his shop. He appeared himself breathless, red, anxious, and asking everyone who was going up the stairs—.

"Why, what's the matter with our interesting strephopode?"

The strephopode was writhing in hideous convulsions, so that the machine in which his leg was enclosed was knocked against the wall enough to break it.

With many precautions, in order not to disturb the position of the limb, the box was removed, and an awful sight presented itself. The outlines of the foot disappeared in such a swelling that the entire skin seemed about to burst, and it was covered with ecchymosis, caused by the famous machine. Hippolyte had already complained of suffering from it. No attention had been paid to him; they had to acknowledge that he had not been altogether wrong, and he was freed for a few hours. But, hardly had the oedema gone down to some extent, than the two savants thought fit to put back the limb in the apparatus, strapping it tighter to hasten matters. At last, three days after, Hippolyte being unable to endure it any longer, they once more removed the machine, and were much surprised at the result they saw. The livid tumefaction spread over the leg, with blisters here and there, whence there oozed a black liquid. Matters were taking a serious turn. Hippolyte began to worry himself, and Mere Lefrancois had him installed in the little room near the kitchen, so that he might at least have some distraction.

But the tax-collector, who dined there every day, complained bitterly of such companionship. Then Hippolyte was removed to the billiard-room. He lay there moaning under his heavy coverings, pale with long beard, sunken eyes, and from time to time turning his perspiring head on the dirty pillow, where the flies alighted. Madame Bovary went to see him. She brought him linen for his poultices; she comforted, and encouraged him. Besides, he did not want for company, especially on market-days, when the peasants were knocking about the billiard-balls round him, fenced with the cues, smoked, drank, sang, and brawled.

"How are you?" they said, clapping him on the shoulder. "Ah! you're not up to much, it seems, but it's your own fault. You should do this! do that!" And then they told him stories of people who had all been cured by other remedies than his. Then by way of consolation they added—.

"You give way too much! Get up! You coddle yourself like a king! All the same, old chap, you don't smell nice!"

Gangrene, in fact, was spreading more and more. Bovary himself turned sick at it. He came every hour, every moment. Hippolyte looked at him with eyes full of terror, sobbing—.

"When shall I get well? Oh, save me! How unfortunate I am! How unfortunate I am!"

And the doctor left, always recommending him to diet himself.

"Don't listen to him, my lad," said Mere Lefrancois, "Haven't they tortured you enough already? You'll grow still weaker. Here! swallow this."

And she gave him some good beef-tea, a slice of mutton, a piece of bacon, and sometimes small glasses of brandy, that he had not the strength to put to his lips.

Abbe Bournisien, hearing that he was growing worse, asked to see him. He began by pitying his sufferings, declaring at the same time that he ought to rejoice at them since it was the will of the Lord, and take advantage of the occasion to reconcile himself to Heaven.

"For," said the ecclesiastic in a paternal tone, "you rather neglected your duties; you were rarely seen at divine worship. How many years is it since you approached the holy table? I understand that your work, that the whirl of the world may have kept you from care for your salvation. But now is the time to reflect. Yet don't despair. I have known great sinners, who, about to appear before God (you are not yet at this point I know), had implored His mercy, and who certainly died in the best frame of mind. Let us hope that, like them, you will set us a good example. Thus, as a precaution, what is to prevent you from saying morning and evening a 'Hail Mary, full of grace,' and 'Our Father which art in heaven'? Yes, do that, for my sake, to oblige me. That won't cost you anything. Will you promise me?"

The poor devil promised. The cure came back day after day. He chatted with the landlady; and even told anecdotes interspersed with jokes and puns that Hippolyte did not understand. Then, as soon as he could, he fell back upon matters of religion, putting on an appropriate expression of face.

His zeal seemed successful, for the club-foot soon manifested a desire to go on a pilgrimage to Bon-Secours if he were cured; to which Monsieur Bournisien replied that he saw no objection; two precautions were better than one; it was no risk anyhow.

Literary Narrative: Chapter Eleven from *Madame Bovary* (1857) by Gustav Flaubert

The druggist was indignant at what he called the manoeuvres of the priest; they were prejudicial, he said, to Hippolyte's convalescence, and he kept repeating to Madame Lefrancois, "Leave him alone! leave him alone! You perturb his morals with your mysticism." But the good woman would no longer listen to him; he was the cause of it all. From a spirit of contradiction she hung up near the bedside of the patient a basin filled with holy-water and a branch of box.

Religion, however, seemed no more able to succour him than surgery, and the invincible gangrene still spread from the extremities towards the stomach. It was all very well to vary the potions and change the poultices; the muscles each day rotted more and more; and at last Charles replied by an affirmative nod of the head when Mere Lefrancois, asked him if she could not, as a forlorn hope, send for Monsieur Canivet of Neufchatel, who was a celebrity.

A doctor of medicine, fifty years of age, enjoying a good position and self-possessed, Charles's colleague did not refrain from laughing disdainfully when he had uncovered the leg, mortified to the knee. Then having flatly declared that it must be amputated, he went off to the chemist's to rail at the asses who could have reduced a poor man to such a state. Shaking Monsieur Homais by the button of his coat, he shouted out in the shop—

"These are the inventions of Paris! These are the ideas of those gentry of the capital! It is like strabismus, chloroform, lithotrity, a heap of monstrosities that the Government ought to prohibit. But they want to do the clever, and they cram you with remedies without, troubling about the consequences. We are not so clever, not we! We are not savants, coxcombs, fops! We are practitioners; we cure people, and we should not dream of operating on anyone who is in perfect health. Straighten club-feet! As if one could straighten club-feet! It is as if one wished, for example, to make a hunchback straight!"

Homais suffered as he listened to this discourse, and he concealed his discomfort beneath a courtier's smile; for he needed to humour Monsier Canivet, whose prescriptions sometimes came as far as Yonville. So he did not take up the defence of Bovary; he did not even make a single remark, and, renouncing his principles, he sacrificed his dignity to the more serious interests of his business.

This amputation of the thigh by Doctor Canivet was a great event in the village. On that day all the inhabitants got up earlier, and the Grande Rue, although full of people, had something lugubrious about it, as if an execution had been expected. At the grocer's they discussed Hippolyte's illness; the shops did no business, and Madame Tuvache, the mayor's wife, did not stir from her window, such was her impatience to see the operator arrive.

He came in his gig, which he drove himself. But the springs of the right side having at length given way beneath the weight of his corpulence, it happened that the carriage as it rolled along leaned over a little, and on the other cushion near him could be seen a large box covered in red sheep-leather, whose three brass clasps shone grandly.

After he had entered like a whirlwind the porch of the "Lion d'Or," the doctor, shouting very loud, ordered them to unharness his horse. Then he went into the stable to see that she was eating her oats all right; for on arriving at a patient's he first of all looked after his mare and his gig. People even said about this—

"Ah! Monsieur Canivet's a character!"

And he was the more esteemed for this imperturbable coolness. The universe to the last man might have died, and he would not have missed the smallest of his habits.

Homais presented himself.

"I count on you," said the doctor. "Are we ready? Come along!"

But the druggist, turning red, confessed that he was too sensitive to assist at such an operation.

"When one is a simple spectator," he said, "the imagination, you know, is impressed. And then I have such a nervous system!"

"Pshaw!" interrupted Canivet; "on the contrary, you seem to me inclined to apoplexy. Besides, that doesn't astonish me, for you chemist fellows are always poking about your kitchens, which must end by spoiling your constitutions. Now just look at me. I get up every day at four o'clock; I shave with cold water (and am never cold). I don't wear flannels, and I never catch cold; my carcass is good enough! I live now in one way, now in another, like a philosopher, taking pot-luck; that is why I am not squeamish like you, and it is as indifferent to me to carve a Christian as the first fowl that turns up. Then, perhaps, you will say, habit! habit!"

Then, without any consideration for Hippolyte, who was sweating with agony between his sheets, these gentlemen entered into a conversation, in which the druggist compared the coolness of a surgeon to that of a general; and this comparison was pleasing to Canivet, who launched out on the exigencies of his art. He looked upon, it as a sacred office, although the ordinary practitioners dishonoured it. At last, coming back to the patient, he examined the bandages brought by Homais, the same that had appeared for the club-foot, and asked for someone to hold the limb for him. Lestiboudois was sent for, and Monsieur Canivet having turned up his sleeves, passed into the billiard-room, while the druggist stayed with Artemise and the landlady, both whiter than their aprons, and with ears strained towards the door.

Bovary during this time did not dare to stir from his house.

He kept downstairs in the sitting-room by the side of the fireless chimney, his chin on his breast, his hands clasped, his eyes staring. "What a mishap!" he thought, "what a mishap!" Perhaps, after all, he had made some slip. He thought it over, but could hit upon nothing. But the most famous surgeons also made mistakes; and that is what no one would ever believe! People, on the contrary, would laugh, jeer! It would spread as far as Forges, as Neufchatel, as Rouen, everywhere! Who could say if his colleagues would not write against him. Polemics would ensue; he would have to answer in the papers. Hippolyte might even prosecute him. He saw himself dishonoured, ruined, lost; and his imagination, assailed by a world of hypotheses, tossed amongst them like an empty cask borne by the sea and floating upon the waves.

Emma, opposite, watched him; she did not share his humiliation; she felt another—that of having supposed such a man was worth anything. As if twenty times already she had not sufficiently perceived his mediocrity.

Charles was walking up and down the room; his boots creaked on the floor.

"Sit down," she said; "you fidget me."

He sat down again.

How was it that she—she, who was so intelligent—could have allowed herself to be deceived again? and through what deplorable madness had she thus ruined her life by continual sacrifices? She recalled all her instincts of luxury, all the privations of her soul, the sordidness of marriage, of the household, her dream sinking into the mire like wounded swallows; all that she had longed for, all that she had denied herself, all that she might have had! And for what? for what?

In the midst of the silence that hung over the village a heart-rending cry rose on the air. Bovary turned white to fainting. She knit her brows with a nervous gesture, then went on. And it was for him, for this creature, for this man, who understood nothing, who felt nothing! For he was there quite quiet, not even suspecting that the ridicule of his name would henceforth sully hers as well as his. She had made efforts to love him, and she had repented with tears for having yielded to another!

"But it was perhaps a valgus!" suddenly exclaimed Bovary, who was meditating.

At the unexpected shock of this phrase falling on her thought like a leaden bullet on a silver plate, Emma, shuddering, raised her head in order to find out what he meant to say; and they looked at the other in silence, almost amazed to see each other, so far sundered were they by their inner thoughts. Charles gazed at her with the dull look of a drunken man, while he listened motionless to the last cries of the sufferer, that followed each other in long-drawn modulations, broken by sharp spasms like the far-off howling of some beast being slaughtered. Emma bit her wan lips, and rolling between her fingers a piece of coral that she had broken, fixed on Charles the burning glance of her eyes like two arrows of fire about to dart forth. Everything in him irritated her now; his face, his dress, what he did not say, his whole person, his existence, in fine. She repented of her past virtue as of a crime, and what still remained of it rumbled away beneath the furious blows of her pride. She revelled in all the evil ironies of triumphant adultery. The memory of her lover came back to her with dazzling attractions; she threw her whole soul into it, borne away towards this image with a fresh enthusiasm; and Charles seemed to her as much removed from her life, as absent forever, as impossible and annihilated, as if he had been about to die and were passing under her eyes.

There was a sound of steps on the pavement. Charles looked up, and through the lowered blinds he saw at the corner of the market in the broad sunshine Dr. Canivet, who was wiping his brow with his handkerchief. Homais, behind him, was carrying a large red box in his hand, and both were going towards the chemist's.

Then with a feeling of sudden tenderness and discouragement Charles turned to his wife saying to her—

"Oh, kiss me, my own!"

"Leave me!" she said, red with anger.

"What is the matter?" he asked, stupefied. "Be calm; compose yourself. You know well enough that I love you. Come!"

"Enough!" she cried with a terrible look.

And escaping from the room, Emma closed the door so violently that the barometer fell from the wall and smashed on the floor.

Charles sank back into his arm-chair overwhelmed, trying to discover what could be wrong with her, fancying some nervous illness, weeping, and vaguely feeling something fatal and incomprehensible whirling round him.

When Rodolphe came to the garden that evening, he found his mistress waiting for him at the foot of the steps on the lowest stair. They threw their arms round one another, and all their rancour melted like snow beneath the warmth of that kiss.

—translated by Eleanor Marx Aveling

Medical Paternalism

One feature of medicine that is often noticed is the "paternalism" of physician practices, especially when in the United States and Europe the vast majority of physicians in the nineteenth and early twentieth centuries were men. Such an attitude towards healthcare might well be a "systematic" or "institutional" mistake a little different from the individual mistakes that Hilfiker catalogues. Paternalism manifests itself when healthcare providers—usually physicians—assume they know best and need not consult in any meaningfully practical way with their patients. In some countries today—Japan is one example—physicians still do not think full forthrightness with patients is the "normal" standard of care. Flaubert's chapter is a subtle example of paternalism, violating the normative situation when patients come to healthcare providers rather than being sought out by them, as in this narrative. (One good example of extreme "paternalism" in medicine is the widely available story by Dr. William Carlos Williams entitled "The Use of Force" [1938: see Bibliography], in which a physician physically fights with a young girl (and bloodies her) in order to make a diagnosis. What is striking about Williams' story is its honesty in focusing on the manners in which the intimacies of patient–physician engagements can give rise to powerful, if inappropriate, emotion, such as we saw in the emotional response—the "mistake"—on the side of the resident in the vignette in Chap. 8, "The Patient with Diabetic Ketoacidosis." A key issue in the narrative from *Madame Bovary* is the "boundary" between the physician's personal ambitions and feelings and his professional responsibilities.

Related Poem

For this chapter, the related poem is the recounting of a "true incident" of terribly imprecise brain surgery. The poem presents—and questions—the behavior of healthcare providers: their truthfulness, their understanding of the medical procedures they undertake, their understanding of the patient in their care. This is a relatively straight-forward, almost factual, narrative, which culminates in a strange aural image in its last lines.

Poem: "In the Theatre" (1983) by Dr. Dannie Abse

> *Author Note*: Dr. Dannie Abse (1923–2014) was a Welsh physician, where he worked in a chest clinic, and a poet, novelist, and playwright. In 2009 he won the Wilfred Owen Poetry Award. He notes of the poem presented here: "My eldest brother is a doctor – I was a schoolboy when he was a medical student and one day he came back from working in the operating theatre in Cardiff when he was a dresser to a well-known brain surgeon by the name of Lambert Rogers. He came back as I say and told us a very strange story, a haunting story, and years passed and it still haunted me and eventually I put down what he said in this poem. […] The operation in question took place in 1938 when they didn't have the scanning devices they now have which can pick out a lesion in the brain very cleverly, whereas in the past sometimes a surgeon, searching for the tumour or whatever it was, broke down more brain tissue than was necessary."

In the Theatre

(A true incident)

> *"Only a local anaesthetic was given because of the blood pressure problem. The patient, thus, was fully awake throughout the operation. But in those days—in 1938, in Cardiff, when I was Lambert Rogers' dresser—they could not locate a brain tumour with precision. Too much normal brain tissue was destroyed as the surgeon searched for it, before he felt the resistance of it…all somewhat hit and miss. One operation I shall never forget…"(Dr. Wilfred Abse)*

Sister saying—"Soon you'll be back in the ward,"
sister thinking—"Only two more on the list,"
the patient saying—"Thank you, I feel fine";
small voices, small lies, nothing untoward,
though, soon, he would blink again and again
because of the fingers of Lambert Rogers,
rash as a blind man's, inside his soft brain.

If items of horror can make a man laugh
then laugh at this: one hour later, the growth
still undiscovered, ticking its own wild time;
more brain mashed because of the probe's braille path;

Lambert Rogers desperate, fingering still;
his dresser thinking, "Christ! Two more on the list,
a cisternal puncture and a neural cyst."

Then, suddenly, the cracked record in the brain,
a ventriloquist voice that cried, "You sod,
leave my soul alone, leave my soul alone,"—
the patient's dummy lips moving to that refrain,
the patient's eyes too wide. And, shocked,
Lambert Rogers drawing out the probe
with nurses, students, sister, petrified.

"Leave my soul alone, leave my soul alone,"
that voice so arctic and that cry so odd
had nowhere else to go—till the antique
gramophone wound down and the words began
to blur and slow, "… leave … my … soul … alone …"
to cease at last when something other died.
And silence matched the silence under snow.

This poem recounts both the lack of knowledge and the lack of skill that Dr. Hilfiker catalogs, but does it in a way that emphasizes the terrible costs of mistakes in medicine.

Lessons for Providers

There are no perfect doctors, nurses, physician assistants. No perfect healthcare providers. We all have to struggle with our mistakes and develop a plan for dealing with them when they occur. They will occur.

Bibliography

Gawande, Atul. 2010. *The Checklist Manifesto: How to Get Things Right*. Amazon Kindle ed. New York: Metropolitan Books.

Hilfiker, David. 1984. Facing Our Mistakes. *New England Journal of Medicine*. http://www.davidhilfiker.com/index.php?option=com_content&view=article&id=51:facing-our-mistakes&Itemid=41

Williams, William Carlos. 1938. The Use of Force. https://bookpdf.services/downloads/the_use_of_force_by_william_carlos_williams.pdf

Section VIII

Death and Dying

Death and Dying 14

The final chapter of *Literature and Medicine* focuses on death and dying. Perhaps the most distinguishing feature of the healthcare professions is the fact that in most of its areas of work, it encounters on a routine and sometimes on a frequent basis the fact of death and dying (and pain and suffering as well). Few other professions—even the law and police work—so *regularly* face the end-of-life and the associated pain and suffering that it entails. This is even more striking in twenty-first-century American culture—and many cultures in advanced developed societies—in which death and dying are not part of everyday life, but relegated to hospitals and other institutions. Many years ago, Ernest Becker was awarded the Pulitzer Prize for General Non-Fiction (1974), two months after he died, for his book *The Denial of Death*, which argued that we can conceive civilization in general, and certainly twentieth-century American civilization, as an elaborate, symbolic defense against the knowledge of our mortality. Such a "denial of death" may not be true of all civilizations: nineteenth-century Euro-American literary culture (including Tolstoy) certainly had a sustained and, to our eyes, a strange fascination with death and dying, and in many non-Western cultures, where death is part of the everyday rhythms of life, death is also a felt aspect of ordinary life. But this "denial of death" is certainly true of twentieth- and twenty-first-century life in America. In fact, the authors of this book have told students for many years, in terms of the intensity of intellectual work, the hours and energy pursuing a PhD in literary studies is no less intense and all-consuming than that of pursuing an MD, PA, or nursing degree—with the powerful exception that, at the heart of training and careers in the health professions, death and dying (and pain and suffering) are almost unavoidable as constant aspects of these professions. This is what Anatole Broyard is getting at when he says "to most physicians, my illness is a routine incident in their rounds, while for me it's the crisis of my life" in "Doctor Talk to Me" (1992). It might be that in healthcare, death and suffering are organized as "routine"—and probably necessarily so—because the "terror and pity" of death and dying (to use Aristotle's description of the effects of tragedy) would otherwise be overwhelming to practicing healthcare providers.

© The Author(s) 2019
R. Schleifer, J. B. Vannatta, *Literature and Medicine*,
https://doi.org/10.1007/978-3-030-19128-3_14

Such feelings are certainly overwhelming to patients and their loved one facing dying and death. For this chapter, then, we have two vignettes from the experience of people—patients and their loved ones—encountering death. This chapter begins with two vignettes of end-of-life: one of a woman and her family who acknowledged mortality over the length of her life—who understood that acknowledging mortality is part of any life we live; and a second vignette of end-of-life where members of the dying patient's family deny death, as Becker notes, and "act out" emotions—fear, anger, bewilderment—because they cannot acknowledge the dying of a loved one.

The Good Death: A Vignette by Dr. Jerry Vannatta

> *Author Note*: Dr. Jerry Burr Vannatta (b. 1948), pursued a career as a medical educator and practicing physician of internal medicine. He served on the faculty of the College of Medicine at the University of Oklahoma from 1975 to 2015, and as its Executive Dean from 1996 to 2002. He retired in 2015, and in 2017, he became Medical Director of the Physician Assistant Program at Oklahoma City University. In 1999 he founded course on "Literature and Medicine" with his friend and colleague Professor Ronald Schleifer, which they taught together at the University of Oklahoma for almost two decades. Together, they published numerous articles and two major projects: *The Chief Concern of Medicine* (2013) and (with Sheila Crow) *Medicine and Humanistic Understanding* (2005). This first vignette describes the "good death" of his wife, Marianne Bea Brown Vannatta, who died in 2009. In 2003 Marianne was awarded the Oklahoma City University Distinguished Philanthropist Award, and in 2004 she established an annual road race, "Race With the Stars," to benefit the Oklahoma City University Kramer School of Nursing Scholarship Fund. Dr. Vannatta is the author of new vignettes in Chaps. 10, 12, and 13 in this volume as well as the two in this chapter. He is also first author (with Ronald Schleifer) of the vignettes taken from *The Chief Concern of Medicine* in earlier chapters.

Five days after she experienced mild abdominal pain she underwent endoscopy. There was a "bulging" in the stomach. The biopsy revealed cancer. "Well, I don't want to die from that," she said.

The phone rang, the gastroenterologist had arranged for her to see a surgeon the next afternoon following a PET scan. The PET scan was positive and she was scheduled for surgery. She smiled. On the way home she turned to her husband, "I get to have surgery on Monday. Let's celebrate." They shared a glass of wine. Their sons were grown, one married with children, the other engaged. On Saturday morning they called both sons and gave them the bad news. She did all the talking and was upbeat about the upcoming surgery. She was optimistic. On Sunday both boys and a granddaughter arrived to be with her during surgery. Her sister likewise was present. Over the next five months the sister and both boys were in and out of town, supportive and helpful. The post-op course was unremarkable and the hospitalization was only five days. She required six weeks to heal before she could take radiation. Every morning she drank a high protein, high calorie shake. It is difficult to get nutrition when one does not have a stomach. She healed well. One week out of the hospital she met daily with her running group for a 2-mile walk. Friends were supportive and stayed connected.

Radiation therapy took six weeks, followed by one month of chemotherapy, which made her weak and nauseated. She and her husband talked weekly about potential outcomes. She stated that if the therapy did not kill the cancer, she did not want to go back to the hospital. At four months she developed abdominal fluids. They were tested, and two days later the oncologist relayed the bad news that the fluid was full of cancer cells. The treatment had failed. When she hung up her face was pale, her voice weak. Her husband held her. After 20 minutes of crying and hugging she announced that she wanted a party – tonight. Friends were invited to bring food and wine. Thirty people arrived; a feast and stories were shared. The next day she arranged to have a valve placed in her abdominal wall so the fluid could be drained at home. Following that procedure they called hospice and enrolled. After a little more than a week the pain in her abdomen was more than she could tolerate so hospice placed a central line and began a morphine drip. The bedroom was a revolving door of family and friends. Sometimes she could visit with them; sometimes they just visited with each other. The family, although sad, were upbeat with her and fully supportive.

The last three days were difficult for everyone, but she and her husband were committed to not return to the hospital and to allow the death to occur at home. Now, family were the only people in the bedroom, playing music, on occasion singing – staying close. At ten in the morning – five months after diagnosis – with her husband holding her on the bed, she took her last few breaths.

The hospice nurse arrived 30 minutes later to pronounce death.

Frenzy Facing Death: A Vignette by Dr. Jerry Vannatta

The patient was transferred because of family demand. She was an 87-year-old widow, suffering from metastatic cancer. Chemo had left her weak and no remission. She developed pneumonia, which was successfully treated in the outside hospital. She had cancer in her lungs, and other places. This patient has four adult children. They seem to disagree about their mother's medical management. She is very debilitated, weighing only 104 pounds. She has stage four cancer, unresponsive to chemotherapy.

This family arrived on day two. There were two women, Helen and Sherry, and two men, Tom and Brady. The attending physician introduced himself and the team: the resident, intern, and two medical students. Tom did not shake hands, and glared at the intern. On the way out of the room Tom aggressively addressed the attending physician, saying his mother was not a Guiney pig, and he did not want student doctors seeing her. The attending suggested they discuss it later. Two hours later Tom presented himself, hostile, upset and adamant that no one but the attending was to see his mother. The attending explained that this was a teaching hospital and that a team of doctors headed by an attending physician sees all patients. Tom rose from his chair and cussing on the way out, slammed the door. The next morning on rounds the resident reported that his attempts to care for the patient this morning were difficult. The two daughters were in the room when he saw her. They were asking why the tumors seem to be growing when their mother had been taking that medicine to make the cancer go away for over a year. Isn't the infection what is causing this illness? Does this older doctor even know what he is doing? The resident was clearly frustrated and had felt attacked by this family. Brady wanted to know how the cancer was responding?

The attending visited with the patient. The conversation revealed that their mother knew she had cancer it wasn't getting better. She revealed that the treatments were making her so sick she didn't want any more of them. When asked why she hadn't stopped them earlier she said it was because the kids would not let her. Tom, as it turns out, has not gotten along with the three siblings most of his life. He is always contrary. "You know, he's just kind of a difficult child," the patient said with a knowing grin. "Sherry seems to always have her head in the sand if you know what I mean," she declared. Helen means well but will not override what the other three want done. "I am so tired, I just want to rest." A discussion then led to the fact that the infection was cleared by the previous hospital. The problem now seems to be the cancer, which has not responded to two types of medicine, and the probability of this cancer getting better was nearly zero, the attending explained.

The patient turned her head, and began to cry. "So I'm not going to get well?" She asked.

"No, I am sorry but you are not going to beat this cancer. It continues to grow in your lungs and down in your abdomen. In fact it is encroaching on your kidneys and they will begin to fail soon."

"Can you keep me comfortable until I pass?" she asked.

"Yes, we can use what we call palliative care." The attending then explained the goals and process of palliative care and the patient stated that is what she wanted.

A family conference was arranged the following morning. Tom refused to come if the "team" was going to be there. Three siblings, the attending, and the team assembled in a room with chairs arranged in a circle. Tom joined them late. The attending opened the discussion by asking each of the siblings what they thought was going on with their mother.

Tom exclaimed, "How the hell am I supposed to know, ain't that what we pay you for?"

Sherry told him to "shut up." Brady asked him to "keep quiet." The attending accepted responsibility and that calmed everyone down.

The attending then explained that the previous hospital had done an excellent job clearing their mother of infection. He explained what the images had shown, that the tumors had grown. He explained that there is no more treatment they could offer for the cancer. He was then quiet; the silence filled the room. Helen spoke up. "So what now?"

The attending retold the conversation he had had with their mother and that she had chosen palliative care. He explained that the goal was comfort not cure. Tom got up and left, Brady followed him out. Sherry asked how he could be so sure and left. Helen stood and thanked the attending and the team for all their help.

The patient was transferred from a treatment room to a palliative care room. Over the next week her kidneys, liver, and respiratory system began to fail. It took an enormous amount of time and energy for the attending to explain each of these issues. It took enormous courage to keep from transferring her to the critical care unit, even when Tom accused him of "killing his mother" and threatened a lawsuit. The DNR, which the patient had signed two weeks ago, was safely and securely settled into the chart, and recorded in the electronic medical record. Less than two weeks after her transfer she died.

These two vignettes together demonstrate the stress and heartbreak of death and dying, but while the first suggests that a "good" death is a function of a "good" life, the second demonstrates the ways in which tensions and unreflected-upon relationships and, perhaps, disappointments, inflect and magnify the terror and pity that death and dying occasion. The first vignette describes people—including the patient herself—who have thought about and acknowledged mortality long before the end-of-life. Family members bring together celebration and sorrow. In the second vignette, family members remain in denial of mortality, and in that denial they cannot trust themselves or one another when facing an inevitable aspect of all our lives.

Ivan Ilych

Ivan Ilych, in Leo Tolstoy's novella, has lived a life of small dishonesties, and he, like the family of the second vignette, has great trouble in acknowledging that death is part of life. *The Death of Ivan Ilych*, written in the late nineteenth century, is an excellent example of literary realism. Tolstoy gives us Ivan, an everyday common man—Ivan Ilych is a Russian name like "John Smith" in English—who in his private life reflects the ideals of the upper middle class of Russia in that century. Tolstoy satirizes the lifestyle and choices made by these people. The style of this novel—its attention to detail—as when he is describing the dead body early in the book, parallels the attention to detail required of the healthcare provider. The book demonstrates many of the characteristics of literary narrative discussed in Chap. 1 of this book. Tolstoy uses the phonic foregrounding where he describes Ivan's visit to the specialist: "The doctor said that so-and-so indicated that there was so-and-so inside the patient, but if the investigation of so-and-so did not confirm this, then he must assume that and that. If he assumed that and that, then . . . and so on." In this short sentence he also demonstrates the specialist's detachment—using medical jargon, what the authors have called "doctor babble" (see *The Chief Concern*: 340–42)—instead of communicating with the patient at their level in the patient's vocabulary.

Throughout the short novel, Ivan constantly is tormented by the inability to get the doctors to address his chief concern: "Is this serious or not?" Ivan's internal dialogue throughout transports the reader into the story and thereby creates empathy for Ivan's character. As discussed in Chap. 1, this transportation and empathy development has the potential to create—as Kidd et al. demonstrated—improved empathy as a general trait in the reader. Ivan's doctors fail altogether in developing a therapeutic patient-physician relationship. They better parallel the doctors in Melville's *White Jacket* (Chap. 11) and the orthopedic surgeon in Tweedy's *Black Man in a Whit Coat* (Chap. 7).

The novella provides the reader with a vicarious experience of not only the patient's view of a bad experience with a set of medical providers but also the vicarious experience of Ivan's poor organization of his private life. This narrative, describing Ivan's life, demonstrates how the patient can wind up at the end-of-life facing death in a frenzy as we saw in the second vignette at the beginning of this chapter.

Tolstoy provides us with a character in this short novel exemplary of the "good practitioner" even though he isn't a practitioner at all—Gerasim. Gerasim is a character who demonstrates the virtues—visited in Chap. 8—that we hope for in our providers. He is compassionate, listens carefully, and on a day to day basis, simply "does his job." It is this doing one's job that represents the behavior providers should be striving to habituate.

At the end of this narrative Tolstoy describes Ivan's frenzied death. He is in severe pain, screaming for hours and for three days on end (the same, although Tolstoy does not explicitly say so, of Jesus's "harrowing" in hell). The family is ignoring his suffering, and not until minutes before his death is Ivan able to turn the stereotype of his life into insight. He suddenly understands—probably for the first time in his life—that he has empathy for others, his son and his wife. (In his *Autobiography*, Dr. William Carlos Williams notes that his career as a physician and as a poet allowed him "to catch the evasive life of the thing… [so that] stereotype will yield a moment of insight" [1967: 359].)

In engaging Tolstoy's literary narrative, the reader can ask—much as a healthcare provider can ask of the story of a patient—what is this story about? Much like the story a patient tells, there is a lot that is unsaid. In reading this short novel, we can imagine that this narrative is about fear, mournfulness, emptiness, and regret. It is this aboutness—or "overall meaning"—which, in both literary narrative and the stories patients present, is a crucial tool for engagement, understanding, and action. This parallel between the purpose of the author of literary narrative and the purpose of patients as they tell their story that makes the study of literature so effective as a pedagogical tool in teaching the breadth of work and responsibility for healthcare providers.

Literary Narrative: "The Death of Ivan Ilych" (1886) by Leo Tolstoy

Author Note: Leo Tolstoy (1828–1910), one of the great novelists in the western literary tradition, was born into a Russian aristocratic family (which might explain his satiric disdain of Ivan Ilych's middle-class social pretensions). He wrote the acclaimed novels *War and Peace* (1869), *Anna Karenina* (1877), and *The Death of Ivan Ilych* (1886), as well as a significant number of short stories and tracts (including *What Is to Be Done?*—a radical anarchist-pacifist manifesto). As noted in Chap. 1 of this book, early twentieth-century Russian literary scholars developed the notion of literary "defamiliarization" in their engagements with his work.

The Death of Ivan Ilych

I

DURING AN INTERVAL in the Melvinski trial in the large building of the Law Courts, the members and public prosecutor met in Ivan Egorovich Shebek's private room, where the conversation turned on the celebrated Krasovski case. Fedor Vasilievich warmly maintained that it was not subject to their jurisdiction, Ivan Egorovich maintained the contrary, while Peter Ivanovich, not having entered into the discussion at the start, took no part in it but looked through the *Gazette* which had just been handed in.

"Gentlemen," he said, "Ivan Ilych has died!"

"You don't say so!"

"Here, read it yourself," replied Peter Ivanovich, handing Fedor Vasilievich the paper still damp from the press. Surrounded by a black border were the words: "Praskovya Fedorovna Golovina, with profound sorrow, informs relatives and friends of the demise of her beloved husband Ivan Ilych Golovin, Member of the Court of Justice, which occurred on February the 4th of this year 1882. The funeral will take place on Friday at one o'clock in the afternoon." Ivan Ilych had been a colleague of the gentlemen present and was liked by them all. He had been ill for some weeks with an illness said to be incurable. His post had been kept open for him, but there had been conjectures that in case of his death Alexeev might receive his appointment, and that either Vinnikov or Shtabel would succeed Alexeev. So on receiving the news of Ivan Ilych's death the first thought of each of the gentlemen in that private room was of the changes and promotions it might occasion among themselves or their acquaintances.

"I shall be sure to get Shtabel's place or Vinnikov's," thought Fedor Vasilievich. "I was promised that long ago, and the promotion means an extra eight hundred rubles a year for me besides the allowance."

"Now I must apply for my brother-in-law's transfer from Kaluga," thought Peter Ivanovich. "My wife will be very glad, and then she won't be able to say that I never do anything for her relations."

"I thought he would never leave his bed again," said Peter Ivanovich aloud. "It's very sad."

"But what really was the matter with him?"

"The doctors couldn't say—at least they could, but each of them said something different. When last I saw him I thought he was getting better."

"And I haven't been to see him since the holidays. I always meant to go."

"Had he any property?"

"I think his wife had a little—but something quiet trifling."

"We shall have to go to see her, but they live so terribly far away."

"Far away from you, you mean. Everything's far away from your place."

"You see, he never can forgive my living on the other side of the river," said Peter Ivanovich, smiling at Shebek. Then, still talking of the distances between different parts of the city, they returned to the Court.

Besides considerations as to the possible transfers and promotions likely to result from Ivan Ilych's death, the mere fact of the death of a near acquaintance aroused, as usual, in all who heard of it the complacent feeling that, "it is he who is dead and not I."

Each one thought or felt, "Well, he's dead but I'm alive!" But the more intimate of Ivan Ilych's acquaintances, his so-called friends, could not help thinking also that they would now have to fulfil the very tiresome demands of propriety by attending the funeral service and paying a visit of condolence to the widow.

Fedor Vasilievich and Peter Ivanovich had been his nearest acquaintances. Peter Ivanovich had studied law with Ivan Ilych and had considered himself to be under obligations to him.

Having told his wife at dinner-time of Ivan Ilych's death, and of his conjecture that it might be possible to get her brother transferred to their circuit, Peter Ivanovich sacrificed his usual nap, put on his evening clothes and drove to Ivan Ilych's house.

At the entrance stood a carriage and two cabs. Leaning against the wall in the hall downstairs near the cloak stand was a coffin-lid covered with cloth of gold, ornamented with gold cord and tassels, that had been polished up with metal powder. Two ladies in black were taking off their fur cloaks. Peter Ivanovich recognized one of them as Ivan Ilych's sister, but the other was a stranger to him. His colleague Schwartz was just coming downstairs, but on seeing Peter Ivanovich enter, he stopped and winked at him, as if to say: "Ivan Ilych has made a mess of things—not like you and me."

Schwartz's face with his Piccadilly whiskers, and his slim figure in evening dress, had as usual an air of elegant solemnity which contrasted with the playfulness of his character and had a special piquancy here, or so it seemed to Peter Ivanovich.

Peter Ivanovich allowed the ladies to precede him and slowly followed them upstairs. Schwartz did not come down but remained where he was, and Peter Ivanovich understood that he wanted to arrange where they should play bridge that evening. The ladies went upstairs to the widow's room, and Schwartz with seriously compressed lips but a playful look in his eyes, indicated by a twist of his eyebrows the room to the right where the body lay.

Peter Ivanovich, like everyone else on such occasions, entered feeling uncertain what he would have to do. All he knew was that at such times it is always safe to cross oneself. But he was not quite sure whether one should make obeisances while doing so. He therefore adopted a middle course. On entering the room, he began crossing himself and made a slight movement resembling a bow. At the same time, as far as the motion of his head and arm allowed, he surveyed the room. Two young men—apparently nephews, one of whom was a high-school pupil—were leaving the room, crossing themselves as they did so. An old woman was standing motionless, and a lady with strangely arched eyebrows was saying something to her in a whisper. A vigorous, resolute Church Reader, in a frock-coat, was reading something in a loud voice with an expression that precluded any contradiction. The butler's assistant, Gerasim, stepping lightly in front of Peter Ivanovich, was strewing something on the floor. Noticing this, Peter Ivanovich was immediately aware of a faint odour of a decomposing body.

The last time he had called on Ivan Ilych, Peter Ivanovich had seen Gerasim in the study. Ivan Ilych had been particularly fond of him and he was performing the duty of a sick nurse.

Peter Ivanovich continued to make the sign of the cross slightly inclining his head in an intermediate direction between the coffin, the Reader, and the icons on the table in a corner of the room. Afterwards, when it seemed to him that this movement of his arm in crossing himself had gone on too long, he stopped and began to look at the corpse.

The dead man lay, as dead men always lie, in a specially heavy way, his rigid limbs sunk in the soft cushions of the coffin, with the head forever bowed on the pillow. His yellow waxen brow with bald patches over his sunken temples was thrust up in the way peculiar to the dead, the protruding nose seeming to press on the upper lip. He was much changed and grown even thinner since Peter Ivanovich had last seen him, but, as is always the case with the dead, his face was handsomer and above all more dignified than when he was alive. The expression on the face said that what was necessary had been accomplished, and accomplished rightly. Besides this there was in that expression a reproach and a warning to the living. This warning seemed to Peter Ivanovich out of place, or at least not applicable to him. He felt a certain discomfort and so he hurriedly crossed himself once more and turned and went out of the door—too hurriedly and too regardless of propriety, as he himself was aware.

Schwartz was waiting for him in the adjoining room with legs spread wide apart and both hands toying with his top-hat behind his back. The mere sight of that playful, well-groomed, and elegant figure refreshed Peter Ivanovich. He felt that Schwartz was above all these happenings and would not surrender to any depressing influences. His very look said that this incident of a church service for Ivan Ilych could not be a sufficient reason for infringing the order of the session—in other words, that it would certainly not prevent his unwrapping a new pack of cards and shuffling them that evening while a footman placed fresh candles on the table: in fact, that there was no reason for supposing that this incident would hinder their spending the evening agreeably. Indeed he said this in a whisper as Peter Ivanovich passed him, proposing that they should meet for a game at Fedor Vasilievich's. But apparently Peter Ivanovich was not destined to play bridge that evening. Praskovya Fedorovna (a short, fat woman who despite all efforts to the contrary had continued to broaden steadily from her shoulders downwards and who had the same extraordinarily arched eyebrows as the lady who had been standing by the coffin), dressed all in black, her head covered with lace, came out of her own room with some other ladies, conducted them to the room where the dead body lay, and said: "The service will begin immediately. Please go in."

Schwartz, making an indefinite bow, stood still, evidently neither accepting nor declining this invitation. Praskovya Fedorovna recognizing Peter Ivanovich, sighed, went close up to him, took his hand, and said: "I know you were a true friend to Ivan Ilych…" and looked at him awaiting some suitable response. And Peter Ivanovich knew that, just as it had been the right thing to cross himself in that room, so what he had to do here was to press her hand, sigh, and say, "Believe me…" So he did all this and as he did it felt that the desired result had been achieved: that both he and she were touched.

"Come with me. I want to speak to you before it begins," said the widow. "Give me your arm."

Peter Ivanovich gave her his arm and they went to the inner rooms, passing Schwartz who winked at Peter Ivanovich compassionately.

"That does for our bridge! Don't object if we find another player. Perhaps you can cut in when you do escape," said his playful look.

Peter Ivanovich sighed still more deeply and despondently, and Praskovya Fedorovna pressed his arm gratefully. When they reached the drawing-room, upholstered in pink cretonne and lighted by a dim lamp, they sat down at the table—she on a sofa and Peter Ivanovich on a low pouffe, the springs of which yielded spasmodically under his weight. Praskovya Fedorovna had been on the point of warning him to take another seat, but felt that such a warning was out of keeping with her present condition and so changed her mind. As he sat down on the pouffe Peter Ivanovich recalled how Ivan Ilych had arranged this room and had consulted him regarding this pink cretonne with green leaves. The whole room was full of furniture and knick-knacks, and on her way to the sofa the lace of the widow's black shawl caught on the edge of the table. Peter Ivanovich rose to detach it, and the springs of the pouffe, relieved of his weight, rose also and gave him a push. The widow began detaching her shawl herself, and Peter Ivanovich again sat down, suppressing the rebellious springs of the pouffe under him. But the widow had not quite freed herself and Peter Ivanovich got up again, and again the pouffe rebelled and even creaked. When this was all over she took out a clean cambric handkerchief and began to weep. The episode with the shawl and the struggle with the pouffe had cooled Peter Ivanovich's emotions and he sat there with a sullen look on his face. This awkward situation was interrupted by Sokolov, Ivan Ilych's butler, who came to report that the plot in the cemetery that Praskovya Fedorovna had chosen would cost two hundred rubles. She stopped weeping, and looking at Peter Ivanovich with the air of a victim, remarked in French that it was very hard for her. Peter Ivanovich made a silent gesture signifying his full conviction that it must indeed be so.

"Please smoke," she said in a magnanimous yet crushed voice, and turned to discuss with Sokolov the price of the plot for the grave.

Peter Ivanovich while lighting his cigarette heard her inquiring very circumstantially into the prices of different plots in the cemetery and finally decide which she would take. When that was done she gave instructions about engaging the choir.

Sokolov then left the room.

"I look after everything myself," she told Peter Ivanovich, shifting the albums that lay on the table; and noticing that the table was endangered by his cigarette-ash, she immediately passed him an ash-tray, saying as she did so: "I consider it an affectation to say that my grief prevents my attending to practical affairs. On the contrary, if anything can—I won't say console me, but—distract me, it is seeing to everything concerning him." She again took out her handkerchief as if preparing to cry, but suddenly, as if mastering her feeling, she shook herself and began to speak calmly. "But there is something I want to talk to you about."

Peter Ivanovich bowed, keeping control of the springs of the pouffe, which immediately began quivering under him.

"He suffered terribly the last few days."

"Did he?" said Peter Ivanovich.

"Oh, terribly! He screamed unceasingly, not for minutes but for hours. For the last three days he screamed incessantly. It was unendurable. I cannot understand how I bore it; you could hear him three rooms off. Oh, what I have suffered!" "Is it possible that he was conscious all that time?" asked Peter Ivanovich.

"Yes," she whispered. "To the last moment. He took leave of us a quarter of an hour before he died, and asked us to take Volodya away."

The thought of the suffering of this man he had known so intimately, first as a merry little boy, then as a schoolmate, and later as a grown-up colleague, suddenly struck Peter Ivanovich with horror, despite an unpleasant consciousness of his own and this woman's dissimulation. He again saw that brow, and that nose pressing down on the lip, and felt afraid for himself.

"Three days of frightful suffering and the death! Why, that might suddenly, at any time, happen to me," he thought, and for a moment felt terrified. But—he did not himself know how—the customary reflection at once occurred to him that this had happened to Ivan Ilych and not to him, and that it should not and could not happen to him, and that to think that it could would be yielding to depression which he ought not to do, as Schwartz's expression plainly showed. After which reflection Peter Ivanovich felt reassured, and began to ask with interest about the details of Ivan Ilych's death, as though death was an accident natural to Ivan Ilych but certainly not to himself.

After many details of the really dreadful physical sufferings Ivan Ilych had endured (which details he learnt only from the effect those sufferings had produced on Praskovya Fedorovna's nerves) the widow apparently found it necessary to get to business.

"Oh, Peter Ivanovich, how hard it is! How terribly, terribly hard!" and she again began to weep.

Peter Ivanovich sighed and waited for her to finish blowing her nose. When she had done so he said, "Believe me..." and she again began talking and brought out what was evidently her chief concern with him—namely, to question him as to how she could obtain a grant of money from the government on the occasion of her husband's death. She made it appear that she was asking Peter Ivanovich's advice about her pension, but he soon saw that she already knew about that to the minutest detail, more even than he did himself. She knew how much could be got out of the government in consequence of her husband's death, but wanted to find out whether she could not possibly extract something more. Peter Ivanovich tried to think of some means of doing so, but after reflecting for a while and, out of propriety, condemning the government for its niggardliness, he said he thought that nothing more could be got. Then she sighed and evidently began to devise means of getting rid of her visitor. Noticing this, he put out his cigarette, rose, pressed her hand, and went out into the anteroom.

In the dining-room where the clock stood that Ivan Ilych had liked so much and had bought at an antique shop, Peter Ivanovich met a priest and a few acquaintances who had come to attend the service, and he recognized Ivan Ilych's daughter, a handsome young woman. She was in black and her slim figure appeared slimmer than ever. She had a gloomy, determined, almost angry expression, and bowed to Peter Ivanovich as though he were in some way to blame. Behind her, with the same offended look, stood a wealthy young man, an examining magistrate, whom Peter Ivanovich also knew and who was her fiancé, as he had heard. He bowed mournfully to them and was about to pass into the death-chamber, when from under the stairs appeared the figure of Ivan Ilych's schoolboy son, who was extremely like his father. He seemed a little Ivan Ilych, such as Peter Ivanovich remembered when they studied law together. His tear-stained eyes had in them the look that is seen in the eyes of boys of thirteen or fourteen who are not pure-minded. When he saw Peter Ivanovich he scowled morosely and shamefacedly. Peter Ivanovich nodded to him and entered the death-chamber. The service began: candles, groans, incense, tears, and sobs. Peter Ivanovich stood looking gloomily down at his feet. He did not look once at the dead man, did not yield to any depressing influence, and was one of the first to leave the room. There was no one in the anteroom, but Gerasim darted out of the dead man's room, rummaged with his strong hands among the fur coats to find Peter Ivanovich's and helped him on with it.

"Well, friend Gerasim," said Peter Ivanovich, so as to say something. "It's a sad affair, isn't it?"

"It's God will. We shall all come to it some day," said Gerasim, displaying his teeth—the even white teeth of a healthy peasant—and, like a man in the thick of urgent work, he briskly opened the front door, called the coachman, helped Peter Ivanovich into the sledge, and sprang back to the porch as if in readiness for what he had to do next.

Peter Ivanovich found the fresh air particularly pleasant after the smell of incense, the dead body, and carbolic acid.

"Where to sir?" asked the coachman.

"It's not too late even now....I'll call round on Fedor Vasilievich."

He accordingly drove there and found them just finishing the first rubber, so that it was quite convenient for him to cut in.

II

IVAN ILYCH'S LIFE had been most simple and most ordinary and therefore most terrible.

He had been a member of the Court of Justice, and died at the age of forty-five. His father had been an official who after serving in various ministries and departments in Petersburg had made the sort of career which brings men to positions from which by reason of their long service they cannot be dismissed, though they are obviously unfit to hold any responsible position, and for whom therefore posts are specially created, which though fictitious carry salaries from six to ten thousand rubles that are not fictitious, and in receipt of which they live on to a great age.

Such was the Privy Councillor and superfluous member of various superfluous institutions, Ilya Epimovich Golovin.

He had three sons, of whom Ivan Ilych was the second. The eldest son was following in his father's footsteps only in another department, and was already approaching that stage in the service at which a similar sinecure would be reached. The third son was a failure. He had ruined his prospects in a number of positions and was now serving in the railway department. His father and brothers, and still more their wives, not merely disliked meeting him, but avoided remembering his existence unless compelled to do so. His sister had married Baron Greff, a Petersburg official of her father's type. Ivan Ilych was *le phenix de la famille* as people said. He was neither as cold and formal as his elder brother nor as wild as the younger, but was a happy mean between them—an intelligent polished, lively and agreeable man. He had studied with his younger brother at the School of Law, but the latter had failed to complete the course and was expelled when he was in the fifth class. Ivan Ilych finished the course well. Even when he was at the School of Law he was just what he remained for the rest of his life: a capable, cheerful, goodnatured, and sociable man, though strict in the fulfillment of what he considered to be his duty: and he considered his duty to be what was so considered by those in authority. Neither as a boy nor as a man was he a toady, but from early youth was by nature attracted to people of high station as a fly is drawn to the light, assimilating their ways and views of life and establishing friendly relations with them. All the enthusiasms of childhood and youth passed without leaving much trace on him; he succumbed to sensuality, to vanity, and latterly among the highest classes to liberalism, but always within limits which his instinct unfailingly indicated to him as correct.

At school he had done things which had formerly seemed to him very horrid and made him feel disgusted with himself when he did them; but when later on he saw that such actions were done by people of good position and that they did not regard them as wrong, he was able not exactly to regard them as right, but to forget about them entirely or not be at all troubled at remembering them.

Having graduated from the School of Law and qualified for the tenth rank of the civil service, and having received money from his father for his equipment, Ivan Ilych ordered himself clothes at Scharmer's, the fashionable tailor, hung a medallion inscribed ∗respice finem∗ on his watch-chain, took leave of his professor and the prince who was patron of the school, had a farewell dinner with his comrades at Donon's first-class restaurant, and with his new and fashionable portmanteau, linen, clothes, shaving and other toilet appliances, and a travelling rug, all purchased at the best shops, he set off for one of the provinces where through his father's influence, he had been attached to the governor as an official for special service.

In the province Ivan Ilych soon arranged as easy and agreeable a position for himself as he had had at the School of Law. He performed his official task, made his career, and at the same time amused himself pleasantly and decorously. Occasionally he paid official visits to country districts where he behaved with dignity both to his superiors and inferiors, and performed the duties entrusted to him, which related chiefly to the sectarians, with an exactness and incorruptible honesty of which he could not but feel proud.

In official matters, despite his youth and taste for frivolous gaiety, he was exceedingly reserved, punctilious, and even severe; but in society he was often amusing and witty, and always good-natured, correct in his manner, and *bon enfant*, as the governor and his wife—with whom he was like one of the family—used to say of him.

In the province he had an affair with a lady who made advances to the elegant young lawyer, and there was also a milliner; and there were carousals with aides-de-camp who visited the district, and after-supper visits to a certain outlying street of doubtful reputation; and there was too some obsequiousness to his chief and even to his chief's wife, but all this was done with such a tone of good breeding that no hard names could be applied to it. It all came under the heading of the French saying: "*Il faut que jeunesse se passe*." It was all done with clean hands, in clean linen, with French phrases, and above all among people of the best society and consequently with the approval of people of rank.

So Ivan Ilych served for five years and then came a change in his official life. The new and reformed judicial institutions were introduced, and new men were needed. Ivan Ilych became such a new man. He was offered the post of examining magistrate, and he accepted it though the post was in another province and obliged him to give up the connexions he had formed and to make new ones. His friends met to give him a send-off; they had a group photograph taken and presented him with a silver cigarette-case, and he set off to his new post.

As examining magistrate Ivan Ilych was just as *comme il faut* and decorous a man, inspiring general respect and capable of separating his official duties from his private life, as he had been when acting as an official on special service. His duties now as examining magistrate were far more interesting and attractive than before. In his former position it had been pleasant to wear an undress uniform made by Scharmer, and to pass through the crowd of petitioners and officials who were timorously awaiting an audience with the governor, and who envied him as with free and easy gait he went straight into his chief's private room to have a cup of tea and a cigarette with him. But not many people had then been directly dependent on him—only police officials and the sectarians when he went on special missions—and he liked to treat them politely, almost as comrades, as if he were letting them feel that he who had the power to crush them was treating them in this simple, friendly way. There were then but few such people. But now, as an examining magistrate, Ivan Ilych felt that everyone without exception, even the most important and self-satisfied, was in his power, and that he need only write a few words on a sheet of paper with a certain heading, and this or that important, self-satisfied person would be brought before him in the role of an accused person or a witness, and if he did not choose to allow him to sit down, would have to stand before him and answer his questions. Ivan Ilych never abused his power; he tried on the contrary to soften its expression, but the consciousness of it and the possibility of softening its effect supplied the chief interest and attraction of his office. In his work itself, especially in his examinations, he very soon acquired a method of eliminating all considerations irrelevant to the legal aspect of the case, and reducing even the most complicated case to a form in which it would be presented on paper only in its externals, completely excluding his personal opinion of the matter, while above all observing every prescribed formality. The work was new and Ivan Ilych was one of the first men to apply the new Code of 1864.

On taking up the post of examining magistrate in a new town, he made new acquaintances and connexions, placed himself on a new footing and assumed a somewhat different tone. He took up an attitude of rather dignified aloofness towards the provincial authorities, but picked out the best circle of legal gentlemen and wealthy gentry living in the town and assumed a tone of slight dissatisfaction with the government, of moderate liberalism, and of enlightened citizenship. At the same time, without at all altering the elegance of his toilet, he ceased shaving his chin and allowed his beard to grow as it pleased.

Ivan Ilych settled down very pleasantly in this new town. The society there, which inclined towards opposition to the governor was friendly, his salary was larger, and he began to play *vint* [a form of bridge], which he found added not a little to the pleasure of life, for he had a capacity for cards, played good-humouredly, and calculated rapidly and astutely, so that he usually won.

After living there for two years he met his future wife, Praskovya Fedorovna Mikhel, who was the most attractive, clever, and brilliant girl of the set in which he moved, and among other amusements and relaxations from his labours as examining magistrate, Ivan Ilych established light and playful relations with her.

While he had been an official on special service he had been accustomed to dance, but now as an examining magistrate it was exceptional for him to do so. If he danced now, he did it as if to show that though he served under the reformed order of things, and had reached the fifth official rank, yet when it came to dancing he could do it better than most people. So at the end of an evening he sometimes danced with Praskovya Fedorovna, and it was chiefly during these dances that he captivated her. She fell in love with him. Ivan Ilych had at first no definite intention of marrying, but when the girl fell in love with him he said to himself: "Really, why shouldn't I marry?"

Praskovya Fedorovna came of a good family, was not bad looking, and had some little property. Ivan Ilych might have aspired to a more brilliant match, but even this was good. He had his salary, and she, he hoped, would have an equal income. She was well connected, and was a sweet, pretty, and thoroughly correct young woman. To say that Ivan Ilych married because he fell in love with Praskovya Fedorovna and found that she sympathized with his views of life would be as incorrect as to say that he married because his social circle approved of the match. He was swayed by both these considerations: the marriage gave him personal satisfaction, and at the same time it was considered the right thing by the most highly placed of his associates.

So Ivan Ilych got married.

The preparations for marriage and the beginning of married life, with its conjugal caresses, the new furniture, new crockery, and new linen, were very pleasant until his wife became pregnant—so that Ivan Ilych had begun to think that marriage would not impair the easy, agreeable, gay and always decorous character of his life, approved of by society and regarded by himself as natural, but would even improve it. But from the first months of his wife's pregnancy, something new, unpleasant, depressing, and unseemly, and from which there was no way of escape, unexpectedly showed itself.

His wife, without any reason—*de gaiete de coeur* as Ivan Ilych expressed it to himself—began to disturb the pleasure and propriety of their life. She began to be jealous without any cause, expected him to devote his whole attention to her, found fault with everything, and made coarse and ill-mannered scenes.

At first Ivan Ilych hoped to escape from the unpleasantness of this state of affairs by the same easy and decorous relation to life that had served him heretofore: he tried to ignore his wife's disagreeable moods, continued to live in his usual easy and pleasant way, invited friends to his house for a game of cards, and also tried going out to his club or spending his evenings with friends. But one day his wife began upbraiding him so vigorously, using such coarse words, and continued to abuse him every time he did not fulfil her demands, so resolutely and with such evident determination not to give way till he submitted—that is, till he stayed at home and was bored just as she was—that he became alarmed. He now realized that matrimony—at any rate with Praskovya Fedorovna—was not always conducive to the pleasures and amenities of life, but on the contrary often infringed both comfort and propriety, and that he must therefore entrench himself against such infringement. And Ivan Ilych began to seek for means of doing so. His official duties were the one thing that imposed upon Praskovya Fedorovna, and by means of his official work and the duties attached to it he began struggling with his wife to secure his own independence.

With the birth of their child, the attempts to feed it and the various failures in doing so, and with the real and imaginary illnesses of mother and child, in which Ivan Ilych's sympathy was demanded but about which he understood nothing, the need of securing for himself an existence outside his family life became still more imperative.

As his wife grew more irritable and exacting and Ivan Ilych transferred the center of gravity of his life more and more to his official work, so did he grow to like his work better and became more ambitious than before.

Very soon, within a year of his wedding, Ivan Ilych had realized that marriage, though it may add some comforts to life, is in fact a very intricate and difficult affair towards which in order to perform one's duty, that is, to lead a decorous life approved of by society, one must adopt a definite attitude just as towards one's official duties.

And Ivan Ilych evolved such an attitude towards married life. He only required of it those conveniences—dinner at home, housewife, and bed—which it could give him, and above all that propriety of external forms required by public opinion. For the rest he looked for lighthearted pleasure and propriety, and was very thankful when he found them, but if he met with antagonism and querulousness he at once retired into his separate fenced-off world of official duties, where he found satisfaction.

Ivan Ilych was esteemed a good official, and after three years was made Assistant Public Prosecutor. His new duties, their importance, the possibility of indicting and imprisoning anyone he chose, the publicity his speeches received, and the success he had in all these things, made his work still more attractive.

More children came. His wife became more and more querulous and ill-tempered, but the attitude Ivan Ilych had adopted towards his home life rendered him almost impervious to her grumbling.

After seven years' service in that town he was transferred to another province as Public Prosecutor. They moved, but were short of money and his wife did not like the place they moved to. Though the salary was higher the cost of living was greater, besides which two of their children died and family life became still more unpleasant for him.

Praskovya Fedorovna blamed her husband for every inconvenience they encountered in their new home. Most of the conversations between husband and wife, especially as to the children's education, led to topics which recalled former disputes, and these disputes were apt to flare up again at any moment. There remained only those rare periods of amorousness which still came to them at times but did not last long. These were islets at which they anchored for a while and then again set out upon that ocean of veiled hostility which showed itself in their aloofness from one another. This aloofness might have grieved Ivan Ilych had he considered that it ought not to exist, but he now regarded the position as normal, and even made it the goal at which he aimed in family life. His aim was to free himself more and more from those unpleasantnesses and to give them a semblance of harmlessness and propriety. He attained this by spending less and less time with his family, and when obliged to be at home he tried to safeguard his position by the presence of outsiders. The chief thing however was that he had his official duties. The whole interest of his life now centered in the official world and that interest absorbed him. The consciousness of his power, being able to ruin anybody he wished to ruin, the importance, even the external dignity of his entry into court, or meetings with his subordinates, his success with superiors and inferiors, and above all his masterly handling of cases, of which he was conscious—all this gave him pleasure and filled his life, together with chats with his colleagues, dinners, and bridge. So that on the whole Ivan Ilych's life continued to flow as he considered it should do—pleasantly and properly.

So things continued for another seven years. His eldest daughter was already sixteen, another child had died, and only one son was left, a schoolboy and a subject of dissension. Ivan Ilych wanted to put him in the School of Law, but to spite him Praskovya Fedorovna entered him at the High School. The daughter had been educated at home and had turned out well: the boy did not learn badly either.

III

SO IVAN ILYCH lived for seventeen years after his marriage. He was already a Public Prosecutor of long standing, and had declined several proposed transfers while awaiting a more desirable post, when an unanticipated and unpleasant occurrence quite upset the peaceful course of his life. He was expecting to be offered the post of presiding judge in a University town, but Happe somehow came to the front and obtained the appointment instead. Ivan Ilych became irritable, reproached Happe, and quarreled both with him and with his immediate superiors—who became colder to him and again passed him over when other appointments were made.

This was in 1880, the hardest year of Ivan Ilych's life. It was then that it became evident on the one hand that his salary was insufficient for them to live on, and on the other that he had been forgotten, and not only this, but that what was for him the greatest and most cruel injustice appeared to others a quite ordinary occurrence. Even his father did not consider it his duty to help him. Ivan Ilych felt himself abandoned by everyone, and that they regarded his position with a salary of 3,500 rubles as quite normal and even fortunate. He alone knew that with the consciousness of the injustices done him, with his wife's incessant nagging, and with the debts he had contracted by living beyond his means, his position was far from normal.

In order to save money that summer he obtained leave of absence and went with his wife to live in the country at her brother's place.

In the country, without his work, he experienced *ennui* for the first time in his life, and not only *ennui* but intolerable depression, and he decided that it was impossible to go on living like that, and that it was necessary to take energetic measures.

Having passed a sleepless night pacing up and down the veranda, he decided to go to Petersburg and bestir himself, in order to punish those who had failed to appreciate him and to get transferred to another ministry.

Next day, despite many protests from his wife and her brother, he started for Petersburg with the sole object of obtaining a post with a salary of five thousand rubles a year. He was no longer bent on any particular department, or tendency, or kind of activity. All he now wanted was an appointment to another post with a salary of five thousand rubles, either in the administration, in the banks, with the railways in one of the Empress Marya's Institutions, or even in the customs—but it had to carry with it a salary of five thousand rubles and be in a ministry other than that in which they had failed to appreciate him.

And this quest of Ivan Ilych's was crowned with remarkable and unexpected success. At Kursk an acquaintance of his, F. I. Ilyin, got into the first-class carriage, sat down beside Ivan Ilych, and told him of a telegram just received by the governor of Kursk announcing that a change was about to take place in the ministry: Peter Ivanovich was to be superseded by Ivan Semonovich.

The proposed change, apart from its significance for Russia, had a special significance for Ivan Ilych, because by bringing forward a new man, Peter Petrovich, and consequently his friend Zachar Ivanovich, it was highly favourable for Ivan Ilych, since Sachar Ivanovich was a friend and colleague of his.

In Moscow this news was confirmed, and on reaching Petersburg Ivan Ilych found Zachar Ivanovich and received a definite promise of an appointment in his former Department of Justice.

A week later he telegraphed to his wife: "Zachar in Miller's place. I shall receive appointment on presentation of report."

Thanks to this change of personnel, Ivan Ilych had unexpectedly obtained an appointment in his former ministry which placed him two states above his former colleagues besides giving him five thousand rubles salary and three thousand five hundred rubles for expenses connected with his removal. All his ill humour towards his former enemies and the whole department vanished, and Ivan Ilych was completely happy.

He returned to the country more cheerful and contented than he had been for a long time. Praskovya Fedorovna also cheered up and a truce was arranged between them. Ivan Ilych told of how he had been feted by everybody in Petersburg, how all those who had been his enemies were put to shame and now fawned on him, how envious they were of his appointment, and how much everybody in Petersburg had liked him.

Praskovya Fedorovna listened to all this and appeared to believe it. She did not contradict anything, but only made plans for their life in the town to which they were going. Ivan Ilych saw with delight that these plans were his plans, that he and his wife agreed, and that, after a stumble, his life was regaining its due and natural character of pleasant lightheartedness and decorum.

Ivan Ilych had come back for a short time only, for he had to take up his new duties on the 10th of September. Moreover, he needed time to settle into the new place, to move all his belongings from the province, and to buy and order many additional things: in a word, to make such arrangements as he had resolved on, which were almost exactly what Praskovya Fedorovna too had decided on.

Now that everything had happened so fortunately, and that he and his wife were at one in their aims and moreover saw so little of one another, they got on together better than they had done since the first years of marriage. Ivan Ilych had thought of taking his family away with him at once, but the insistence of his wife's brother and her sister-in-law, who had suddenly become particularly amiable and friendly to him and his family, induced him to depart alone.

So he departed, and the cheerful state of mind induced by his success and by the harmony between his wife and himself, the one intensifying the other, did not leave him. He found a delightful house, just the thing both he and his wife had dreamt of. Spacious, lofty reception rooms in the old style, a convenient and dignified study, rooms for his wife and daughter, a study for his son—it might have been specially built for them. Ivan Ilych himself superintended the arrangements, chose the wallpapers, supplemented the furniture (preferably with antiques which he considered particularly *comme il faut*), and supervised the upholstering. Everything progressed and progressed and approached the ideal he had set himself: even when things were only half completed they exceeded his expectations. He saw what a refined and elegant character, free from vulgarity, it would all have when it was ready. On falling asleep he pictured to himself how the reception room would look. Looking at the yet unfinished drawing room he could see the fireplace, the screen, the what-not, the little chairs dotted here and there, the dishes and plates on the walls, and the bronzes, as they would be when everything was in place. He was pleased by the thought of how his wife and daughter, who shared his taste in this matter, would be impressed by it. They were certainly not expecting as much. He had been particularly successful in finding, and buying cheaply, antiques which gave a particularly aristocratic character to the whole place. But in his letters he intentionally understated everything in order to be able to surprise them. All this so absorbed him that his new duties—though he liked his official work—interested him less than he had expected. Sometimes he even had moments of absent-mindedness during the court sessions and would consider whether he should have straight or curved cornices for his curtains. He was so interested in it all that he often did things himself, rearranging the furniture, or rehanging the curtains. Once when mounting a step-ladder to show the upholsterer, who did not understand, how he wanted the hangings draped, he made a false step and slipped, but being a strong and agile man he clung on and only knocked his side against the knob of the window frame. The bruised place was painful but the pain soon passed, and he felt particularly bright and well just then. He wrote: "I feel fifteen years younger." He thought he would have everything ready by September, but it dragged on till mid-October. But the result was charming not only in his eyes but to everyone who saw it.

In reality it was just what is usually seen in the houses of people of moderate means who want to appear rich, and therefore succeed only in resembling others like themselves: there are damasks, dark wood, plants, rugs, and dull and polished bronzes—all the things people of a certain class have in order to resemble other people of that class. His house was so like the others that it would never have been noticed, but to him it all seemed to be quite exceptional. He was very happy when he met his family at the station and brought them to the newly furnished house all lit up, where a footman in a white tie opened the door into the hall decorated with plants, and when they went on into the drawing-room and the study uttering exclamations of delight. He conducted them everywhere, drank in their praises eagerly, and beamed with pleasure. At tea that evening, when Praskovya Fedorovna among others things asked him about his fall, he laughed, and showed them how he had gone flying and had frightened the upholsterer.

"It's a good thing I'm a bit of an athlete. Another man might have been killed, but I merely knocked myself, just here; it hurts when it's touched, but it's passing off already—it's only a bruise."

So they began living in their new home—in which, as always happens, when they got thoroughly settled in they found they were just one room short—and with the increased income, which as always was just a little (some five hundred rubles) too little, but it was all very nice. Things went particularly well at first, before everything was finally arranged and while something had still to be done: this thing bought, that thing ordered, another thing moved, and something else adjusted. Though there were some disputes between husband and wife, they were both so well satisfied and had so much to do that it all passed off without any serious quarrels. When nothing was left to arrange it became rather dull and something seemed to be lacking, but they were then making acquaintances, forming habits, and life was growing fuller.

Ivan Ilych spent his mornings at the law court and came home to dinner, and at first he was generally in a good humour, though he occasionally became irritable just on account of his house. (Every spot on the tablecloth or the upholstery, and every broken windowblind string, irritated him. He had devoted so much trouble to arranging it all that every disturbance of it distressed him.) But on the whole his life ran its course as he believed life should do: easily, pleasantly, and decorously.

He got up at nine, drank his coffee, read the paper, and then put on his undress uniform and went to the law courts. There the harness in which he worked had already been stretched to fit him and he donned it without a hitch: petitioners, inquiries at the chancery, the chancery itself, and the sittings public and administrative. In all this the thing was to exclude everything fresh and vital, which always disturbs the regular course of official business, and to admit only official relations with people, and then only on official grounds. A man would come, for instance, wanting some information. Ivan Ilych, as one in whose sphere the matter did not lie, would have nothing to do with him: but if the man had some business with him in his official capacity, something that could be expressed on officially stamped paper, he would do everything, positively everything he could within the limits of such relations, and in doing so would maintain the semblance of friendly human relations, that is, would observe the courtesies of life. As soon as the official relations ended, so did everything else. Ivan Ilych possessed this capacity to separate his real life from the official side of affairs and not mix the two, in the highest degree, and by long practice and natural aptitude had brought it to such a pitch that sometimes, in the manner of a virtuoso, he would even allow himself to let the human and official relations mingle. He let himself do this just because he felt that he could at any time he chose resume the strictly official attitude again and drop the human relation. And he did it all easily, pleasantly, correctly, and even artistically. In the intervals between the sessions he smoked, drank tea, chatted a little about politics, a little about general topics, a little about cards, but most of all about official appointments. Tired, but with the feelings of a virtuoso—one of the first violins who has played his part in an orchestra with precision—he would return home to find that his wife and daughter had been out paying calls, or had a visitor, and that his son had been to school, had done his homework with his tutor, and was surely learning what is taught at High Schools. Everything was as it should be. After dinner, if they had no visitors, Ivan Ilych sometimes read a book that was being much discussed at the time, and in the evening settled down to work, that is, read official papers, compared the depositions of witnesses, and noted paragraphs of the Code applying to them. This was neither dull nor amusing. It was dull when he might have been playing bridge, but if no bridge was available it was at any rate better than doing nothing or sitting with his wife. Ivan Ilych's chief pleasure was giving little dinners to which he invited men and women of good social position, and just as his drawing-room resembled all other drawing-rooms so did his enjoyable little parties resemble all other such parties.

Once they even gave a dance. Ivan Ilych enjoyed it and everything went off well, except that it led to a violent quarrel with his wife about the cakes and sweets. Praskovya Fedorovna had made her own plans, but Ivan Ilych insisted on getting everything from an expensive confectioner and ordered too many cakes, and the quarrel occurred because some of those cakes were left over and the confectioner's bill came to forty-five rubles. It was a great and disagreeable quarrel. Praskovya Fedorovna called him "a fool and an imbecile," and he clutched at his head and made angry allusions to divorce.

But the dance itself had been enjoyable. The best people were there, and Ivan Ilych had danced with Princess Trufonova, a sister of the distinguished founder of the Society "Bear My Burden".

The pleasures connected with his work were pleasures of ambition; his social pleasures were those of vanity; but Ivan Ilych's greatest pleasure was playing bridge. He acknowledged that whatever disagreeable incident happened in his life, the pleasure that beamed like a ray of light above everything else was to sit down to bridge with good players, not noisy partners, and of course to four-handed bridge (with five players it was annoying to have to stand out, though one pretended not to mind), to play a clever and serious game (when the cards allowed it) and then to have supper and drink a glass of wine. After a game of bridge, especially if he had won a little (to win a large sum was unpleasant), Ivan Ilych went to bed in a specially good humour.

So they lived. They formed a circle of acquaintances among the best people and were visited by people of importance and by young folk. In their views as to their acquaintances, husband, wife and daughter were entirely agreed, and tacitly and unanimously kept at arm's length and shook off the various shabby friends and relations who, with much show of affection, gushed into the drawing-room with its Japanese plates on the walls. Soon these shabby friends ceased to obtrude themselves and only the best people remained in the Golovins' set.

Young men made up to Lisa, and Petrishchev, an examining magistrate and Dmitri Ivanovich Petrishchev's son and sole heir, began to be so attentive to her that Ivan Ilych had already spoken to Praskovya Fedorovna about it, and considered whether they should not arrange a party for them, or get up some private theatricals.

So they lived, and all went well, without change, and life flowed pleasantly.

IV

THEY WERE ALL in good health. It could not be called ill health if Ivan Ilych sometimes said that he had a queer taste in his mouth and felt some discomfort in his left side.

But this discomfort increased and, though not exactly painful, grew into a sense of pressure in his side accompanied by ill humour. And his irritability became worse and worse and began to mar the agreeable, easy, and correct life that had established itself in the Golovin family. Quarrels between husband and wife became more and more frequent, and soon the ease and amenity disappeared and even the decorum was barely maintained. Scenes again became frequent, and very few of those islets remained on which husband and wife could meet without an explosion. Praskovya Fedorovna now had good reason to say that her husband's temper was trying. With characteristic exaggeration she said he had always had a dreadful temper, and that it had needed all her good nature to put up with it for twenty years. It was true that now the quarrels were started by him. His bursts of temper always came just before dinner, often just as he began to eat his soup. Sometimes he noticed that a plate or dish

IV

was chipped, or the food was not right, or his son put his elbow on the table, or his daughter's hair was not done as he liked it, and for all this he blamed Praskovya Fedorovna. At first she retorted and said disagreeable things to him, but once or twice he fell into such a rage at the beginning of dinner that she realized it was due to some physical derangement brought on by taking food, and so she restrained herself and did not answer, but only hurried to get the dinner over. She regarded this self-restraint as highly praiseworthy. Having come to the conclusion that her husband had a dreadful temper and made her life miserable, she began to feel sorry for herself, and the more she pitied herself the more she hated her husband. She began to wish he would die; yet she did not want him to die because then his salary would cease. And this irritated her against him still more. She considered herself dreadfully unhappy just because not even his death could save her, and though she concealed her exasperation, that hidden exasperation of hers increased his irritation also.

After one scene in which Ivan Ilych had been particularly unfair and after which he had said in explanation that he certainly was irritable but that it was due to his not being well, she said that he was ill it should be attended to, and insisted on his going to see a celebrated doctor.

He went. Everything took place as he had expected and as it always does. There was the usual waiting and the important air assumed by the doctor, with which he was so familiar (resembling that which he himself assumed in court), and the sounding and listening, and the questions which called for answers that were foregone conclusions and were evidently unnecessary, and the look of importance which implied that "if only you put yourself in our hands we will arrange everything—we know indubitably how it has to be done, always in the same way for everybody alike." It was all just as it was in the law courts. The doctor put on just the same air towards him as he himself put on towards an accused person.

The doctor said that so-and-so indicated that there was so-and-so inside the patient, but if the investigation of so-and-so did not confirm this, then he must assume that and that. If he assumed that and that, then…and so on. To Ivan Ilych only one question was important: was his case serious or not? But the doctor ignored that inappropriate question. From his point of view it was not the one under consideration, the real question was to decide between a floating kidney, chronic catarrh, or appendicitis. It was not a question the doctor solved brilliantly, as it seemed to Ivan Ilych, in favour of the appendix, with the reservation that should an examination of the urine give fresh indications the matter would be reconsidered. All this was just what Ivan Ilych had himself brilliantly accomplished a thousand times in dealing with men on trial. The doctor summed up just as brilliantly, looking over his spectacles triumphantly and even gaily at the accused. From the doctor's summing up Ivan Ilych concluded that things were bad, but that for the doctor, and perhaps for everybody else, it was a matter of indifference, though for him it was bad. And this conclusion struck him painfully, arousing in him a great feeling of pity for himself and of bitterness towards the doctor's indifference to a matter of such importance.

He said nothing of this, but rose, placed the doctor's fee on the table, and remarked with a sigh: "We sick people probably often put inappropriate questions. But tell me, in general, is this complaint dangerous, or not?…"

The doctor looked at him sternly over his spectacles with one eye, as if to say: "Prisoner, if you will not keep to the questions put to you, I shall be obliged to have you removed from the court."

"I have already told you what I consider necessary and proper. The analysis may show something more." And the doctor bowed.

Ivan Ilych went out slowly, seated himself disconsolately in his sledge, and drove home. All the way home he was going over what the doctor had said, trying to translate those complicated, obscure, scientific phrases into plain language and find in them an answer to the question: "Is my condition bad? Is it very bad? Or is there as yet nothing much wrong?" And it seemed to him that the meaning of what the doctor had said was that it was very bad. Everything in the streets seemed depressing. The cabmen, the houses, the passersby, and the shops, were dismal. His ache, this dull gnawing ache that never ceased for a moment, seemed to have acquired a new and more serious significance from the doctor's dubious remarks. Ivan Ilych now watched it with a new and oppressive feeling.

He reached home and began to tell his wife about it. She listened, but in the middle of his account his daughter came in with her hat on, ready to go out with her mother. She sat down reluctantly to listen to this tedious story, but could not stand it long, and her mother too did not hear him to the end.

"Well, I am very glad," she said. "Mind now to take your medicine regularly. Give me the prescription and I'll send Gerasim to the chemist's." And she went to get ready to go out.

While she was in the room Ivan Ilych had hardly taken time to breathe, but he sighed deeply when she left it.

"Well," he thought, "perhaps it isn't so bad after all."

He began taking his medicine and following the doctor's directions, which had been altered after the examination of the urine. But then it happened that there was a contradiction between the indications drawn from the examination of the urine and the symptoms that showed themselves. It turned out that what was happening differed from what the doctor had told him, and that he had either forgotten or blundered, or hidden something from him. He could not, however, be blamed for that, and Ivan Ilych still obeyed his orders implicitly and at first derived some comfort from doing so.

From the time of his visit to the doctor, Ivan Ilych's chief occupation was the exact fulfillment of the doctor's instructions regarding hygiene and the taking of medicine, and the observation of his pain and his excretions. His chief interest came to be people's ailments and people's health. When sickness, deaths, or recoveries were mentioned in his presence, especially when the illness resembled his own, he listened with agitation which he tried to hide, asked questions, and applied what he heard to his own case.

The pain did not grow less, but Ivan Ilych made efforts to force himself to think that he was better. And he could do this so long as nothing agitated him. But as soon as he had any unpleasantness with his wife, any lack of success in his official work, or held bad cards at bridge, he was at once acutely sensible of his disease. He had formerly borne such mischances, hoping soon to adjust what was wrong, to master it and attain success, or make a grand slam. But now every mischance upset him and plunged him into despair. He would say to himself: "there now, just as I was beginning to get better and the medicine had begun to take effect, comes this accursed misfortune, or unpleasantness…" And he was furious with the mishap, or with the people who were causing the unpleasantness and killing him, for he felt that this fury was killing him but he could not restrain it. One would have thought that it should have been clear to him that this exasperation with circumstances and people aggravated his illness, and that he ought therefore to ignore unpleasant occurrences. But he drew the very opposite conclusion: he said that he needed peace, and he watched for everything that might disturb it and became irritable at the slightest infringement of it.

His condition was rendered worse by the fact that he read medical books and consulted doctors. The progress of his disease was so gradual that he could deceive himself when comparing one day with another—the difference was so slight. But when he consulted the doctors it seemed to him that he was getting worse, and even very rapidly. Yet despite this he was continually consulting them.

That month he went to see another celebrity, who told him almost the same as the first had done but put his questions rather differently, and the interview with this celebrity only increased Ivan Ilych's doubts and fears. A friend of a friend of his, a very good doctor, diagnosed his illness again quite differently from the others, and though he predicted recovery, his questions and suppositions bewildered Ivan Ilych still more and increased his doubts. A homeopathist diagnosed the disease in yet another way, and prescribed medicine which Ivan Ilych took secretly for a week. But after a week, not feeling any improvement and having lost confidence both in the former doctor's treatment and in this one's, he became still more despondent. One day, a lady acquaintance mentioned a cure effected by a wonder-working icon. Ivan Ilych caught himself listening attentively and beginning to believe that it had occurred. This incident alarmed him. "Has my mind really weakened to such an extent?" he asked himself. "Nonsense! It's all rubbish. I mustn't give way to nervous fears but having chosen a doctor must keep strictly to his treatment. That is what I will do. Now it's all settled. I won't think about it, but will follow the treatment seriously till summer, and then we shall see. From now there must be no more of this wavering!" This was easy to say but impossible to carry out. The pain in his side oppressed him and seemed to grow worse and more incessant, while the taste in his mouth grew stranger and stranger. It seemed to him that his breath had a disgusting smell, and he was conscious of a loss of appetite and strength. There was no deceiving himself: something terrible, new, and more important than anything before in his life, was taking place within him of which he alone was aware. Those about him did not understand or would not understand it, but thought everything in the world was going on as usual. That tormented Ivan Ilych more than anything. He saw that his household, especially his wife and daughter who were in a perfect whirl of visiting, did not understand anything of it and were annoyed that he was so depressed and so exacting, as if he were to blame for it. Though they tried to disguise it he saw that he was an obstacle in their path, and that his wife had adopted a definite line in regard to his illness and kept to it regardless of anything he said or did. Her attitude was this: "You know," she would say to her friends, "Ivan Ilych can't do as other people do, and keep to the treatment prescribed for him. One day he'll take his drops and keep strictly to his diet and go to bed in good time, but the next day unless I watch him he'll suddenly forget his medicine, eat sturgeon—which is forbidden—and sit up playing cards till one o'clock in the morning."

"Oh, come, when was that?" Ivan Ilych would ask in vexation. "Only once at Peter Ivanovich's."

"And yesterday with Shebek."

"Well, even if I hadn't stayed up, this pain would have kept me awake."

"Be that as it may you'll never get well like that, but will always make us wretched."

Praskovya Fedorovna's attitude to Ivan Ilych's illness, as she expressed it both to others and to him, was that it was his own fault and was another of the annoyances he caused her. Ivan Ilych felt that this opinion escaped her involuntarily—but that did not make it easier for him.

At the law courts too, Ivan Ilych noticed, or thought he noticed, a strange attitude towards himself. It sometimes seemed to him that people were watching him inquisitively as a man whose place might soon be vacant. Then again, his friends would suddenly begin to chaff him in a friendly way about his low spirits, as if the awful, horrible, and unheard-of thing that was going on within him, incessantly gnawing at him and irresistibly drawing him away, was a very agreeable subject for jests. Schwartz in particular irritated him by his jocularity, vivacity, and *savoir-faire*, which reminded him of what he himself had been ten years ago.

Friends came to make up a set and they sat down to cards. They dealt, bending the new cards to soften them, and he sorted the diamonds in his hand and found he had seven. His partner said "No trumps" and supported him with two diamonds. What more could be wished for? It ought to be jolly and lively. They would make a grand slam. But suddenly Ivan Ilych was conscious of that gnawing pain, that taste in his mouth, and it seemed ridiculous that in such circumstances he should be pleased to make a grand slam.

He looked at his partner Mikhail Mikhaylovich, who rapped the table with his strong hand and instead of snatching up the tricks pushed the cards courteously and indulgently towards Ivan Ilych that he might have the pleasure of gathering them up without the trouble of stretching out his hand for them. "Does he think I am too weak to stretch out my arm?" thought Ivan Ilych, and forgetting what he was doing he over-trumped his partner, missing the grand slam by three tricks. And what was most awful of all was that he saw how upset Mikhail Mikhaylovich was about it but did not himself care. And it was dreadful to realize why he did not care.

They all saw that he was suffering, and said: "We can stop if you are tired. Take a rest." Lie down? No, he was not at all tired, and he finished the rubber. All were gloomy and silent. Ivan Ilych felt that he had diffused this gloom over them and could not dispel it. They had supper and went away, and Ivan Ilych was left alone with the consciousness that his life was poisoned and was poisoning the lives of others, and that this poison did not weaken but penetrated more and more deeply into his whole being.

With this consciousness, and with physical pain besides the terror, he must go to bed, often to lie awake the greater part of the night. Next morning he had to get up again, dress, go to the law courts, speak, and write; or if he did not go out, spend at home those twenty-four hours a day each of which was a torture. And he had to live thus all alone on the brink of an abyss, with no one who understood or pitied him.

V

SO ONE MONTH passed and then another. Just before the New Year his brother-in-law came to town and stayed at their house. Ivan Ilych was at the law courts and Praskovya Fedorovna had gone shopping. When Ivan Ilych came home and entered his study he found his brother-in-law there—a healthy, florid man—unpacking his portmanteau himself. He raised his head on hearing Ivan Ilych's footsteps and looked up at him for a moment without a word. That stare told Ivan Ilych everything. His brother-in-law opened his mouth to utter an exclamation of surprise but checked himself, and that action confirmed it all.

"I have changed, eh?"

"Yes, there is a change."

And after that, try as he would to get his brother-in-law to return to the subject of his looks, the latter would say nothing about it. Praskovya Fedorovna came home and her brother went out to her. Ivan Ilych locked to door and began to examine himself in the glass, first full face, then in profile. He took up a portrait of himself taken with his wife, and compared it with what he saw in the glass. The change in him was immense. Then he bared his arms to the elbow, looked at them, drew the sleeves down again, sat down on an ottoman, and grew blacker than night.

"No, no, this won't do!" he said to himself, and jumped up, went to the table, took up some law papers and began to read them, but could not continue. He unlocked the door and went into the reception-room. The door leading to the drawing-room was shut. He approached it on tiptoe and listened.

"No, you are exaggerating!" Praskovya Fedorovna was saying.

"Exaggerating! Don't you see it? Why, he's a dead man! Look at his eyes—there's no life in them. But what is it that is wrong with him?"

"No one knows. Nikolaevich [that was another doctor] said something, but I don't know what. And Seshchetitsky [this was the celebrated specialist] said quite the contrary…"

Ivan Ilych walked away, went to his own room, lay down, and began musing; "The kidney, a floating kidney." He recalled all the doctors had told him of how it detached itself and swayed about. And by an effort of imagination he tried to catch that kidney and arrest it and support it. So little was needed for this, it seemed to him. "No, I'll go to see Peter Ivanovich again." [That was the friend whose friend was a doctor.] He rang, ordered the carriage, and got ready to go.

"Where are you going, Jean?" asked his wife with a specially sad and exceptionally kind look.

This exceptionally kind look irritated him. He looked morosely at her.

"I must go to see Peter Ivanovich."

He went to see Peter Ivanovich, and together they went to see his friend, the doctor. He was in, and Ivan Ilych had a long talk with him.

Reviewing the anatomical and physiological details of what in the doctor's opinion was going on inside him, he understood it all.

There was something, a small thing, in the vermiform appendix. It might all come right. Only stimulate the energy of one organ and check the activity of another, then absorption would take place and everything would come right. He got home rather late for dinner, ate his dinner, and conversed cheerfully, but could not for a long time bring himself to go back to work in his room. At last, however, he went to his study and did what was necessary, but the consciousness that he had put something aside—an important, intimate matter which he would revert to when his work was done—never left him. When he had finished his work he remembered that this intimate matter was the thought of his vermiform appendix. But he did not give himself up to it, and went to the drawing-room for tea. There were callers there, including the examining magistrate who was a desirable match for his daughter, and they were conversing, playing the piano, and singing.

Ivan Ilych, as Praskovya Fedorovna remarked, spent that evening more cheerfully than usual, but he never for a moment forgot that he had postponed the important matter of the appendix. At eleven o'clock he said goodnight and went to his bedroom. Since his illness he had slept alone in a small room next to his study. He undressed and took up a novel by Zola, but instead of reading it he fell into thought, and in his imagination that desired improvement in the vermiform appendix occurred. There was the absorption and evacuation and the re-establishment of normal activity. "Yes, that's it!" he said to himself. "One need only assist nature, that's all." He remembered his medicine, rose, took it, and lay down on his back watching for the beneficent action of the medicine and for it to lessen the pain. "I need only take it regularly and avoid all injurious influences. I am already feeling better, much better." He began touching his side: it was not painful to the touch. "There, I really don't feel it. It's much better already." He put out the light and turned on his side ... "The appendix is getting better, absorption is occurring." Suddenly he felt the old, familiar, dull, gnawing pain, stubborn and serious. There was the same familiar loathsome taste in his mouth.

His heart sank and he felt dazed. "My God! My God!" he muttered. "Again, again! And it will never cease." And suddenly, the matter presented itself in a quite different aspect. "Vermiform appendix! Kidney!" he said to himself. "It's not a question of appendix or kidney, but of life and…death. Yes, life was there and now it is going, going and I cannot stop it. Yes. Why deceive myself? Isn't it obvious to everyone but me that I'm dying, and that it's only a question of weeks, days…it may happen this moment. There was light and now there is darkness. I was here and now I'm going there! Where?" A chill came over him, his breathing ceased, and he felt only the throbbing of his heart.

"When I am not, what will there be? There will be nothing. Then where shall I be when I am no more? Can this be dying? No, I don't want to!" He jumped up and tried to light the candle, felt for it with trembling hands, dropped candle and candlestick on the floor, and fell back on his pillow.

"What's the use? It makes no difference," he said to himself, staring with wide-open eyes into the darkness. "Death. Yes, death. And none of them knows or wishes to know it, and they have no pity for me. Now they are playing." (He heard through the door the distant sound of a song and its accompaniment.) "It's all the same to them, but they will die too! Fools! I first, and they later, but it will be the same for them. And now they are merry…the beasts!"

Anger choked him and he was agonizingly, unbearably miserable. "It is impossible that all men have been doomed to suffer this awful horror!" He raised himself.

"Something must be wrong. I must calm myself—must think it all over from the beginning." And he again began thinking. "Yes, the beginning of my illness: I knocked my side, but I was still quite well that day and the next. It hurt a little, then rather more. I saw the doctors, then followed despondency and anguish, more doctors, and I drew nearer to the abyss. My strength grew less and I kept coming nearer and nearer, and now I have wasted away and there is no light in my eyes. I think of the appendix—but this is death! I think of mending the appendix, and all the while here is death! Can it really be death?" Again terror seized him and he gasped for breath. He leant down and began feeling for the matches, pressing with his elbow on the stand beside the bed. It was in his way and hurt him, he grew furious with it, pressed on it still harder, and upset it. Breathless and in despair he fell on his back, expecting death to come immediately.

Meanwhile the visitors were leaving. Praskovya Fedorovna was seeing them off. She heard something fall and came in.

"What has happened?"

"Nothing. I knocked it over accidentally."

She went out and returned with a candle. He lay there panting heavily, like a man who has run a thousand yards, and stared upwards at her with a fixed look.

"What is it, Jean?"

"No...o...thing. I upset it." ("Why speak of it? She won't understand," he thought.)

And in truth she did not understand. She picked up the stand, lit his candle, and hurried away to see another visitor off. When she came back he still lay on his back, looking upwards.

"What is it? Do you feel worse?"

"Yes."

She shook her head and sat down.

"Do you know, Jean, I think we must ask Leshchetitsky to come and see you here."

This meant calling in the famous specialist, regardless of expense. He smiled malignantly and said "No." She remained a little longer and then went up to him and kissed his forehead.

While she was kissing him he hated her from the bottom of his soul and with difficulty refrained from pushing her away.

"Good night. Please God you'll sleep."

"Yes."

VI

IVAN ILYCH SAW that he was dying, and he was in continual despair.

In the depth of his heart he knew he was dying, but not only was he not accustomed to the thought, he simply did not and could not grasp it.

The syllogism he had learnt from Kiesewetter's Logic: "Caius is a man, men are mortal, therefore Caius is mortal," had always seemed to him correct as applied to Caius, but certainly not as applied to himself. That Caius—man in the abstract—was mortal, was perfectly correct, but he was not Caius, not an abstract man, but a creature quite, quite separate from all others. He had been little Vanya, with a mamma and a papa, with Mitya and Volodya, with the toys, a coachman and a nurse, afterwards with Katenka and will all the joys, griefs, and delights of childhood, boyhood, and youth. What did Caius know of the smell of that striped leather ball Vanya had been so fond of? Had Caius kissed his mother's hand like that, and did the silk of her dress rustle so for Caius? Had he rioted like that at school when the pastry was bad? Had Caius been in love like that? Could Caius preside at a session as he did? "Caius really was mortal, and it was right for him to die; but for me, little Vanya, Ivan Ilych, with all my thoughts and emotions, it's altogether a different matter. It cannot be that I ought to die. That would be too terrible."

Such was his feeling.

"If I had to die like Caius I would have known it was so. An inner voice would have told me so, but there was nothing of the sort in me and I and all my friends felt that our case was quite different from that of Caius. And now here it is!" he said to himself. "It can't be. It's impossible! But here it is. How is this? How is one to understand it?"

He could not understand it, and tried to drive this false, incorrect, morbid thought away and to replace it by other proper and healthy thoughts. But that thought, and not the thought only but the reality itself, seemed to come and confront him.

And to replace that thought he called up a succession of others, hoping to find in them some support. He tried to get back into the former current of thoughts that had once screened the thought of death from him. But strange to say, all that had formerly shut off, hidden, and destroyed his consciousness of death, no longer had that effect. Ivan Ilych now spent most of his time in attempting to re-establish that old current. He would say to himself: "I will take up my duties again—after all I used to live by them." And banishing all doubts he would go to the law courts, enter into conversation with his colleagues, and sit carelessly as was his wont, scanning the crowd with a thoughtful look and leaning both his emaciated arms on the arms of his oak chair; bending over as usual to a colleague and drawing his papers nearer he would interchange whispers with him, and then suddenly raising his eyes and sitting erect would pronounce certain words and open the proceedings. But suddenly in the midst of those proceedings the pain in his side, regardless of the stage the proceedings had reached, would begin its own gnawing work. Ivan Ilych would turn his attention to it and try to drive the thought of it away, but without success. *It* would come and stand before him and look at him, and he would be petrified and the light would die out of his eyes, and he would again begin asking himself whether *It* alone was true. And his colleagues and subordinates would see with surprise and distress that he, the brilliant and subtle judge, was becoming confused and making mistakes. He would shake himself, try to pull himself together, manage somehow to bring the sitting to a close, and return home with the sorrowful consciousness that his judicial labours could not as formerly hide from him what he wanted them to hide, and could not deliver him from *It*. And what was worst of all was that *It* drew his attention to itself not in order to make him take some action but only that he should look at *It*, look it straight in the face: look at it and without doing anything, suffer inexpressibly.

And to save himself from this condition Ivan Ilych looked for consolations—new screens—and new screens were found and for a while seemed to save him, but then they immediately fell to pieces or rather became transparent, as if *It* penetrated them and nothing could veil *It*.

In these latter days he would go into the drawing-room he had arranged—that drawing-room where he had fallen and for the sake of which (how bitterly ridiculous it seemed) he had sacrificed his life—for he knew that his illness originated with that knock. He would enter and see that something had scratched the polished table. He would look for the cause of this and find that it was the bronze ornamentation of an album, that had got bent. He would take up the expensive album which he had lovingly arranged, and feel vexed with his daughter and her friends for their untidiness—for the album was torn here and there and some of the photographs turned upside down. He would put it carefully in order and bend the ornamentation back into position. Then it would occur to him to place all those things in another corner of the room, near the plants. He would call the footman, but his daughter or wife would come to help him. They would not agree, and his wife would contradict him, and he would dispute and grow angry. But that was all right, for then he did not think about *It*. *It* was invisible.

But then, when he was moving something himself, his wife would say: "Let the servants do it. You will hurt yourself again." And suddenly *It* would flash through the screen and he would see it. It was just a flash, and he hoped it would disappear, but he would involuntarily pay attention to his side. "It sits there as before, gnawing just the same!" And he could no longer forget *It*, but could distinctly see it looking at him from behind the flowers. "What is it all for?" "It really is so! I lost my life over that curtain as I might have done when storming a fort. Is that possible? How terrible and how stupid. It can't be true! It can't, but it is."

He would go to his study, lie down, and again be alone with *It*: face to face with *It*. And nothing could be done with *It* except to look at it and shudder.

VII

HOW IT HAPPENED it is impossible to say because it came about step by step, unnoticed, but in the third month of Ivan Ilych's illness, his wife, his daughter, his son, his acquaintances, the doctors, the servants, and above all he himself, were aware that the whole interest he had for other people was whether he would soon vacate his place, and at last release the living from the discomfort caused by his presence and be himself released from his sufferings.

He slept less and less. He was given opium and hypodermic injections of morphine, but this did not relieve him. The dull depression he experienced in a somnolent condition at first gave him a little relief, but only as something new, afterwards it became as distressing as the pain itself or even more so.

Special foods were prepared for him by the doctors' orders, but all those foods became increasingly distasteful and disgusting to him.

For his excretions also special arrangements had to be made, and this was a torment to him every time—a torment from the uncleanliness, the unseemliness, and the smell, and from knowing that another person had to take part in it.

But just through his most unpleasant matter, Ivan Ilych obtained comfort. Gerasim, the butler's young assistant, always came in to carry the things out. Gerasim was a clean, fresh peasant lad, grown stout on town food and always cheerful and bright. At first the sight of him, in his clean Russian peasant costume, engaged on that disgusting task embarrassed Ivan Ilych.

Once when he got up from the commode too weak to draw up his trousers, he dropped into a soft armchair and looked with horror at his bare, enfeebled thighs with the muscles so sharply marked on them.

Gerasim with a firm light tread, his heavy boots emitting a pleasant smell of tar and fresh winter air, came in wearing a clean Hessian apron, the sleeves of his print shirt tucked up over his strong bare young arms; and refraining from looking at his sick master out of consideration for his feelings, and restraining the joy of life that beamed from his face, he went up to the commode.

"Gerasim!" said Ivan Ilych in a weak voice.

Gerasim started, evidently afraid he might have committed some blunder, and with a rapid movement turned his fresh, kind, simple young face which just showed the first downy signs of a beard.

"Yes, sir?"

"That must be very unpleasant for you. You must forgive me. I am helpless."

"Oh, why, sir," and Gerasim's eyes beamed and he showed his glistening white teeth, "what's a little trouble? It's a case of illness with you, sir."

And his deft strong hands did their accustomed task, and he went out of the room stepping lightly. five minutes later he as lightly returned.

Ivan Ilych was still sitting in the same position in the armchair.

"Gerasim," he said when the latter had replaced the freshlywashed utensil. "Please come here and help me." Gerasim went up to him. "Lift me up. It is hard for me to get up, and I have sent Dmitri away."

Gerasim went up to him, grasped his master with his strong arms deftly but gently, in the same way that he stepped—lifted him, supported him with one hand, and with the other drew up his trousers and would have set him down again, but Ivan Ilych asked to be led to the sofa. Gerasim, without an effort and without apparent pressure, led him, almost lifting him, to the sofa and placed him on it.

"That you. How easily and well you do it all!"

Gerasim smiled again and turned to leave the room. But Ivan Ilych felt his presence such a comfort that he did not want to let him go.

"One thing more, please move up that chair. No, the other one—under my feet. It is easier for me when my feet are raised."

VII

Gerasim brought the chair, set it down gently in place, and raised Ivan Ilych's legs on it. It seemed to Ivan Ilych that he felt better while Gerasim was holding up his legs.

"It's better when my legs are higher," he said. "Place that cushion under them."

Gerasim did so. He again lifted the legs and placed them, and again Ivan Ilych felt better while Gerasim held his legs.

When he set them down Ivan Ilych fancied he felt worse.

"Gerasim," he said. "Are you busy now?"

"Not at all, sir," said Gerasim, who had learnt from the townsfolk how to speak to gentlefolk.

"What have you still to do?"

"What have I to do? I've done everything except chopping the logs for tomorrow."

"Then hold my legs up a bit higher, can you?"

"Of course I can. Why not?" and Gerasim raised his master's legs higher and Ivan Ilych thought that in that position he did not feel any pain at all.

"And how about the logs?"

"Don't trouble about that, sir. There's plenty of time."

Ivan Ilych told Gerasim to sit down and hold his legs, and began to talk to him. And strange to say it seemed to him that he felt better while Gerasim held his legs up.

After that Ivan Ilych would sometimes call Gerasim and get him to hold his legs on his shoulders, and he liked talking to him. Gerasim did it all easily, willingly, simply, and with a good nature that touched Ivan Ilych. Health, strength, and vitality in other people were offensive to him, but Gerasim's strength and vitality did not mortify but soothed him.

What tormented Ivan Ilych most was the deception, the lie, which for some reason they all accepted, that he was not dying but was simply ill, and he only need keep quiet and undergo a treatment and then something very good would result. He however knew that do what they would nothing would come of it, only still more agonizing suffering and death. This deception tortured him—their not wishing to admit what they all knew and what he knew, but wanting to lie to him concerning his terrible condition, and wishing and forcing him to participate in that lie. Those lies—lies enacted over him on the eve of his death and destined to degrade this awful, solemn act to the level of their visitings, their curtains, their sturgeon for dinner—were a terrible agony for Ivan Ilych. And strangely enough, many times when they were going through their antics over him he had been within a hairbreadth of calling out to them: "Stop lying! You know and I know that I am dying. Then at least stop lying about it!" But he had never had the spirit to do it. The awful, terrible act of his dying was, he could see, reduced by those about him to the level of a casual, unpleasant, and almost indecorous incident (as if someone entered a drawing room defusing an unpleasant odour) and this was done by that very decorum which he had served all his life long. He saw that no one felt for him, because no one even wished to grasp his position. Only Gerasim recognized it and pitied him. And so Ivan Ilych felt at ease only with him. He felt comforted when Gerasim supported his legs (sometimes all night long) and refused to go to bed, saying: "Don't you worry, Ivan Ilych. I'll get sleep enough later on," or when he suddenly became familiar and exclaimed: "If you weren't sick it would be another matter, but as it is, why should I grudge a little trouble?" Gerasim alone did not lie; everything showed that he alone understood the facts of the case and did not consider it necessary to disguise them, but simply felt sorry for his emaciated and enfeebled master. Once when Ivan Ilych was sending him away he even said straight out: "We shall all of us die, so why should I grudge a little trouble?"—expressing the fact that he did not think his work burdensome, because he was doing it for a dying man and hoped someone would do the same for him when his time came.

Apart from this lying, or because of it, what most tormented Ivan Ilych was that no one pitied him as he wished to be pitied. At certain moments after prolonged suffering he wished most of all (though he would have been ashamed to confess it) for someone to pity him as a sick child is pitied. He longed to be petted and comforted. He knew he was an important functionary, that he had a beard turning grey, and that therefore what he longed for was impossible, but still he longed for it. And in Gerasim's attitude towards him there was something akin to what he wished for, and so that attitude comforted him. Ivan Ilych wanted to weep, wanted to be petted and cried over, and then his colleague Shebek would come, and instead of weeping and being petted, Ivan Ilych would assume a serious, severe, and profound air, and by force of habit would express his opinion on a decision of the Court of Cassation and would stubbornly insist on that view. This falsity around him and within him did more than anything else to poison his last days.

VIII

IT WAS MORNING. He knew it was morning because Gerasim had gone, and Peter the footman had come and put out the candles, drawn back one of the curtains, and begun quietly to tidy up. Whether it was morning or evening, Friday or Sunday, made no difference, it was all just the same: the gnawing, unmitigated, agonizing pain, never ceasing for an instant, the consciousness of life inexorably waning but not yet extinguished, the approach of that ever dreaded and hateful Death which was the only reality, and always the same falsity. What were days, weeks, hours, in such a case?

"Will you have some tea, sir?"

"He wants things to be regular, and wishes the gentlefolk to drink tea in the morning," thought Ivan Ilych, and only said "No."

"Wouldn't you like to move onto the sofa, sir?"

"He wants to tidy up the room, and I'm in the way. I am uncleanliness and disorder," he thought, and said only:

"No, leave me alone."

The man went on bustling about. Ivan Ilych stretched out his hand. Peter came up, ready to help.

"What is it, sir?"

"My watch."

Peter took the watch which was close at hand and gave it to his master.

"Half-past eight. Are they up?"

"No sir, except Vladimir Ivanovich" (the son) "who has gone to school. Praskovya Fedorovna ordered me to wake her if you asked for her. Shall I do so?"

"No, there's no need to."

"Perhaps I'd better have some tea," he thought, and added aloud: "Yes, bring me some tea."

Peter went to the door, but Ivan Ilych dreaded being left alone. "How can I keep him here? Oh yes, my medicine."

"Peter, give me my medicine."

"Why not? Perhaps it may still do some good." He took a spoonful and swallowed it. "No, it won't help. It's all tomfoolery, all deception," he decided as soon as he became aware of the familiar, sickly, hopeless taste. "No, I can't believe in it any longer. But the pain, why this pain? If it would only cease just for a moment!" And he moaned. Peter turned towards him. "It's all right. Go and fetch me some tea."

Peter went out. Left alone Ivan Ilych groaned not so much with pain, terrible though that was, as from mental anguish. Always and for ever the same, always these endless days and nights.

If only it would come quicker! If only *what* would come quicker? Death, darkness?... No, no! anything rather than death!

When Peter returned with the tea on a tray, Ivan Ilych stared at him for a time in perplexity, not realizing who and what he was. Peter was disconcerted by that look and his embarrassment brought Ivan Ilych to himself.

"Oh, tea! All right, put it down. Only help me to wash and put on a clean shirt."

And Ivan Ilych began to wash. With pauses for rest, he washed his hands and then his face, cleaned his teeth, brushed his hair, looked in the glass. He was terrified by what he saw, especially by the limp way in which his hair clung to his pallid forehead.

While his shirt was being changed he knew that he would be still more frightened at the sight of his body, so he avoided looking at it. Finally he was ready. He drew on a dressing-gown, wrapped himself in a plaid, and sat down in the armchair to take his tea. For a moment he felt refreshed, but as soon as he began to drink the tea he was again aware of the same taste, and the pain also returned. He finished it with an effort, and then lay down stretching out his legs, and dismissed Peter.

Always the same. Now a spark of hope flashes up, then a sea of despair rages, and always pain; always pain, always despair, and always the same. When alone he had a dreadful and distressing desire to call someone, but he knew beforehand that with others present it would be still worse. "Another dose of morphine—to lose consciousness. I will tell him, the doctor, that he must think of something else. It's impossible, impossible, to go on like this."

An hour and another pass like that. But now there is a ring at the door bell. Perhaps it's the doctor? It is. He comes in fresh, hearty, plump, and cheerful, with that look on his face that seems to say: "There now, you're in a panic about something, but we'll arrange it all for you directly!" The doctor knows this expression is out of place here, but he has put it on once for all and can't take it off—like a man who has put on a frock-coat in the morning to pay a round of calls.

The doctor rubs his hands vigorously and reassuringly.

"Brr! How cold it is! There's such a sharp frost; just let me warm myself!" he says, as if it were only a matter of waiting till he was warm, and then he would put everything right.

"Well now, how are you?"

Ivan Ilych feels that the doctor would like to say: "Well, how are our affairs?" but that even he feels that this would not do, and says instead: "What sort of a night have you had?"

Ivan Ilych looks at him as much as to say: "Are you really never ashamed of lying?" But the doctor does not wish to understand this question, and Ivan Ilych says: "Just as terrible as ever. The pain never leaves me and never subsides. If only something …"

"Yes, you sick people are always like that…. There, now I think I am warm enough. Even Praskovya Fedorovna, who is so particular, could find no fault with my temperature. Well, now I can say good-morning," and the doctor presses his patient's hand.

Then dropping his former playfulness, he begins with a most serious face to examine the patient, feeling his pulse and taking his temperature, and then begins the sounding and auscultation.

Ivan Ilych knows quite well and definitely that all this is nonsense and pure deception, but when the doctor, getting down on his knee, leans over him, putting his ear first higher then lower, and performs various gymnastic movements over him with a significant expression on his face, Ivan Ilych submits to it all as he used to submit to the speeches of the lawyers, though he knew very well that they were all lying and why they were lying.

The doctor, kneeling on the sofa, is still sounding him when Praskovya Fedorovna's silk dress rustles at the door and she is heard scolding Peter for not having let her know of the doctor's arrival.

She comes in, kisses her husband, and at once proceeds to prove that she has been up a long time already, and only owing to a misunderstanding failed to be there when the doctor arrived.

Ivan Ilych looks at her, scans her all over, sets against her the whiteness and plumpness and cleanness of her hands and neck, the gloss of her hair, and the sparkle of her vivacious eyes. He hates her with his whole soul. And the thrill of hatred he feels for her makes him suffer from her touch.

Her attitude towards him and his diseases is still the same. Just as the doctor had adopted a certain relation to his patient which he could not abandon, so had she formed one towards him—that he was not doing something he ought to do and was himself to blame, and that she reproached him lovingly for this—and she could not now change that attitude.

"You see he doesn't listen to me and doesn't take his medicine at the proper time. And above all he lies in a position that is no doubt bad for him—with his legs up."

She described how he made Gerasim hold his legs up.

The doctor smiled with a contemptuous affability that said: "What's to be done? These sick people do have foolish fancies of that kind, but we must forgive them."

When the examination was over the doctor looked at his watch, and then Praskovya Fedorovna announced to Ivan Ilych that it was of course as he pleased, but she had sent today for a celebrated specialist who would examine him and have a consultation with Michael Danilovich (their regular doctor). "Please don't raise any objections. I am doing this for my own sake," she said ironically, letting it be felt that she was doing it all for his sake and only said this to leave him no right to refuse. He remained silent, knitting his brows. He felt that he was surrounded and involved in a mesh of falsity that it was hard to unravel anything.

Everything she did for him was entirely for her own sake, and she told him she was doing for herself what she actually was doing for herself, as if that was so incredible that he must understand the opposite.

At half-past eleven the celebrated specialist arrived. Again the sounding began and the significant conversations in his presence and in another room, about the kidneys and the appendix, and the questions and answers, with such an air of importance that again, instead of the real question of life and death which now alone confronted him, the question arose of the kidney and appendix which were not behaving as they ought to and would now be attached by Michael Danilovich and the specialist and forced to amend their ways.

The celebrated specialist took leave of him with a serious though not hopeless look, and in reply to the timid question Ivan Ilych, with eyes glistening with fear and hope, put to him as to whether there was a chance of recovery, said that he could not vouch for it but there was a possibility. The look of hope with which Ivan Ilych watched the doctor out was so pathetic that Praskovya Fedorovna, seeing it, even wept as she left the room to hand the doctor his fee.

The gleam of hope kindled by the doctor's encouragement did not last long. The same room, the same pictures, curtains, wall-paper, medicine bottles, were all there, and the same aching suffering body, and Ivan Ilych began to moan. They gave him a subcutaneous injection and he sank into oblivion.

It was twilight when he came to. They brought him his dinner and he swallowed some beef tea with difficulty, and then everything was the same again and night was coming on.

After dinner, at seven o'clock, Praskovya Fedorovna came into the room in evening dress, her full bosom pushed up by her corset, and with traces of powder on her face. She had reminded him in the morning that they were going to the theatre. Sarah Bernhardt was visiting the town and they had a box, which he had insisted on their taking. Now he had forgotten about it and her toilet offended him, but he concealed his vexation when he remembered that he had himself insisted on their securing a box and going because it would be an instructive and aesthetic pleasure for the children.

Praskovya Fedorovna came in, self-satisfied but yet with a rather guilty air. She sat down and asked how he was, but, as he saw, only for the sake of asking and not in order to learn about it, knowing that there was nothing to learn—and then went on to what she really wanted to say: that she would not on any account have gone but that the box had been taken and Helen and their daughter were going, as well as Petrishchev (the examining magistrate, their daughter's fiancé) and that it was out of the question to let them go alone; but that she would have much preferred to sit with him for a while; and he must be sure to follow the doctor's orders while she was away.

"Oh, and Fedor Petrovich" (the fiancé) "would like to come in. May he? And Lisa?"

"All right."

Their daughter came in in full evening dress, her fresh young flesh exposed (making a show of that very flesh which in his own case caused so much suffering), strong, healthy, evidently in love, and impatient with illness, suffering, and death, because they interfered with her happiness.

Fedor Petrovich came in too, in evening dress, his hair curled *a la Capoul*, a tight stiff collar round his long sinewy neck, an enormous white shirt-front and narrow black trousers tightly stretched over his strong thighs. He had one white glove tightly drawn on, and was holding his opera hat in his hand.

Following him the schoolboy crept in unnoticed, in a new uniform, poor little fellow, and wearing gloves. Terribly dark shadows showed under his eyes, the meaning of which Ivan Ilych knew well.

His son had always seemed pathetic to him, and now it was dreadful to see the boy's frightened look of pity. It seemed to Ivan Ilych that Vasya was the only one besides Gerasim who understood and pitied him.

They all sat down and again asked how he was. A silence followed. Lisa asked her mother about the opera glasses, and there was an altercation between mother and daughter as to who had taken them and where they had been put. This occasioned some unpleasantness.

Fedor Petrovich inquired of Ivan Ilych whether he had ever seen Sarah Bernhardt. Ivan Ilych did not at first catch the question, but then replied: "No, have you seen her before?"

"Yes, in *Adrienne Lecouvreur*."

Praskovya Fedorovna mentioned some roles in which Sarah Bernhardt was particularly good. Her daughter disagreed. Conversation sprang up as to the elegance and realism of her acting—the sort of conversation that is always repeated and is always the same.

In the midst of the conversation Fedor Petrovich glanced at Ivan Ilych and became silent. The others also looked at him and grew silent. Ivan Ilych was staring with glittering eyes straight before him, evidently indignant with them. This had to be rectified, but it was impossible to do so. The silence had to be broken, but for a time no one dared to break it and they all became afraid that the conventional deception would suddenly become obvious and the truth become plain to all. Lisa was the first to pluck up courage and break that silence, but by trying to hide what everybody was feeling, she betrayed it.

"Well, if we are going it's time to start," she said, looking at her watch, a present from her father, and with a faint and significant smile at Fedor Petrovich relating to something known only to them. She got up with a rustle of her dress.

They all rose, said good-night, and went away.

When they had gone it seemed to Ivan Ilych that he felt better; the falsity had gone with them. But the pain remained—that same pain and that same fear that made everything monotonously alike, nothing harder and nothing easier. Everything was worse.

Again minute followed minute and hour followed hour. Everything remained the same and there was no cessation. And the inevitable end of it all became more and more terrible.

"Yes, send Gerasim here," he replied to a question Peter asked.

IX

HIS WIFE RETURNED late at night. She came in on tiptoe, but he heard her, opened his eyes, and made haste to close them again. She wished to send Gerasim away and to sit with him herself, but he opened his eyes and said: "No, go away."

"Are you in great pain?"

"Always the same."

"Take some opium."

He agreed and took some. She went away.

Till about three in the morning he was in a state of stupefied misery. It seemed to him that he and his pain were being thrust into a narrow, deep black sack, but though they were pushed further and further in they could not be pushed to the bottom. And this, terrible enough in itself, was accompanied by suffering. He was frightened yet wanted to fall through the sack, he struggled but yet co-operated. And suddenly he broke through, fell, and regained consciousness. Gerasim was sitting at the foot of the bed dozing quietly and patiently, while he himself lay with his emaciated stockinged legs resting on Gerasim's shoulders; the same shaded candle was there and the same unceasing pain.

"Go away, Gerasim," he whispered.

"It's all right, sir. I'll stay a while."

"No. Go away."

He removed his legs from Gerasim's shoulders, turned sideways onto his arm, and felt sorry for himself. He only waited till Gerasim had gone into the next room and then restrained himself no longer but wept like a child. He wept on account of his helplessness, his terrible loneliness, the cruelty of man, the cruelty of God, and the absence of God.

"Why hast Thou done all this? Why hast Thou brought me here? Why, why dost Thou torment me so terribly?"

He did not expect an answer and yet wept because there was no answer and could be none. The pain again grew more acute, but he did not stir and did not call. He said to himself: "Go on! Strike me! But what is it for? What have I done to Thee? What is it for?"

Then he grew quiet and not only ceased weeping but even held his breath and became all attention. It was as though he were listening not to an audible voice but to the voice of his soul, to the current of thoughts arising within him.

"What is it you want?" was the first clear conception capable of expression in words, that he heard.

"What do you want? What do you want?" he repeated to himself.

"What do I want? To live and not to suffer," he answered.

And again he listened with such concentrated attention that even his pain did not distract him.

"To live? How?" asked his inner voice.

"Why, to live as I used to—well and pleasantly."

"As you lived before, well and pleasantly?" the voice repeated.

And in imagination he began to recall the best moments of his pleasant life. But strange to say none of those best moments of his pleasant life now seemed at all what they had then seemed—none of them except the first recollections of childhood. There, in childhood, there had been something really pleasant with which it would be possible to live if it could return. But the child who had experienced that happiness existed no longer, it was like a reminiscence of somebody else.

As soon as the period began which had produced the present Ivan Ilych, all that had then seemed joys now melted before his sight and turned into something trivial and often nasty.

And the further he departed from childhood and the nearer he came to the present the more worthless and doubtful were the joys. This began with the School of Law. A little that was really good was still found there—there was light-heartedness, friendship, and hope. But in the upper classes there had already been fewer of such good moments. Then during the first years of his official career, when he was in the service of the governor, some pleasant moments again occurred: they were the memories of love for a woman. Then all became confused and there was still less of what was good; later on again there was still less that was good, and the further he went the less there was. His marriage, a mere accident, then the disenchantment that followed it, his wife's bad breath and the sensuality and hypocrisy: then that deadly official life and those preoccupations about money, a year of it, and two, and ten, and twenty, and always the same thing. And the longer it lasted the more deadly it became. "It is as if I had been going downhill while I imagined I was going up. And that is really what it was. I was going up in public opinion, but to the same extent life was ebbing away from me. And now it is all done and there is only death.

"Then what does it mean? Why? It can't be that life is so senseless and horrible. But if it really has been so horrible and senseless, why must I die and die in agony? There is something wrong!

"Maybe I did not live as I ought to have done," it suddenly occurred to him. "But how could that be, when I did everything properly?" he replied, and immediately dismissed from his mind this, the sole solution of all the riddles of life and death, as something quite impossible.

"Then what do you want now? To live? Live how? Live as you lived in the law courts when the usher proclaimed 'The judge is coming!' The judge is coming, the judge!" he repeated to himself. "Here he is, the judge. But I am not guilty!" he exclaimed angrily. "What is it for?" And he ceased crying, but turning his face to the wall continued to ponder on the same question: Why, and for what purpose, is there all this horror? But however much he pondered he found no answer. And whenever the thought occurred to him, as it often did, that it all resulted from his not having lived as he ought to have done, he at once recalled the correctness of his whole life and dismissed so strange an idea.

X

ANOTHER FORTNIGHT PASSED. Ivan Ilych now no longer left his sofa. He would not lie in bed but lay on the sofa, facing the wall nearly all the time. He suffered ever the same unceasing agonies and in his loneliness pondered always on the same insoluble question: "What is this? Can it be that it is Death?" And the inner voice answered: "Yes, it is Death."

"Why these sufferings?" And the voice answered, "For no reason—they just are so." Beyond and besides this there was nothing.

From the very beginning of his illness, ever since he had first been to see the doctor, Ivan Ilych's life had been divided between two contrary and alternating moods: now it was despair and the expectation of this uncomprehended and terrible death, and now hope and an intently interested observation of the functioning of his organs. Now before his eyes there was only a kidney or an intestine that temporarily evaded its duty, and now only that incomprehensible and dreadful death from which it was impossible to escape.

These two states of mind had alternated from the very beginning of his illness, but the further it progressed the more doubtful and fantastic became the conception of the kidney, and the more real the sense of impending death.

He had but to call to mind what he had been three months before and what he was now, to call to mind with what regularity he had been going downhill, for every possibility of hope to be shattered.

Latterly during the loneliness in which he found himself as he lay facing the back of the sofa, a loneliness in the midst of a populous town and surrounded by numerous acquaintances and relations but that yet could not have been more complete anywhere—either at the bottom of the sea or under the earth—during that terrible loneliness Ivan Ilych had lived only in memories of the past. Pictures of his past rose before him one after another. They always began with what was nearest in time and then went back to what was most remote—to his childhood—and rested there. If he thought of the stewed prunes that had been offered him that day, his mind went back to the raw shrivelled French plums of his childhood, their peculiar flavour and the flow of saliva when he sucked their stones, and along with the memory of that taste came a whole series of memories of those days: his nurse, his brother, and their toys. "No, I mustn't think of that....It is too painful," Ivan Ilych said to himself, and brought himself back to the present—to the button on the back of the sofa and the creases in its morocco. "Morocco is expensive, but it does not wear well: there had been a quarrel about it. It was a different kind of quarrel and a different kind of morocco that time when we tore father's portfolio and were punished, and mamma brought us some tarts...." And again his thoughts dwelt on his childhood, and again it was painful and he tried to banish them and fix his mind on something else.

Then again together with that chain of memories another series passed through his mind—of how his illness had progressed and grown worse. There also the further back he looked the more life there had been. There had been more of what was good in life and more of life itself. The two merged together. "Just as the pain went on getting worse and worse, so my life grew worse and worse," he thought. "There is one bright spot there at the back, at the beginning of life, and afterwards all becomes blacker and blacker and proceeds more and more rapidly—in inverse ration to the square of the distance from death," thought Ivan Ilych. And the example of a stone falling downwards with increasing velocity entered his mind. Life, a series of increasing sufferings, flies further and further towards its end—the most terrible suffering. "I am flying...." He shuddered, shifted himself, and tried to resist, but was already aware that resistance was impossible, and again with eyes weary of gazing but unable to cease seeing what was before them, he stared at the back of the sofa and waited—awaiting that dreadful fall and shock and destruction.

"Resistance is impossible!" he said to himself. "If I could only understand what it is all for! But that too is impossible. An explanation would be possible if it could be said that I have not lived as I ought to. But it is impossible to say that," and he remembered all the legality, correctitude, and propriety of his life. "That at any rate can certainly not be admitted," he thought, and his lips smiled ironically as if someone could see that smile and be taken in by it. "There is no explanation! Agony, death....What for?"

XI

ANOTHER TWO WEEKS went by in this way and during that fortnight an even occurred that Ivan Ilych and his wife had desired. Petrishchev formally proposed. It happened in the evening. The next day Praskovya Fedorovna came into her husband's room considering how best to inform him of it, but that very night there had been a fresh change for the worse in his condition. She found him still lying on the sofa but in a different position. He lay on his back, groaning and staring fixedly straight in front of him.

She began to remind him of his medicines, but he turned his eyes towards her with such a look that she did not finish what she was saying; so great an animosity, to her in particular, did that look express.

"For Christ's sake let me die in peace!" he said.

She would have gone away, but just then their daughter came in and went up to say good morning. He looked at her as he had done at his wife, and in reply to her inquiry about his health said dryly that he would soon free them all of himself. They were both silent and after sitting with him for a while went away.

"Is it our fault?" Lisa said to her mother. "It's as if we were to blame! I am sorry for papa, but why should we be tortured?"

The doctor came at his usual time. Ivan Ilych answered "Yes" and "No," never taking his angry eyes from him, and at last said: "You know you can do nothing for me, so leave me alone."

"We can ease your sufferings."

"You can't even do that. Let me be."

The doctor went into the drawing room and told Praskovya Fedorovna that the case was very serious and that the only resource left was opium to allay her husband's sufferings, which must be terrible.

It was true, as the doctor said, that Ivan Ilych's physical sufferings were terrible, but worse than the physical sufferings were his mental sufferings which were his chief torture.

His mental sufferings were due to the fact that that night, as he looked at Gerasim's sleepy, good-natured face with it prominent cheek-bones, the question suddenly occurred to him: "What if my whole life has been wrong?"

It occurred to him that what had appeared perfectly impossible before, namely that he had not spent his life as he should have done, might after all be true. It occurred to him that his scarcely perceptible attempts to struggle against what was considered good by the most highly placed people, those scarcely noticeable impulses which he had immediately suppressed, might have been the real thing, and all the rest false. And his professional duties and the whole arrangement of his life and of his family, and all his social and official interests, might all have been false. He tried to defend all those things to himself and suddenly felt the weakness of what he was defending. There was nothing to defend.

"But if that is so," he said to himself, "and I am leaving this life with the consciousness that I have lost all that was given me and it is impossible to rectify it—what then?"

He lay on his back and began to pass his life in review in quite a new way. In the morning when he saw first his footman, then his wife, then his daughter, and then the doctor, their every word and movement confirmed to him the awful truth that had been revealed to him during the night. In them he saw himself—all that for which he had lived—and saw clearly that it was not real at all, but a terrible and huge deception which had hidden both life and death. This consciousness intensified his physical suffering tenfold. He groaned and tossed about, and pulled at his clothing which choked and stifled him. And he hated them on that account.

He was given a large dose of opium and became unconscious, but at noon his sufferings began again. He drove everybody away and tossed from side to side.

His wife came to him and said:

"Jean, my dear, do this for me. It can't do any harm and often helps. Healthy people often do it." He opened his eyes wide.

"What? Take communion? Why? It's unnecessary! However…"

She began to cry.

"Yes, do, my dear. I'll send for our priest. He is such a nice man."

"All right. Very well," he muttered.

When the priest came and heard his confession, Ivan Ilych was softened and seemed to feel a relief from his doubts and consequently from his sufferings, and for a moment there came a ray of hope. He again began to think of the vermiform appendix and the possibility of correcting it. He received the sacrament with tears in his eyes.

When they laid him down again afterwards he felt a moment's ease, and the hope that he might live awoke in him again. He began to think of the operation that had been suggested to him. "To live! I want to live!" he said to himself.

His wife came in to congratulate him after his communion, and when uttering the usual conventional words she added:

"You feel better, don't you?"

Without looking at her he said "Yes."

Her dress, her figure, the expression of her face, the tone of her voice, all revealed the same thing. "This is wrong, it is not as it should be. All you have lived for and still live for is falsehood and deception, hiding life and death from you." And as soon as he admitted that thought, his hatred and his agonizing physical suffering again sprang up, and with that suffering a consciousness of the unavoidable, approaching end. And to this was added a new sensation of grinding shooting pain and a feeling of suffocation.

The expression of his face when he uttered that "Yes" was dreadful. Having uttered it, he looked her straight in the eyes, turned on his face with a rapidity extraordinary in his weak state and shouted:

"Go away! Go away and leave me alone!"

XII

FROM THAT MOMENT the screaming began that continued for three days, and was so terrible that one could not hear it through two closed doors without horror. At the moment he answered his wife realized that he was lost, that there was no return, that the end had come, the very end, and his doubts were still unsolved and remained doubts.

"Oh! Oh! Oh!" he cried in various intonations. He had begun by screaming "I won't!" and continued screaming on the letter "O".

For three whole days, during which time did not exist for him, he struggled in that black sack into which he was being thrust by an invisible, resistless force. He struggled as a man condemned to death struggles in the hands of the executioner, knowing that he cannot save himself. And every moment he felt that despite all his efforts he was drawing nearer and nearer to what terrified him. He felt that his agony was due to his being thrust into that black hole and still more to his not being able to get right into it. He was hindered from getting into it by his conviction that his life had been a good one. That very justification of his life held him fast and prevented his moving forward, and it caused him most torment of all.

Suddenly some force struck him in the chest and side, making it still harder to breathe, and he fell through the hole and there at the bottom was a light. What had happened to him was like the sensation one sometimes experiences in a railway carriage when one thinks one is going backwards while one is really going forwards and suddenly becomes aware of the real direction.

"Yes, it was not the right thing," he said to himself, "but that's no matter. It can be done. But what *is* the right thing?" he asked himself, and suddenly grew quiet.

This occurred at the end of the third day, two hours before his death. Just then his schoolboy son had crept softly in and gone up to the bedside. The dying man was still screaming desperately and waving his arms. His hand fell on the boy's head, and the boy caught it, pressed it to his lips, and began to cry.

At that very moment Ivan Ilych fell through and caught sight of the light, and it was revealed to him that though his life had not been what it should have been, this could still be rectified. He asked himself, "What *is* the right thing?" and grew still, listening. Then he felt that someone was kissing his hand. He opened his eyes, looked at his son, and felt sorry for him. His wife came up to him and he glanced at her. She was gazing at him open-mouthed, with undried tears on her nose and cheek and a despairing look on her face. He felt sorry for her too.

"Yes, I am making them wretched," he thought. "They are sorry, but it will be better for them when I die." He wished to say this but had not the strength to utter it. "Besides, why speak? I must act," he thought. With a look at his wife he indicated his son and said: "Take him away...sorry for him... sorry for you too...." He tried to add, "Forgive me," but said "Forego" and waved his hand, knowing that He whose understanding mattered would understand.

And suddenly it grew clear to him that what had been oppressing him and would not leave him was all dropping away at once from two sides, from ten sides, and from all sides. He was sorry for them, he must act so as not to hurt them: release them and free himself from these sufferings. "How good and how simple!" he thought. "And the pain?" he asked himself. "What has become of it? Where are you, pain?"

He turned his attention to it.

"Yes, here it is. Well, what of it? Let the pain be."

"And death...where is it?"

He sought his former accustomed fear of death and did not find it. "Where is it? What death?" There was no fear because there was no death.

In place of death there was light.

"So that's what it is!" he suddenly exclaimed aloud. "What joy!"

To him all this happened in a single instant, and the meaning of that instant did not change. For those present his agony continued for another two hours. Something rattled in his throat, his emaciated body twitched, then the gasping and rattle became less and less frequent.

"It is finished!" said someone near him.

He heard these words and repeated them in his soul.

"Death is finished," he said to himself. "It is no more!"

He drew in a breath, stopped in the midst of a sigh, stretched out, and died.

— translated by Louise and Aylmer Maude

Reading Tolstoy

One might profitably try to isolate in Tolstoy's novella scenes and moments of "defamiliarlization," where Tolstoy forces his readers to rethink what they already imagine they know. In a similar fashion, one can examine the degree to which *The Death of Ivan Ilych* is a "twice-told story." Finally, one might wonder about the relationship between the personal emotions that the narrative provokes and the less personal satire of middle-class and professional values and behavior Tolstoy seems to present in his narrative as well.

Related Poem

For this chapter, the related poem written by John Donne, "Death Be Not Proud." The form of the poem is a sonnet—it joins sonnets by Shakespeare and Yeats in *Literature and Medicine*, which we described in Chaps. 2 and 6. (Donne's poem takes the form of a Shakespearean rather than a Petrarchan sonnet.) This formal design contrasts powerfully with the seeming "primal cry" of a fear of death. In a way, the formality of artistic form can be imagined to temper, so to speak, the terror and pity of death, just as dying woman, friends, and family in the first vignette of this chapter tempers sorrow with celebration—something that neither the frenzied family of the second vignette nor Ivan Ilych throughout much of his life, can do. This poem is a central focus of the potent play *Wit* by Margaret Edson (1999) that examines the terminal care of Vivian Bearing, an English Professor, dying of ovarian cancer, who spent her career studying Donne's poetry. Like Ivan's work as a judge, Professor Bearing exhibited the same nonempathetic engagement with students that she finds in her physicians. (There is also a first-rate film adaptation of the play.)

Poem: "Death Be Not Proud" (1609) by John Donne

> *Author Note*: John Donne (1572–1631) was an English poet and cleric in the early seventeenth century, a younger contemporary of Shakespeare. He is considered a pre-eminent representative of the "metaphysical poets," whose work focuses on wit and paradox; his poems have been hailed as some of the most important poetry of the English Renaissance. His powerful uses of images, sound, and unusual syntax are hypnotic and captivating. His *Holy Sonnets*—a series of 19 poems, also known as *The Divine Meditations*—were published posthumously in 1633.

Holy Sonnet X: Death Be Not Proud

Death be not proud, though some have callèd thee
Mighty and dreadfull, for, thou art not soe,
For, those, whom thou think'st thou dost overthrow,
Die not, poore death, nor yet canst thou kill mee;
From rest and sleepe, which but they pictures bee,
Much pleasure, then from thee, much more must flow;
And soonest our best men with thee doe goe,
Rest of their bones, and soules deliverie.
Thou'art slave to Fate, chance, kings and desperate men,
And dost with poyson, warre, and sicknesse dwell,
And poppie,' or charmes can make us sleepe as well,
And better than thy stroake; why swell'st thou then?

One short sleepe past, we wake eternally,
And death shall be no more, Death thou shalt die.

Lessons for Providers
1. Learning to identify the "tipping point" of a patient's illness, where all further treatment is futile, is difficult and requires experience. But the provider who learns to do this and turns her attention to caring and palliative measures at the right point in time will help the patient through a frightening time and into a dignified death.
2. As we go through a medical education and begin a professional life, reflecting on how we organize our personal lives will benefit the care we provide our patients.
3. Mindfully attending to our private and professional lives—to do the right things for the right reasons—will avoid our own death experience paralleling Ivan's, whose career as a judge offers parallels to a career in healthcare.

Bibliography

Broyard, Anatole. 1992. Doctor Talk to Me. *New York Times*. https://www.nytimes.com/1990/08/26/magazine/doctor-talk-to-me.html

Edson, Margaret. 1999. *Wit*. New York: Faber and Faber. The film version by Mike Nichols, staring Emma Thompson, is also available.

Schleifer, Ronald, and Jerry Vannatta. 2013. *The Chief Concern of Medicine: The Integration of the Medical Humanities and Narrative Knowledge into Medical Practices*. Ann Arbor: University of Michigan Press.

Williams, William Carlos. 1967. *The Autobiography of William Carlos Williams*. New York: New Directions.

Section IX

Postscript: The Fulfillments of Healthcare

Afterword

15

Throughout this book, we have presented vignettes and literary works that touch upon the difficult—and often heart-rending—aspects of healthcare. But we want to conclude with a postscript that emphasizes the great fulfillments of relieving people faced with great crises, caring for them, and sometimes simply providing them solace. Many of the literary texts of this book—from the first, where Grace Paley and her father find connections to one another even in the face of death, to the last, where Ivan Ilych comes to understand value beyond selfishness—offer images and stories of the goodness of life even faced with the problems and limitations of pain, suffering, and mortality. Anton Chekhov offers us sympathy and goodwill, as he often does in his stories; William Carlos Williams underlines how much depends on our engagements with the world; and William Blake extols the virtues of honest friendship, even as he narrates the creation of enmity. For this reason we ended our book, *The Chief Concern of Medicine*, by describing how the connection between literature and medicine

> offers us all, patients and physicians alike, a richer sense of the vocation of healthcare and also a richer sense of our shared human lives. Many of the writers, physicians, scholars, and even patients we have encountered in *The Chief Concern of Medicine* telling stories and analyzing experience have suggested, as we noted in the beginning, that to be a healthcare worker is an especially privileged position in our and in any society. Like many other professions, healthcare has the potential for great social, intellectual, and – as we mentioned in the Introduction – spiritual rewards in its engagements with the great crises of health and illness, well being and suffering, life and death that face all people. That is, in its encounters with ailing human beings, the profession of medicine also entails interactions with people that touch upon the vital centers of human life in general…. [T]here are few professions that call upon the intimacies, the emotions, the potentiality of honest and heartfelt interchange that characterize the best part of our private lives as does a profession in medicine. Medicine and doctoring are built around this human relationship between patient and physician; they are grounded in storytelling, good listening, and the sense – which can always be improved and shared – of how stories work; and because they touch on the great crises of our shared lives, they are always, in their smallest gestures as well as largest decisions, a profoundly ethical enterprise. But saying these things is a way of saying that the privilege of doctoring simply underscores the fact that the work of medicine in the face of suffering and also in its

© The Author(s) 2019
R. Schleifer, J. B. Vannatta, *Literature and Medicine*,
https://doi.org/10.1007/978-3-030-19128-3_15

restorations of health is, more generally, something we all share in as we bring and encounter care and care-taking in our lives. To make us mindful of this – again as both patients and physicians – is an important aspect of the significance of narrative in medical practices. (*The Chief Concern*: 355–56)

Related Poem

In order to spell out the fulfillments of the work of healthcare, we conclude this volume with a poem about the "good news" of medicine, the consolation of reassurance which is also a significant part of the work of healthcare. For this postscript, the related poem, written by Derek Mahon, "Everything Is Going To Be All Right," speaks outright the good news that healthcare providers often bring to their patients. Talking about this poem, Dr. John Stone notes that

> one of the great functions of the physician is to say those few words ["everything is going to be all right"] to the patient from time to time. When I first came to Atlanta, I was assigned to a congenital heart disease clinic,…and I would go and see these little kids [and] I would hear, as the stethoscope went to the left lower sternal border, … a groaning, musical, short systolic murmur like "mmm, mmm." This is an innocent murmur, and … one of the great things was to be able to say to the mother, usually, "Everything is going to be all right." Hearing the good news is at least as important as hearing the bad news. (Vannatta et al. 2005: Chap. 4, p. 50, hypertext [video] 3)

Poem: "Everything Is Going to Be All Right" (1979) by Derek Mahon

> *Author Note*: Derek Mahon (b. 1941) is an Irish poet, playwright, and translator. His award-winning books include *Harbour Lights* (2006), *Somewhere the Wave* (2007), and *Life on Earth* (2008), all of which won the *Irish Times* Poetry Now Award.

Everything Is Going To Be All Right

> How should I not be glad to contemplate
> the clouds clearing beyond the dormer window
> and a high tide reflected on the ceiling?
> There will be dying, there will be dying,
> but there is no need to go into that.
> The poems flow from the hand unbidden
> and the hidden source is the watchful heart.
> The sun rises in spite of everything
> and the far cities are beautiful and bright.
> I lie here in a riot of sunlight
> watching the day break and the clouds flying.
> Everything is going to be all right.

Lessons for Providers
1. Embrace the opportunities to inform the patient that they are "all right." A positive response to the patient's "chief concern" is as healing for providers as it is for the patients themselves.
2. When you deliver good news to a patient, be sure to empathize with their joy. Make sure they know how joyful you feel that they are going to be all right. Empathy for patient's joy is as important as empathy for their suffering.

Bibliography

Schleifer, Ronald, and Jerry Vannatta. 2013. *The Chief Concern of Medicine: The Integration of the Medical Humanities and Narrative Knowledge into Medical Practices*. Ann Arbor: University of Michigan Press.

Vannatta, Jerry, Ronald Schleifer, and Sheila Crow. 2005. *Medicine and Humanistic Understanding: The Significance of Narrative in Medical Practices*. Philadelphia: University of Pennsylvania Press. A DVD-Rom publication.

Appendix 1: Experimental Results: The Cognitive Science of Literary Reading

As we mention in Chap. 1, in the last twenty-five years, there have been many studies, pursued with scientific rigor, that demonstrate the fact that reading literary fiction effects changes in people that increase particular forms of cognition and fellow-feeling. How these effects are brought about has until recently been analyzed and discussed, usually in literary studies, but rarely by means of scientific protocols. Over the past twenty-five years, however, data has been published, derived from rigorous testing, which shed light on this phenomenon and indeed allowed us to discern a causal relationship between reading fiction and increased empathy and vigorous enactments of Theory of Mind (ToM). Some of these studies have shown how—by means of the analysis of literary *features*—literary fiction goes about effecting these changes. But besides these "interpretative" analyses, many other studies have demonstrated by means of empirical, quantitative and qualitative research techniques that reading literary fiction leads to enhanced ToM, measurable transportation states ("vicarious experiences"), and increased empathy. As we suggested in Part I of the Introduction, these phenomena are desirable for people in the healthcare field. The exciting development of this data has come about through a confluence of studies involving a number of disciplines including cognitive and social psychology, narratology (including stylistics and linguistics), neuro-imaging, and, as we note throughout this book, literary semiotics and medical pedagogy. ("Semiotics" is the systematic study of "signs" and the manner in which meaning is generated. It grew out of linguistics and logic in the early twentieth century. For a historical and cultural account of why this is so, see Schleifer 2018a). In this appendix, we set forth a short summary of this research by focusing on a small number of these studies that are representative of the wider empirical work of the last two decades. Taken as a whole, this work in cognitive and affective science describes how reading literary fiction (variously defined as "writerly," "polyphonic," and "stylistically sophisticated discourse"), creates in the reader the desirable effects of enhanced empathy, more rigorous ToM, and the vicarious experiences that Transportation Theory analyzes.

In 1994, D. S. Miall, an English professor, and Don Kuiken, a psychology professor, both from Alberta, Canada, demonstrated that "foregrounding" is systematically correlated with increased reading times and changes in affect (emotional response), and it is also correlated with readers' judgment of "strikingness" in a series of experiments utilizing student-readers. The term "foregrounding" was coined in the 1930s by Jan Mukařovský, a semiotician and member of the Prague School of Linguistics. By "foregrounding" he means "the range of stylistic variations that occur in literature, whether at the phonetic level (e.g., alliteration, rhyme), the grammatical level (e.g., inversion, ellipsis), or the semantic level (e.g., metaphor, irony)" (Miall and Kuiken 1994: 390). In Part II of Chap. 1, we describe and examine thirteen of these "features" in relation to Grace Paley's short story, "A Conversation with My Father." These features, we suggest, are useful in developing the careful engagement with the literary narratives and poems of our text-anthology—and, importantly, in the careful engagement with patients' stories in the clinic. While these features can occur in all language uses, Miall and Kuiken argue (following Mukařovský) that they are systematically present in literary texts: foregrounding, they argue, "enables literature to present meaning with an intricacy and complexity that ordinary language does not normally allow" (1994: 390). One such measure—using a term we discuss in Part II of Chap. 1—is the ability of literature to "defamiliarize" experience and make it new. "Defamiliarization," as we mention in Chap. 1, is a term developed by Russian scholars in the early twentieth century to allow for the systematic study of the ways that discursive art—literary narrative—provokes effects and responses in readers/listeners. Miall and Kuiken measured the effects of foregrounding in four formal studies of readers that measure the "strikingness" of literature (i.e., the attention it arrests by means of defamiliarization), the provocation of feeling (affect), and the ways that foregrounding increases reading time.

Much work has been done on the emotive responses to reading literary fiction (Appendix 5 refers to some of this material). In 2002, Miall and Kuiken published an innovative research paper that showed that readers of literary fiction were moved emotionally by certain passages, and when they reflected on that emotion they discovered that the passages and attendant emotion had stimulated reflections in their real world lives or in other texts. (This result corresponds with the features describing the borderline between everyday life and literary fiction in Part II of Chap. 1.) Furthermore, they found that the reflections stimulated "boundary crossing." Specifically, they demonstrate that "the experience of feelings in one situation leads to the re-experiencing of those feelings in situations that are similar" (2002: 226). This phenomenon, as we note in the vignette in Chap. 4, precisely occurred in Dr. Vannatta's practice when he re-experienced feelings provoked by Toni Morrison's novel *Beloved* in his interchange with a patient. Miall and Kuiken test this with the systematic study of students reading Sean O'Faolain's short story "The Trout." In a review article, in 2011 Raymond Mar et al. reviewed the literature on emotion and narrative fiction in which he and his colleagues examine in fine detail *empirical* studies that demonstrate the evocation and transformation of readers' emotions, how these emotions affect readers' experiences of narrative, and, finally, the consequences of these experiences in readers' subsequent lives well after closing the book.

Much of the work on how literary narrative does its work to enhance empathy focuses on Transportation Theory. These studies demonstrate that literary fiction is more effective in producing its cognitive and affective responses when the reader is "transported" into the story. This transportation is an integrative melding of attention, imagery, and feelings, such as Miall and Kuiken describe under the category of "foregrounding" and we describe in Part II of Chap. 1 in relation to "features" of literary narrative. In 2010, Melanie Green used a well-validated measure of transportation to demonstrate that being transported into a story was correlated with perceived realism, and perceived realism mediates the effect of transportation on beliefs in another version of "boundary crossing" from the literary fiction to everyday life. She also demonstrated that emotional changes, discussed in the studies mentioned above and many additional studies, are correlated with the degree of transportation.

In the last decade, studies have appeared that looked at reading literary fiction and its effect on Theory of Mind (ToM). In 2013, David Kidd and Emanuele Castano reported in *Science* a randomized control trial of the effects on ToM of reading fiction vs. non-fiction. They found that literary fiction was statistically more effective at increasing performance on advanced ToM tests. They also found a difference in ToM testing when comparing literary fiction with popular fiction. Similarly, Raymond Mar et al. showed that exposure to literary fiction predicts performance on an empathy task, controlling for age, gender, English fluency, personality trait openness, and transportation. Finally, in this short survey, in their 2013 article "How Does Fiction Reading Influence Empathy? An Experimental Investigation on the role of Emotional Transportation," Matthijs Bal and Martijn Veltkamp reported in two experiments that the emotional effects of reading fiction vs. non-fiction increase over time. They made measurements immediately following reading and again after one week. They also found that when the readers were transported into the story, they performed higher on empathy tests than when they had been exposed to newspaper articles.

Appendix 2: Discussion Questions for the Chapters

In this appendix, we offer a number of general questions for class discussion appropriate for all (or most) of the chapters of *Literature and Medicine* and then present specific questions for each chapter. It is our hope that these questions will stimulate instructors and readers to devise their own questions that will stimulate reflection on the usefulness of narrative to medical practice.

General Questions

These questions can be addressed to the literary narratives and, if necessary, to the poems in almost every chapter of *Literature and Medicine*. By and large, they are linked to Chap. 1. In presenting these questions, we will give examples from James Joyce's story, "Araby" in Chap. 5, "Listening to Patients."

1. In the literary narrative and/or poem you have read, what phrase/sentence/paragraph generated the most emotion?
    In "Araby" readers are usually struck when the young-boy narrator says: "I imagined that I bore my chalice safely through a throng of foes. Her name sprang to my lips at moments in strange prayers and praises which I myself did not understand. My eyes were often full of tears (I could not tell why) and at times a flood from my heart seemed to pour itself out into my bosom. I thought little of the future. I did not know whether I would ever speak to her or not or, if I spoke to her, how I could tell her of my confused adoration. But my body was like a harp and her words and gestures were like fingers running upon the wires."
2. In the literary narrative and/or poem you have read, what phrase/sentence/paragraph generated the most reflection?
    In "Araby" readers usually spend time pondering and discussing the fact of the boy's "anguish and anger" at the end of the story.
3. In the literary narrative and/or poem you have read, what phrase/sentence/paragraph is the most striking? (The term "strikingness" is used by Miall and Kuiken in their discussion of the "foregrounding" strategies in literary fiction that demonstrate the effectiveness of literary fiction to promote empathy and "transportation.")

© The Author(s) 2019
R. Schleifer, J. B. Vannatta, *Literature and Medicine*,
https://doi.org/10.1007/978-3-030-19128-3

In "Araby" readers are struck by the seeming obsessiveness of the boy-narrator's feelings for Mangan's sister. (This suggests that "strikingness" is an overall narrative feature and overlaps with the first two questions.)

4. In the literary narrative and/or poem you have read, what phrase/sentence/paragraph generated the most incomprehension? That is, what is the greatest anomaly the text presents? How does the author/text create this anomaly?

    In "Araby" perhaps the greatest anomaly is when the boy-narrator says, "I could interpret these signs," when his uncle comes home, but then fails to narrate his interpretation.

5. In the literary narrative you have read, describe the way the action of ignoring symptoms (or other phenomena) by the healthcare provider (or by some other character in narratives without healthcare professionals) is self-serving. That is, what particular ends are served by the act of ignoring evidence. (Quite often, the act of ignoring information is "unsaid" insofar as the narrative does not explicitly note it is a self-serving act.) This "negative" action by physicians is particularly pronounced in Leo Tolstoy's *The Death of Ivan Ilych*.

    In "Araby," the boy-narrator seems to be having experiences—of "love," infatuation, desire—that he hasn't experienced before, but instead of questioning those experiences, he narrates them as "confused adoration." (In this, Joyce's narrator is like many patients, who present bodily symptom which they can only explain by falling back to "familiar" vocabularies.)

6. In the literary narrative and/or poem you have read, consult the catalogue of narrative elements in the article reproduced in Appendix 5 (p. 266 below) and describe the possible *narrative roles* of the characters portrayed.

    In "Araby," the boy-narrator conceives of himself as the "hero," his uncle as the hero's "helper," and the object of desire (sometimes the "heroine") as Mangan's sister. But at the end of the story, the hero-narrator confesses the "vanity" of his heroic role. Thus, it is also possible to conceive of the boy-narrator as "ironic" rather than heroic.

7. In the literary narrative and/or poem you have read, consult Appendix 5 (p. 267 below) and describe the possible *narrative genres* of the characters portrayed.

    The catalogue of narrative genres in the article reproduced in Appendix 5 suggests that one way of understanding the genre of a narrative is to discover who is left with the sought-for good at the end of a narrative. Thus, when the "hero" gains the sought-for good, we have a melodrama; when the "helper" does we have a tragedy; when the heroine does we have comedy. Finally, when the sought-for good is gained by the opponent, we have an ironic narrative. In "Araby," the clerks at the bazaar—who have English accents in Ireland ruled by British colonial power—who flirt with one another, seem to possess the sought-for good that the boy-narrator seeks.

8. Describe the witness who learns in relation to this narrative as a whole. Note: the witness can be a character in the narrative or the reader or even the teller. What is learned?

    In "Araby" the witness who learns is the young boy who experiences disillusionment at the end of the story. What he learns is complicated: the "anguish and anger" he feels might seem to adults reading this story itself a childish response to commonplace experience. But the story as a whole (the author's/text's "overall meaning") might also suggest that as a grown person the boy-narrator is haunted by this story. If readers are haunted as well, it is because they are "transported" into the situation Joyce narrates.

9. Describe the "overall meaning" of the narrative or poem in this chapter.
   The analysis of the conclusion of "Araby" discussed in the preceding question suggests the story's "overall meaning."
10. In what ways does the narrative/poem we have read shed light on healthcare?
    In "Araby," the narrative provokes examination of the private "cultural" vocabularies of people and patients. The boy-narrator is having an experience he never had before, and the only vocabulary he has to describe it is not medical—neither he nor the author conceives of his "puppy love" as a function of puberty, hormones, or other medically-related terms—but rather the language of the Catholic Church that permeates his life and experience.

Chapter Questions

The following questions are designed for discussion and reflection for each of the chapters of *Literature and Medicine*.

Chapter 2: "The Narrative Structure of Diagnosis." The major literary narrative in this chapter is Dr. Arthur Conan Doyle's Sherlock Holmes story, "The Resident Patient." Since this chapter examines the *process* of abduction, most of the discussion questions focus on interpretative processes rather than understanding of elements of the narrative. (See also Appendix 4 for teaching aids for this chapter.)

1. How are the police and the detective, Sherlock Holmes, like doctors?
2. Why is the crime situation narrated twice, once in more or less fragmented form by Dr. Trevelyan and once by Sherlock Holmes?
3. Why does Doyle begin by referring to Edgar Allan Poe?
4. Using the narrative as an example, discuss how abduction differs from induction.
5. Doyle has Holmes describes his method as "deduction." Is this correct? Why does he do so?
6. In describing abduction, Charles Sanders Peirce notes that in encountering a "surprising fact," one can see that it is a matter of course in the context of a particular hypothesis. What is the "surprising fact" in "The Resident Patient"?
7. Is the police failure to take into account all the evidentiary facts an instance of self-serving ignorance? What ends are served?

In addition to Doyle's story, Chap. 2 presents William Shakespeare's sonnet, "That Time of Year," with its lines mixed out of order. One can ask students (or oneself) to re-arrange the lines in correct order. Such an exercise should focus on the language (e.g., the rhymes), structure (e.g., the Shakespearean sonnet form), and discursive logic (e.g., the semantic meaning of the poem).

1. Are there any parallels between the focus on phonic "facts," cultural conventions, and the logic of meaning between this exercise and the abductive work of Sherlock Holmes in "The Resident Patient"?

Chapter 3: "Literature and Professionalism in Medicine." The major literary narrative in this chapter is Richard Selzer's "Imelda." The following questions are related to that story and to the poem by Dr. Audrey Shafer at the end of the chapter.

1. How do healthcare providers decide the definition of "truth-telling"? and does this mean a healthcare provider should never withhold information?
2. Why does Dr. Fransiscus perform the operation on a corpse?
3. How does Dr. Audrey Shafer's poem, "Monday Morning," speak to the relationship between personal/family life and professional life? What do you make of the fact that her son is naked when she leaves her home and her patient "arrives/ Naked under hospital issue/Ready to sleep"?

Chapter 4: "Rapport and Empathy in Medicine." The literary piece in this chapter is Dr. Anton Chekhov's "The Doctor's Visit." The questions are related to that story.

1. What does the factory represent to Korolyov?
2. Why does Korolyov "connect" with Lisa on his second encounter with her?
3. What role does the "roundabout" conversation play in Korolyov's interaction with Lisa?
4. What is the insight Korolyov has that leads to engagement on the second visit as opposed to the detachment on the first?

Chapter 5: "Listening to Patients." The major literary narrative in this chapter is James Joyce's short story, "Araby." The following are questions related to that story.

1. How old is the boy in the story?
2. Why does he articulate "strange prayers and praises which I myself did not understand"?
3. Why does his uncle come home late?
4. Discuss the implicit dichotomy between experience and understanding. How might this dichotomy affect the patient–provider interaction?
5. Discuss the end of the story. What happens at the bazaar to provoke his final reaction? What, if anything, does he learn from this experience?

In addition to Joyce's story, Chap. 5 presents a poem by Dr. William Carlos Williams, "The Red Wheelbarrow."

1. What do you make of the opening line of the poem?
2. Is this poem presenting an implicit narrative or simply describing a scene? Does any "action" take place?
3. Why would anyone speak this sentence?

Appendix 2: Discussion Questions for the Chapters   251

Chapter 6: "The Patient." The literary narrative in this chapter is "The Yellow Wallpaper." The vignette is by Audre Lorde, and the poem is Dr. Rafael Campo's "The Couple." The questions refer to those three narrative/literary pieces.

1. In the vignette by Lorde, she states "As women we are raised to fear." What does Lorde mean by that?
2. In "The Yellow Wallpaper," what is the author hoping we will understand about the patient's husband/doctor?
3. How would you describe the patient's relationship with the husband/doctor in this story?
4. In Dr. Campo's poem "The Couple," what is the "awful light" mentioned in the last line?

Chapter 7: "The Doctor." The literary narrative in this chapter is "The Lynching of Jube Benson" by Paul Laurence Dunbar. The vignette is from Dr. Damon Tweedy's *Black Man in a White Coat*, and the poem is the Slave Spiritual "Sometimes I Feel like a Motherless Child." The questions are related to all three of these pieces.

1. What is meant by "unconscious" or "implicit" bias?
2. How can we bring unconscious bias to consciousness?
3. In Dr. Tweedy's vignette, why does his doctor brush him off early in the story?
4. What was your primary emotion as you read "The Lynching of Jube Benson"?
5. How does that primary emotion work to help you remember the story?
6. In what ways, if any, does the Slave Spiritual in this chapter relate to its theme of the role of the doctor?

Chapter 8: "Everyday Ethics of Medical Practices." The literary narrative in this chapter is Dr. Anton Chekhov's "Enemies." The questions will relate to that story and its discussion.

1. How do virtue ethics differ from normative ethics?
2. How should healthcare providers develop the virtues mentioned in this chapter?
3. To what degree are the enmities of both Kirilov and Abogin examples of arrogance?
4. Contrast the qualities of empathy and arrogance.
5. What is the origin of arrogance in healthcare providers?

Chapter 9: "Culture." The literary narrative in this chapter is "The Annunciation: Lupe" by Demetria Martinez; the poem is "Making Tortillas" by Alicia Gaspar de Alba. The questions relate to these literary narratives.

1. What is the effect of telling the story in the second person in "The Annunciation"?
2. Why does the narrator talk to her unborn child?
3. Why does she talk to her neighbor?

4. Name some of the different assumptions held by the narrator and the larger American society in which she lives.
5. Is the narrator's pregnancy a medical condition?
6. Why should there be a parallel between making tortillas and love making in "Making Tortillas"?
7. What does the poem suggest about the "hum" of culture mentioned in the chapter?

Chapter 10: "Sexual and Domestic Abuse." This chapter focuses on domestic abuse and violence more generally in its vignette, literary narrative (Edgar Allan Poe's "Berenice"), and poem (William Butler Yeats's "Leda and the Swan"). In so doing, it also discusses Roddy Doyle's full-length novelistic representation of domestic abuse, *The Woman Who Walked into Doors*.

1. In Roddy Doyle's *The Woman Who Walked into Doors*, physicians repeatedly overlook and dismiss possible causes of Paula's injuries. How does this affect Dr. Vannatta's engagement with his patient?
2. What is Dr. Vannatta's patient's "chief concern"?
3. What is the "chief concern" of Egaeus, the narrator of "Berenice"?
4. In what way might we see that Poe's story is "twice-told"?
5. Discuss what emotions Poe's text provokes and the ways the narration seems to provoke them.
6. Does Yeats's poem "romanticize" violence? (What might the phrase "romanticize violence" mean?)
7. Is the violence of the poem as "graphic" as that of the Poe story? If not, what aspects of the different languages these literary works use might account for the difference? If it is as graphic, are the similarities produced by similar language uses?
8. Why does Egaeus narrate his story? Why would anyone speak Yeats's poem?

Chapter 11: "Pain." The literary narrative in this chapter is the chapter, "The Operation," from Herman Melville's novel *White Jacket*; the chapter also presents Lous Heshusius's harrowing experience of chronic pain and Emily Dickinson's meditation on the nature of pain.

1. Why does Melville give the physicians and doctors such strange names?
2. How does someone learn to "honor" the patient's story of fear and pain since it cannot be corroborated?
3. What can a healthcare provider do to ensure the patient with chronic pain feels "heard" and "understood"?
4. What is missing—which virtues—in Dr. Cuticle's behavior as he interacts with his patient in "The Operation"?
5. Why is the acknowledgment of her pain by others so important to Lous Heshusius?
6. What does Dickinson mean by "element of blank"?

Appendix 2: Discussion Questions for the Chapters

Chapter 12: "Ageing." The literary narrative in this chapter is the second chapter of Nathaniel Hawthorne's novel, *The House of Seven Gables* entitled "The Little Shop Window." These questions are related to this story as well as to the vignette, "Treating a Very Old Woman." The related poem is Thomas Hardy's "I Look into My Glass."

1. What virtue could the provider habituate in the vignette "Treating a Very Old Woman" to improve his care of future patients?
2. What is the "overall meaning" of Hawthorne's chapter "The Little Shop Window"?
3. What can we learn about human hope and hopelessness from Hawthorne's chapter?
4. What must the healthcare provider be conscious of when caring for the elderly that is not so important with young patients?
5. What is the greatest difficulty of aging in Hardy's poem?

Chapter 13: "Mistakes in Medicine." The literary narrative in this chapter is Gustav Flaubert's representation of an operation in *Madame Bovary*. The vignette describes a horrible mistake in medicine in relation to Dr. David Hilfiker's systematic analyses of medical mistakes in "Facing Our Mistakes," a widely available text. Questions will relate to the narrative vignette and short story as well as Dr. Dannie Abse's related poem, "In the Theatre."

1. How should mistakes in health care be dealt with?
2. What is the connection between mistakes in medicine and the virtue of truth-telling?
3. What was the "main" mistake in Dr. Vannatta's narrative?
4. What was the "main" mistake in the operation Flaubert describes.
5. What is the relationship between Dr. Bovary's practice and Madame Bovary's adulterous love affair?
6. Does Dr. Abse's poem describe a failure of skill or a failure of knowledge? Should a failure based upon lack of technology be considered a "mistake"? Does this poem suggest another "kind" of mistake?
7. What is the emotion provoked by each of these texts, the vignette, fictional narrative, and poem? Are there any connections among the emotions provoked by each of these texts?

Chapter 14: "Death and Dying." The literary narrative in this chapter is *The Death of Ivan Ilych* by Leo Tolstoy. The questions will relate to that story.

1. What is the big lie that tormented Ivan?
2. Why did Ivan's family fail to connect with his suffering?
3. What was Ivan's epiphany just before death?
4. What was the parallel Tolstoy made between Ivan's life as a judge and Ivan's doctors?
5. Describe the quality of the differences in facing death in the two vignettes.
6. Why does the speaker in John Donne's poem address "death" directly?

Postscript: "The Fulfillment of Healthcare." The literary text in the postscript is Derek Mahon's poem "Everything is Going to be All Right."

1. What do we make of the poem's assertion "there will be dying"?
2. What does it mean by "the watchful heart"?

Appendix 3: Daily Writing Assignment

[Here is a sample description of daily writings for a course on "Literature and Medicine." Needless to say, it is a sample, which could be modified to "weekly writing assignments" or "occasional writing assignments."]

There will be a one-page daily writing assignment (single space, up to about 500 words) that should be prepared for each class meeting. We will talk about the daily writing the first day of class. The assignments must focus on literary texts: a story, excerpt, or poem, *not on a vignette*. Each assignment should have a thesis statement underlined in the paper. The daily writing will be collected during each class meeting and returned in the subsequent meeting. Grades on papers are based on several factors. The most important requirement for the daily writings is a clear and *arguable* thesis based upon an aspect of the day's reading set forth in the daily writing topics. Failure to present such a thesis, while engaging with our readings, will earn a minimum grade. A *thesis* by definition is something that can be argued against: "'The Yellow Wallpaper' describes progressive psychosis" is not a strong thesis because it is almost impossible to disagree with it. (The opposite of a thesis-driven essay is a descriptive essay, which this weak thesis enacts.) "Richard Selzer's narrator in 'Imelda' uses three narrative techniques to represent ambivalence" is a strong thesis. The more *specific* the focus/argument is, the more likely that the paper presents a strong thesis and argument.

The following 15 topics could be the focus of the daily writing. As we mentioned, these writings should present a thesis associated with the topics. (The exception to this rule is the "parody" assignment, where a thesis—the student-writer's claim for the "most characteristic" feature of the author being parodied—will necessarily remain implicit in the turned-in assignment). Here are some general rules to be followed.

- Each daily writing assignment must explicitly name the topic examined (e.g., "Professionalism"; "empathy").
- The odd topic of "parody" must be one of the course's assignment (even while students can choose other topics and thereby in short courses leave out some topics).
- In the topics, you will see terms such as "how it works," "importance," "power and meaning." These refer to the ways narrative fiction creates meaning and

© The Author(s) 2019
R. Schleifer, J. B. Vannatta, *Literature and Medicine*,
https://doi.org/10.1007/978-3-030-19128-3

exerts rhetorical power (e.g., changing someone's mind, making you notice something not noticed before, focusing attention and expectation, suggesting a moral judgment, etc.). Arguments can, and often should, be organized in relation to these larger concerns.
- In pursuing these topics, you may look for the assumptions, values, perspectives, overall meaning, beliefs, ideas, and fantasies, both explicit and implicit, that play a role in the narrative. Attention to such (often "unsaid") aspects of texts is the work of critical thinking and critical writing.
- Finally, on days when there is more than one author (e.g., a story and a poem), you can decide to focus on one text or examine the topic in relation to more than one author.

Daily Writing Topics

Here is the list of the 15 topics. Although they are somewhat ordered here, you can choose any one for any daily reading insofar as no topic is repeated.

- **Title**. Argue for a particular way the title of a narrative functions. You can decide the particular function and describe how it *works* in the narrative.
- **Beginnings**. Argue for a particular way the opening of a narrative functions. You can decide the particular function and describe how it *works* in the narrative.
- **Endings**. Argue for a particular way the conclusion of a narrative functions. You can decide the particular function and describe how it *works* in the narrative.
- **Detail**. Choose some small detail in the text—for instance, the syllogism in Leo Tolstoy's *The Death of Ivan Ilych*—and argue how it *works* in the narrative. (When the novelist Vladimir Nabokov taught fiction at Cornell University *to graduate students in literature*, he would give them quizzes about small details, such as the color of one character's shoes in Marcel Proust's enormously long novel, *Remembrance of Things Past*. Students weren't particularly happy with these quizzes affecting their grades, but they got into the habit of paying attention to minute details.)
- **Action**. Choose a particular action that takes place in a text—for instance, Ivan's behavior as a judge in *Ivan Ilych*—and argue for its importance to the narrative as a whole.
- **Idea**. Choose a particular idea that is presented in a text—for instance, the idea of loneliness in *Ivan Ilych*—and argue for its importance to the narrative as a whole. (You can, but do not have to, equate "idea" with "theme.")
- **Language**. Choose a notable use of language in a text—a phrase, metaphor, colloquialism, the tense, or simply a well- or ill-formed paragraph—and argue for the ways it is important within that text. One example is to argue for the particular ways a text such as Thomas Hardy's poem creates a sense of music out of language by means of manipulations of words/sounds.

- **Represented Emotion**. Argue that a particular emotion *felt by a character* that is represented in a narrative, and argue for the particular way the author creates that representation. (Notice this is closely related to the preceding topic: one example is to argue for the particular ways a text creates a sense of anger.)
- **Provoked Emotion**. Argue that a particular emotion is provoked *in the reader* of a narrative, and argue for the particular way the author creates that emotion in the reader.
- **Repetition/Pattern**. Describe the presentations of repetitions of patterns in a text—of action, locutions, details, etc.—and argue how this pattern/repetition contributes to the power and meaning of the narrative.
- **Roles**. Argue that a particular *narrative role* is assumed by a character or an object in a narrative. (Note that the catalogue of narrative elements in Appendix 5 (p. 266) presents, among other things, a small number of narrative roles that some believe inhabit all narrative storytelling. You may disagree with the nature of these roles and argue that a particular narrative exhibits others or you may agree with it, but in any case this should present a model for discussing narrative roles.)
- **Genre**. Argue that a particular *narrative genre* that characterizes a particularly literary text. (Note that Appendix 5 (p. 267) presents, among other things, a small number of narrative genres that some believe organize all narrative storytelling. You may disagree with the nature of these genres and argue that a particular narrative exhibits others or you may agree with it, but in any case this should present a model for discussing narrative roles.)
- **History/Politics**. Argue for the importance to a narrative of the particular moment in history in which it takes place.
- **Voice**. Argue for the distinctness of a particular "voice" in a narrative—the characteristic modes of speech in a character or the narrator—and its importance within the narrative.
- **Parody**. Write a parody of the author (or one of the authors) in the daily assignment.

Appendix 4: Guide for Discussing Diagnosis and Diagnosis Errors

A class session on "The Logic of Diagnosis" is presented with four primary goals in mind.

"Goals of Session"
1) To introduce beginning students to the logic of making a diagnosis.
2) To introduce the logical method of ***Abduction*** (or "inference to the best explanation") as described in Chap. 2.
3) To demonstrate how physicians can "***miss the diagnosis***" by making *systematic* errors in method.
4) To demonstrate the usefulness and fun of the use of literary works, in this case detective stories, in the learning and practice of medicine.

The Method of Making a Diagnosis

The diagnostic method always begins with a detailed history and physical examination. Most experts state that the most important diagnostic information the physician will get is the History of Present Illness (HPI). The HPI is the patient's narrative, a story. This story must be told by the patient, fully and artfully facilitated by the healthcare provider; carefully apprehended by the caretaker (listening carefully to the said and the unsaid); studied acutely for the body language and emotive content. The past medical history, review of systems, etc. is acquired to make sure the "story" is as complete as possible.

A physical examination is then performed to complete the act of reading the "text" of the patient. Once this data base is completed, and possibly some laboratory or imaging done, the physician uses this information to make his or her differential diagnosis and or best guess diagnosis. The diagnosis is always a ***best guess***—an educated guess. It is a hypothesis; in the language of C. S. Pierce, an abduction. (Note these detailed procedures are present in the excerpt from Dr. Damon Tweedy's *Black Man in a White Coat* in Chap. 7.)

Physicians commonly think they are using "induction" when making a diagnosis. Dr. Arthur Conan Doyle, writing in the late nineteenth century, called Sherlock Holmes's method "deduction." Both are incorrect, since the detective of Doyle and the present-day physician both actually use a logical process more closely resembling "abduction." Abduction is a logical process, sometimes called "hypothesis generation" or "inference to the best explanation." The following comparison will be used.

Induction	Abduction
Classifies objects or facts	Begins with characteristics or qualities; these run in categories
Observation of facts only	Imagining what might be
Tests a Hypothesis	Forms a Hypothesis

Chart 1

Another way of thinking about abduction is the following:

> A surprising fact C is discovered, such that, if A were true, then C would be a matter of course. (C. S. Peirce)

This is a formal way of stating a common piece of diagnostic wisdom handed down through generations. It is not clear where it originates, but can be stated another way as well:

> **Look for the unusual, or the piece of evidence that doesn't fit. Follow that piece of evidence and you will most commonly make the correct diagnosis**

Chart 2

The following charts use texts from Chap. 2 of *Literature and Medicine* to clarify this systematic procedure.

Woman with Hyponatremia

Induction (performed by the Resident)	Abduction (performed by C. G. Gunn, M.D.)
Classifies objects of facts: 1. patient is hyponatremic 2. denies all possible causes 3. physical exam unremarkable except pos tilt	**Begins with Characteristics** (qualities) that run in categories 1. Quality of resident's report probably not accurate or complete. (He investigated all known causes of hypo Na and found none) 2. Quality and characteristics of patient's story as taken by Dr. Gunn—pt's demeanor, her glancing at the bedside table. The **"unsaid"**, the **overall meaning** of the story
Observation of Facts only 1. Patient denies 1, 2, 3, etc. 2. Physical exam shows ...	**Imagining what might be true** 1. What if the glance at the bedside table is meaningful? What if the meaningfulness is that her purse is in there? What if she is lying to us?
Tests a Hypothesis 1. She has hyponatremia, all known causes are not present so the case is **abstruse** or this is a **brand new disease**!!!	**Form a Hypothesis** The patient has chlorthalidone in her purse which is in the night stand. The diuretic made her hyponatremic. *We do not know why she is lying. But the fact that she is lying makes the surprising fact (hyponatremia without a cause) a matter of course*

Chart 3

Here, Dr. Gunn knew more about hyponatremia than the resident. However, he was also better at reading the text of the patient, and he knew to follow the evidence that did not fit. In this case, that she is lying.

Appendix 4: Guide for Discussing Diagnosis and Diagnosis Errors

"The Resident Patient"

Induction (performed by the police)	Abduction (performed by Sherlock Holmes)
Classifies objects or facts 1. Time of death 2. Door locked from the inside 3. List number of cigars	**Begins with characteristics** (Qualities) run in categories 1. characteristics of the cigars 2. size of the shoe prints 3. Blessington is lying (attended to the body language)
Observation of Facts Only 1. time of death 2. door locked 3. number of cigars	**Imagine what might be true** 1. Size of shoe prints and characteristics of the cigars make it likely that more than one person was in the room
Tests a Hypothesis The hypothesis in many cases is sometimes already biased. In this case, it was biased by the time of death. The hypothesis was that it was suicide. 1. Blessington was a nervous guy who smoked a lot. He died by hanging at 5:00 in the morning (the most common time for suicide). Therefore this is a suicide	**Forms a Hypothesis** 2. Evidence points to multiple people in the room—foot prints, different kinds of cigars—therefore, this was most likely not a suicide but a different category of death—murder. The **surprising fact** was the **multiple kinds of cigars** left in the room. It becomes a matter of course if there were other people in the room

Chart 4

In this case, the detective knew more than the police (i.e., knowledge about cigars). He read the text of the patient better (he noticed the lying of Blessington); and he used the issue of categories to his advantage in "abducting" a cause.

Common Errors in Diagnostic Medicine

1. Error of incomplete Data Base

    This occurs when the healthcare provider does not obtain a complete history, perform a complete physical examination, listen to the unsaid as well as the said, and appropriately apprehend the body language of the patient. This category contains the following sub-species of error:

    a. Taking all patient responses at face value, thus ignoring the effect of denial, repression, and lying
    b. Failing to apprehend the "meaningful whole" of the patient as text. This includes **gender, ethnic, cultural, linguistic,** and other meaningful ways storytelling is inflected
    c. Failing to have technical skills for the physical examination and its interpretation
    d. Being too tired, hungry, or distracted when evaluating the patient

2. Failure to consider all **categories of illness** that might result from the collection of symptoms, physical findings, and other aspects of the data base

    a. Examples of categories of illness include congenital, behavioral, environmental, infectious, immune and autoimmune, idiopathic (we just don't know enough yet), cancer. And others

3. Errors of Confusing the Unusual with the Abstruse (these are the terms of Edgar Allan Poe's detective, Auguste Dupin)

    a. This occurs when the physician is confronted with a common illness with a very unusual set of presenting signs and symptoms. Since she cannot figure it out, she throws up her hands and says "it's just not solvable"

(*continued*)

(continued)

4. Error of lack of Knowledge
   a. One cannot diagnose a disease one does not know about
5. Error of Semiotics (Interpretation of signs)
   a. An example might be "misreading" the presence of edema in a malnourished patient as due to congestive heart failure
   b. Another might be "misinterpretation" of the low white count in the first patient with AIDS
6. Error of "Worshiping at the Altar of Technology"
   a. This error is made when the laboratory or imaging results do not correlate well with the *most important diagnostic information*—the **History and Physical Examination.** (For a striking example read David Hilfiker's "Facing Our Mistakes" [1984].)

Chart 5
Class Format:

1. Discuss the readings: (30 minutes)
   a. Please facilitate the discussion so that you bring out the important topics that students will need to know to discuss the difference between Induction and Abduction.
   b. Please facilitate the discussion so that it is clear in the learner's minds *HOW THE DIAGNOSTIC ERROR WAS MADE*.
   c. Our educational goal is for the learner to discover the above during their discussion. This will be best accomplished if the facilitation is done well.
2. Distribute charts 1 and 2 of this appendix as a class handout.
   a. Discuss the method of diagnosis and the difference between induction and abduction (approximately 15 minutes)
3. Distribute charts 3 and 4 of this appendix.
   a. Discuss the differences between induction and abduction as found in the two readings.
   b. Encourage questions, discussion, etc. (10–15 min)
4. Distribute chart 5 and discuss the errors of making a diagnosis.
   a. Encourage the students to think of other possible types of errors. Encourage their discussion of the anxiety associated with making an error in diagnosis.
   b. Offer ideas on how to deal with making a diagnostic error from your own experience, or from sources like Hilfiker's essay and his catalogue of "mistakes." (As mentioned earlier, the whole of this article should be available online.)

Appendix 5: Medical Professionalism: Using Literary Narrative to Explore and Evaluate Medical Professionalism[1]

Medical Professionalism

Medical Professionalism is a central tenet to the practice of medicine and has been described by D. T. Stern in his book, *Measuring Medical Professionalism*, in the following manner:

> Professionalism is demonstrated through a foundation of clinical competence, communication skills, and ethical understanding, upon which is built the aspiration to and wise application of the principles of professionalism: excellence, humanism, accountability, and altruism. (19)

The Accreditation Council of Graduate Medical Education (ACGME) incorporates professionalism as one of the six "core competencies" that are required to be assessed by graduate medical education training programs for all trainees (see *Advancing*).[2] Resident and fellow physicians in all specialties must demonstrate competency in professionalism appropriate to their training level in order to progress to the next training level and ultimately to graduate to become an accredited independently practicing physician.

Beginning in 2013, the ACGME asked each medical specialty to define "Milestones" for each of the six competency areas, including professionalism. Milestones are defined as competency-based developmental outcomes (i.e., knowledge, skills, attitudes, and performance) that can be demonstrated progressively by residents and fellows from the beginning of their education through graduation to the unsupervised practice of their specialties (available at *Milestones*). The Milestones are meant to be observable activities, with specific behaviors described for each level from beginning (novice) through master. A review of the ACGME Professional Milestones available on the ACGME website for each specialty reveals a wide variation in how each specialty defines professionalism by attributing differing attitudes and behaviors to the term (see Accreditation: *Milestones.*). Some

---

[1] By Casey Hester, MD, Jerry Vannatta, MD, and Ronald Schleifer, Ph.D., reproduced from *New Directions in Literature and Medicine Studies*, ed. Stephanie Hilger (New York: Palgrave, 2017), pp. 99–116.

[2] The other 5 Competencies are: Patient Care, Medical Knowledge, Interpersonal and Communication Skills, Systems Based Practice, and Practice-Based Learning and Improvement.

specialties use as few as three to four attributes, while others use up to six to ten. A highly representative but not exhaustive list includes such professional behaviors as demonstrating, through discernable and measurable behavior:

awareness of personal/professional boundaries,
compassion,
cultural sensitivity,
empathy,
honesty,
integrity,
professional duty,
respect,
self-awareness,
sensitivity to ambiguity,
trustworthiness.

In the workshop we describe in this essay, we have medical learners assess the behavioral manifestation of the six bold-faced professional behaviors on this list, which are the "milestones" of professionalism for the specialty of pediatrics.

These general behavioral qualities are somewhat vague. In significant part they coincide with Aristotle's conception of "virtue ethics" mentioned later in this essay. The experience of two of the authors (CH and JV) as physician educators is that it is difficult to get faculty to define professionalism in a consistent manner. Because of this, it can also be difficult for these same faculty to arrive at accurate and meaningful assessments of residents' professionalism, since teaching faculty find it difficult, if not impossible, to assess the learner on professional objectives they may not be able to define, let alone measure. The Milestones are discrete, observable behaviors that can be situated and related to one another along a developmental trajectory; thus, they are a step towards eliminating the ambiguity of inconsistent definitions of professional objectives and the resulting ambiguity in assessment. It was similarly our goal in developing the workshop—using literary narrative to define and evaluate medical professionalism—to create a practical and experiential method of exploring ambiguous conceptions of professionalism. Our hope was to create a method that residents and faculty alike can feel comfortable using, and one that can move any community of medical learners toward profound understanding and agreement of what medical professionalism is. Because two of the authors (RS and JV) have sixteen years' experience engaging literary narrative to teach medical themes—professionalism among them—and because literary narrative studies lends itself to exploring issues that are ambiguous, we turned to the use of literary narrative to develop a workshop for residents and faculty that builds competency in defining and evaluating medical professionalism. As a Pediatric Residency Program Director, one author (CH) is charged with helping the residents in her program understand, define, and evaluate medical professionalism. She must also shepherd the faculty in her department through the same process so that the evaluation of the residents' attitudes and behaviors can be reliably and accurately observed, developed, and assessed.

Narrative Medicine

Using literary narratives for the purpose of exploring medical issues has become known as "Narrative Medicine." This is a term coined by medical and literary scholar Rita Charon M.D., PhD. She states that Narrative Medicine provides healthcare professionals with practical wisdom in comprehending what patients endure in illness and what physicians themselves undergo in the care of the sick. She further discusses in her book *Narrative Medicine: Honoring the Stories of Illness* that one of the goals of studying literary narrative for doctors is to become competent at recognizing, absorbing, interpreting, and being moved by the stories—the medical histories—that patients tell doctors (vii). The absorption and interpretation of narrative has been labeled "narrative knowledge." This knowledge differs from bio-scientific knowledge in that it is organized such that the whole is greater than the sum of its parts whereas bio-scientific knowledge is organized such that the whole is equal to the sum of its parts. It is also knowledge that allows value to enter its understandings, whereas the knowledge of science that physicians use to diagnose and prescribe therapies tends to eschew value judgments in favor of quantification (i.e., the whole being *equal* to the sum of its parts). Narrative knowledge engages ambiguity, not so much to consistently "resolve" it as to take ambiguity into account in its understandings. This is important because even though medicine is often taught as if it were unambiguous, the practice of medicine is blanketed in ambiguity. Therefore, it has been said by many writers and thinkers in the field of narrative medicine that physicians should be taught to think in the ways of narrative in addition to the bio-scientific ways that medicine is primarily taught in order to comprehend (rather than dismiss) the ambiguities that arise in its practice.

The strategies of narrative medicine, which Dr. Charon analyzes, entail narrative knowledge, gained through the study of literature. The resulting knowledge—and, indeed, the resulting wisdom—that follows from narrative medicine can be thought of as the Aristotelean concept of *phronesis*. *Phronesis*—often translated into English as "practical knowledge" or "practical wisdom"—is one of the virtues that Aristotle lists in his discussion of virtues. (His chief examples of people who achieve *phronesis* were physicians and navigators.) He believed it was necessary for individuals to habituate these virtues—including *phronesis*—in order to live a good life. Aristotle argued that the virtues necessary to live a good life needed to be habituated since people were not born with these attitudes and behaviors. He also argued that these virtues facilitated the development of good character in particular people and that building good character led to achieving a good life. One can think of the same process as being necessary for the medical professional—that the specific attributes and behaviors of professionalism are the "virtues" that need to be habituated for the physician to live the good professional life. Engaging literary narrative provides an excellent vehicle for exploring, defining, and teaching these virtues, so that they can be brought forward into consciousness where the learner can begin to habituate them and the faculty can more clearly identify them (see Aristotle for his account of *phronesis*; and Chap. 2 of Schleifer and Vannatta for an extended discussion).

Although Aristotle suggests that *phronesis* was the result of long practice, fictional stories provide an efficient and safe way to do the work of achieving *phronesis*. The fictional story provides a medical narrative—different from the residents' own experience—that allows them to explore the physician's attitudes and behaviors from a distance, eliminating the barrier of self-consciousness and shyness. Literary narratives also provide a medium which is similar to what Schleifer and Vannatta describe as the "medical drama" of every day practice of medicine (262–74). In medical practice, there are characters—specifically a patient and a physician—and there is a plot such as moving toward a diagnosis or therapy. Fictional (or sometimes autobiographical) medical narratives provide themes, growing out of plot and characters, that can be explored. Moreover, depending on the behavior of the characters in the story (the patient and the physician), as well as how the plot plays out, a genre can be assigned to the story by the readers in the workshop. Asking the medical learners to use concepts usually only found in literary education is troublesome to many medical educators. Although the ideas and concepts of narrative medicine are becoming more common, there remains a minority of academic physicians who feel competent to facilitate discussions involving these issues.

With that potential barrier in mind, Schleifer and Vannatta, in their book *The Chief Concern of Medicine: The Integration of the Medical Humanities and Narrative Medicine into Medical Practices,* introduced schemas of narrative, based upon work in narratology over the past fifty years, that they believe can help non-experts begin to approach medical education using literary narrative. The use of these schemas can aid the non-expert in facilitating the discussion of the literary piece in the workshop. The following are the schemas of narrative as they appear in their book (383–84). These schemas set forth six "elements" of narrative (A); four character "roles" in narrative based upon the analogy between the structure of the sentence and the structure of narrative (B); and four basic genres of narrative based upon the interaction of the events and characters of narrative (C):

A. Narrative Structure

   Narrative Possesses

   1. A sequence of events;
   2. An end; and
   3. Recognizable agents.

   Narrative also possesses

   4. A teller and a listener (i.e., narrative is both articulated and received);
   5. A *witness who learns*—who is "concerned"—about the end; and
   6. Its witness learns *from experience.*

B. Roles in Narrative

Narrative	Sentence	Medical Roles
Hero	subject	***patient*** (hero)
Desired object	object	***health*** (desired object/condition)
[Action	verb	*to purge* (to remove the disease)
		to purify (to achieve well-being)
		to clarify (to figure out whatever works)]
Helper	adverb	***physician*** (helper)
Opponent	adverb	***illness*** (opponent)

For medicine, Schleifer and Vannatta name the four "medical roles" as *patient, health, physician, illness* corresponding to the more general narrative roles of *hero, desired object, helper, opponent*. The three action verbs of medicine are taken from different understandings/translations of the Greek word *catharsis*, the medical term Aristotle uses in his analysis of tragedy.

C. The Genres of narrative
   Heroic Melodrama (epic):
   a heroic narrative, where the hero receives the wished-for goods (in myth and tradition, the bride and the kingdom). The hero conquers the opponent in the process.
   Tragedy:
   a tragic narrative, where the helper receives the wished-for goods (both the storied knowledge of what has taken place on the level of the individual destruction of the hero and the promised reconstruction of the community on the brink of collapse with the destruction of the hero, which is often accomplished by the helper).
   Comedy:
   a comic narrative, where the heroine receives the wished-for goods (in myth and tradition, the hero as husband and the estate of marriage).
   Irony:
   a more or less "modern" narrative, where the opponent receives the wished-for goods (to destroy them on the level of the individual and to transform them on the level of general value).

It is helpful for the purposes of teaching this workshop to point out that the genre is generally defined by what happens to the hero or in the case of comedy to the heroine. Since in the medical drama or story there is always a patient and a physician, it is important for the learners in the workshop to identify these two characters as enacting particular narrative roles (i.e., the general roles of hero [patient] and helper [physician]) so that the genre can be explored in some detail.

An Exemplary Medical Narrative

We will provide an example of applying these schemas to one of the stories—the one we have the most experience with when running the workshop—namely "Imelda" (1982) by Dr. Richard Selzer. In this short story, the setting is a medical school in the U.S. in which the chief of plastic surgery is preparing to take a group to Honduras for a "mission trip." He finds a third-year medical student who speaks fluent Spanish and invites the student along. The student is the narrator of the story. Although the surgeon is brilliant and competent with a scalpel, early in the story he is depicted as curt with patients and less than compassionate. Upon arrival in Honduras, they meet a young girl, Imelda, who presents for evaluation and surgery on a cleft palate and lip. The surgeon once again is impatient with his young, embarrassed patient, Imelda, and rushes through the evaluation. She is scheduled for surgery the next morning. During the induction of anesthesia, Imelda experiences malignant hyperthermia and dies. Following the failed attempt at resuscitation the surgeon goes to tell Imelda's mother. After the surgeon informs her that Imelda died, the mother replies, "at least she will go to heaven beautiful as God intended." The surgeon does not clarify that the death had occurred before the operation could be done. That night the surgeon enters the morgue, locks the door behind himself and under light from a candle completes the operation on the dead girl. The next day the student notices that Imelda's body is out in front of the clinic being readied to travel back to her village, and the student approaches the mother with money for flowers. The mother thanks the student for making her daughter beautiful. The student peeks under the sheet covering Imelda's body and discovers an Imelda with a repaired cleft lip and palate.

Upon return home the proceedings of the mission trip are being presented at grand rounds by the surgeon. The student is managing the slide projector. The surgeon calls for the next slide and sees the image of Imelda. He mentions her name, but says no more. The picture is of Imelda with the disfiguring cleft lip. The student cannot figure out what the surgeon is doing and does not project the next two slides, which show Imelda repaired. The ending of Selzer's story is quite ambiguous insofar as the author does not let us know exactly what the student-narrator or the surgeon was thinking at the time. The last paragraphs are narrated after much time has passed—the student-narrator is much older and still fascinated, if not obsessed, by the occurrences on the mission—and the student's final meditation on these events are highly metaphorical: "I, too, have not been entirely free of [Imelda]. Now and then, in the years that have passed, I see that donkey-cart cortège, or [the surgeon's] face bent over hers in the morgue. I would like to have told him what I now know, that his unrealistic act was one of goodness, one of those small, persevering acts done, perhaps, to ward off madness. Like lighting a lamp, boiling water for tea, washing a shirt. But, of course, it's too late now" (35–6).

In this story, there is obviously a plot which is the subject of Schema A. It has a narrator, the student, and an audience that learns. What exactly we learn is one of the aims of the workshop and work that the learners in the workshop must do. We must wonder about the surgeon's motives, whether he was compassionate or whether this behavior was in self-interest. To understand the genre of this story we must assign roles to the surgeon, to the cleft palate (his patient's condition[3]), and to the patient herself. If the learners assign the surgeon the role of melodramatic hero, which is commonly the case, then, insofar as the hero does not receive the desired object (which would suggest a "melodramatic" narrative), the surgeon must lose something important, die or experience exile (in a kind of failed melodrama). They can also assign the surgeon the role of tragic hero, in which case the student-narrator

---

[3] Animate people are not necessarily the only parts of narratives that can perform narrative "roles": the study of narrative suggests that inanimate objects, such as the ring in *Lord of the Rings*, function like "characters" in stories insofar as they perform character-defining roles.

would be his "helper," who "reconstructs" the community on the brink of collapse. If, on the other hand, they assign the hero role to Imelda, her death is exemplary of a tragedy in the traditional sense, with the physician assigned the role of helper and the cleft palate being the opponent. (In Schema B above we have assigned the patient the role of hero in all medical dramas and the doctor the role of helper.) These various assignments of roles allow learners to understand in concrete terms the *ambiguity* of narrative knowledge, and it allows them to confront the ambiguity inherent in narrative—and in professional as well as fictional encounters between patients and physicians—rather than to dismiss it. Such self-conscious encounters can help define and understand real-life attributes of medical professionalism.

The Workshop

Objectives, Structure, Process and Lessons Learned

Our workshop can be taught in one of two ways. One is to provide the learners with a short story (a medical story) well before the workshop and then explore it at the workshop. The other is to use a developed modification of the short story found in Savitt's *Medical Readers' Theater*—in which the story has been transformed into a drama that can be read by participants during the workshop. This last format has worked best for us because of time constraints for the residents and the faculty which often preclude ability to read and prepare beforehand.

Objectives
By the end of the workshop, participants will be able to:

1) Define Professionalism Milestones specific to their specialty
2) Assess Professionalism Milestone levels for the physician from the story, based on observable attitudes and behaviors
3) Assign a specific narrative genre from Schema C to the literary narrative literature according to the roles—themselves defined by action—of characters in a story

Structure
Like the objectives, the structure of the Workshop is best set forth in terms of a working list of elements.

The participants are divided into working groups of three to six individuals.

1. A pre-assessment of learners' knowledge and understanding of Professionalism Milestones is taken by written survey (Likert scales).
2. Professionalism Milestones are distributed for the chosen Specialty. In the Workshop described here, the six Milestones of Professionalism for Pediatrics mentioned earlier—behaviors demonstrating: empathy, professional duty, an awareness of personal/professional boundaries, self-awareness, trustworthiness,

sensitivity to ambiguity, (Accreditation: *Pediatrics*)—were described in relation to five levels of accomplishment for each professional behavior. (See Appendix B for these Milestones and the descriptions of their levels of accomplishment.)
3. A reading of the story occurs before the workshop or the *Medical Readers' Theater* approach is used to read it during the workshop.
4. The attitudes and behaviors of the physician in the story are examined within the context of each Professionalism Milestone. In Pediatrics, as Appendix B sets forth, each of the six Milestones focuses respectively on: empathy, duty, (enforcing) boundaries, self-awareness, trustworthiness, and (acceptance of) ambiguity.
5. Specific examples of the physician's displayed behaviors and attitudes are taken from the story to appraise the physician's level for each of the six Milestones on a scale of 1 (novice) to 5 (Master).
6. Milestone levels for the physician are initially assigned either individually or in small groups, depending on number of workshop participants. A large group discussion is then led by the facilitator, as individuals and small groups attempt to justify and reconcile any discrepant opinions on appraised Milestone levels. During this discussion, professionalism terms are disambiguated by the facilitator in an attempt for the group to reach consensus on a single level (1–5) for each of the Milestones for the behaviors occurring in the literary narrative.
7. A post-assessment of learners' knowledge and understanding of Professionalism Milestones is taken by written survey (Likert scales).

Process

The participants in the workshop are divided into groups. They are given the milestones of professionalism for a particular specialty. (We generally use pediatrics because it is one of the most robust with respect to both the quantity of Professionalism Milestones and clear descriptions of the behaviors that are expected to be exhibited for each level along the novice-mastery continuum.) We first ask the participants to read through each of the Professionalism Milestones and write down the "mastery" level physician they know personally for each Milestone. This allows them to "anchor" the behavior through previous observation.

In our workshop, several members of the group assumed the "role" of characters in the story and read the dramatic version of "Imelda" from the *Medical Readers' Theater*. We asked the participants serving as the audience to make notes during the reading describing and assessing the attitudes and the behaviors of the doctor in the story. This is analogous to direct observation of a resident in a clinic with a patient and using this first-hand knowledge as data for evaluation.

Following the reading of the drama, each group is asked to discuss the story and the doctor's behavior and evaluate the professionalism of the chief of plastic surgery in "Imelda." Once the groups are finished evaluating the doctor, each group is asked to state what level of achievement they assigned the doctor on each of the professional Milestones listed. (See Appendix B for the specific evaluation criteria for professionalism in pediatrics.) The facilitator then leads a discussion of the story, the characters' attitudes and behaviors, and helps the participants explore the story in terms of attitudes, behaviors, genre, and character assignment. This group discussion, if facilitated appropriately, should demonstrate that unlike scientific discussions where terms are clearly defined and answers are more concrete, the discussion of a story is more ambiguous. As is the case in our joint analysis of the professional Milestones for Pediatrics in relation to "Imelda," there will be many perspectives on the behavior of the patient and the doctor, the plot and the genre. Getting the participants more comfortable with the ambiguity of the process is one of the goals of the workshop.

Following the discussion and the reporting of all groups, the participants are asked to reflect upon and journal how they may use literary narrative or the *Medical Readers' Theater* in their home institution to help their own trainees and faculty more precisely understand and articulate the Milestones of medical professionalism by participating in a careful discussion of actions and their assessment performed by physicians in literary narrative.

Outcomes and Lessons Learned

Average levels assigned for the Surgeon by workshop participants (four workshops, with 54 total participants) were as follows for each of the six Pediatric Professionalism Milestones (1 = Novice, 5 = Master).

- **Empathy**—*consensus level*: 1.5. *Examples*: Did not seem to feel or display empathy; would not touch his patients, was dismissive of Spanish-speaking man with leg wound, ripped rag away from Imelda's face. He did seem to show more emotion towards end of story.
- **Duty**—*consensus level*: 4.5. *Examples*: Always reading; high sense of duty to the profession.
- **Boundaries**—*consensus level*: 1. *Examples*: Repaired her face without consent after she died.
- **Self-Awareness**—*consensus level*: 1 at beginning of story, 3 at end. *Examples*: Could not accept less than perfect role, but then after Imelda changed his practice, he was: "quieter, softer."
- **Trustworthiness**—*consensus level*: 3.5. *Examples*: clinically conscientious but could not be trusted if his self-interest superseded patient interest.
- **Ambiguity**—*consensus level*: 1.5. *Examples*: Did not advise mother or Imelda on risks/benefits of operation, did not consider patient input early in story; "rigid and authoritarian"; could not accept that Imelda had died without repair after mother thanked him for fixing her cleft palate.

For some of the Professionalism Milestones (duty, trustworthiness), the surgeon ranked quite high, whereas for others he ranked quite low (empathy, boundaries). This parallels what happen in real life as well—people are not all good or all bad, and similarly the surgeon is neither all professional nor all unprofessional.

Our experience is that participants are often initially uncomfortable with the discussion of professionalism primarily because of the ambiguity of the terms. They report being able to recognize professionalism when they see it, but putting it into words and thus being able to offer formative feedback and meaningful assessment can be elusive. We have found the Milestones, set up as observable behaviors along a developmental continuum, assuage some of this uncertainty and allow for more objective assessment, but only after participants are walked through the process in the workshop. It is, therefore, very helpful for participants to go through the process of seeing that in some areas the surgeon is highly professional, but in others he falls short. Being able to cite specific examples from the story allows participants to sort through the components of what professional behavior is and what it is not—in this case, defined by the six Professionalism Milestones for Pediatrics. This helps participants gain clarity in the process of defining professionalism; and it also helps participants in realizing that professionalism is not a dichotomous concept (e.g., one that lends itself to complete disambiguation). We have also found that the participants are initially uncomfortable with the teaching of literary terms and concepts, but by the end of the workshop they are a bit more comfortable. One of the largest barriers we have discovered is that they are worried about finding a competent facilitator to run workshops at their home institutions. The purpose of this paper is to set forth a procedure that can structure the work of facilitating the achievement of well-defined professional Milestones by means of the shared experience of a literary narrative.

Conclusions

In our experience medical learners are nearly always predisposed to assign the role of hero to the physician. When we explain the physician should always play the role of helper in the medical drama, it often comes as a surprise. However, once it is explained that the physician's role in real life should also be helper instead of hero, participants usually begin to understand. We point out several reasons why the physician should be playing helper in a medical drama, including that the role of helper is an easier place from which to adopt and indeed habituate the virtues that have been defined at the "mastery" end of the Professionalism Milestones for each medical specialty. Further, by adopting the role of helper, a physician is potentially much less liable to "burn out" in his or her career. For example, if we as physicians assign ourselves the role of hero, then we must either always win by defeating the illness (which we know cannot always happen) as in the case of Melodrama, or we must lose something of ourselves, die, or become exiled in the case of Tragedy. This assignment of roles for the characters in conjunction with the assignment of genre to the narrative as a whole allows the participants to learn narrative knowledge, which is critical because the practice of medicine is primarily narrative in nature. Our patients tell us stories. We re-story them in a biomedical narrative (the history of present Illness). Daily, we, as physicians, use narrative to tease out a history, negotiate a diagnosis, and communicate good as well as difficult news. The more our physicians in training know about stories, specifically how narratives function and are structured within the context of the intertwined roles of doctor, patient, and disease, the better helpers they will become, and thus better physicians.

Further, by using narrative structure and roles in a schematized fashion we have given physician-educators sophisticated tools with which to demonstrate a profound nexus between literary narratology and the practice of medicine. The process always allows us to clarify what professional behaviors look like when carried out in everyday practice. By critically examining and assessing the professional—and unprofessional—behaviors that physicians demonstrate as the characters in literary works, workshop attendees can reflect on their own behaviors, with the goal of striving towards the Mastery end for each of the Professionalism Milestones. For those who are already narratively competent, the Workshop we are describing still allows them the tools of schematization of elements of narrative that offer an efficient way to convey what they already know for the benefit of workshop participants. Moreover, those who have not yet reflected on their professional experience in relation to narrative structures can discover—both as participants in and even facilitators of Workshops—another framework in which to understand their work and a more precise sense of professional Milestones. Thus, by integrating the Professional Milestones into Narrative Medicine, we have provided the learners and trainers new ways of thinking about physician attitudes and behaviors, and have taught them something about how stories work as well.

Appendix A: Suggested Texts

The following is a list of stories that work well to explore the professionalism of a medical provider.

Anton Chekhov, "Enemies"
Arthur Conan Doyle, "The Doctors of Hoyland"
David Hilfiker, "Mistakes"
Jerome Groopman, "Unprepared"
Stories included in the *Medical Readers' Theater:*
William Carlos Williams
"A Face of Stone"
"The Girl with a Pimply Face"
"The Use of Force" (annotated by Felice Aull and by Pamela Moore and Jack Coulehan)
"Old Doc Rivers"
Richard Selzer
"Fetishes"
"Imelda"
"Whither Thou Goest"
Susan Onthank Mates
"Ambulance"
"Laundry"
Pearl S. Buck, "The Enemy"
Arthur Conan Doyle, "Round the Red Lamp"
Katherine Anne Porter, "He"
Mary E. Wilkins Freeman, "A Mistaken Charity"
Margaret Lamb, "Management"

Appendix B: Pediatric Professionalism Milestones (Accreditation: *Pediatric Milestones*)

	Level 1	Level 2	Level 3	Level 4	Level 5
Empathy	\multicolumn{5}{c}{PROF1: Humanism, compassion, integrity, and respect for others; based on the characteristics of an empathetic practitioner}				
	Sees the patients in a "we versus they" framework and is detached and not sensitive to the human needs of the patient and family	Demonstrates compassion for patients in selected situations (e.g., tragic circumstances, such as unexpected death), but has a pattern of conduct that demonstrates a lack of sensitivity to many of the needs of others	Demonstrates consistent understanding of patient and family expressed needs and a desire to meet those needs on a regular basis; is responsive in demonstrating kindness and compassion	Is altruistic and goes beyond responding to expressed needs of patients and families; anticipates the human needs of patients and families and works to meet those needs as part of his skills in daily practice	Is a proactive advocate on behalf of individual patients, families, and groups of children in need

	Level 1	Level 2	Level 3	Level 4	Level 5
Duty	\multicolumn{5}{c}{PROF2: Professionalization: A sense of duty and accountability to patients, society, and the profession}				
	Appears to be interested in learning pediatrics but not fully engaged and involved as a professional, which results in an observational or passive role	Although the learner appreciates her role in providing care and being a professional, at times has difficulty in seeing self as a professional, which may result in not taking appropriate primary responsibility	Demonstrates understanding and appreciation of the professional role and the gravity of being the "doctor" by becoming fully engaged in patient care activities; has a sense of duty; has rare lapses into behaviors that do not reflect a professional self-view	Has internalized and accepts full responsibility of the professional role and develops fluency with patient care and professional relationships in caring for a broad range of patients and team members	Extends professional role beyond the care of patients and sees self as a professional who is contributing to something larger (e.g., a community, a specialty, or the medical profession)

	Level 1	Level 2	Level 3	Level 4	Level 5
Boundaries	\multicolumn{5}{PROF3. Professional Conduct: High standards of ethical behavior which includes maintaining appropriate professional boundaries}				
	Has repeated lapses in professional conduct wherein responsibility to patients, peers, and/or the program are not met. These lapses may be due to an apparent lack of insight about the professional role and expected behaviors or other conditions or causes (e.g., depression, substance use, poor health)	Under conditions of stress or fatigue, has documented lapses in professional conduct that lead others to remind, enforce, and resolve conflicts; may have some insight into behavior, but an inability to modify behavior when placed in stressful situations	In nearly all circumstances, conducts interactions with a professional mindset, sense of duty, and accountability; has insight into his or her own behavior, as well as likely triggers for professionalism lapses, and is able to use this information to remain professional	Demonstrates an in-depth understanding of professionalism that allows her to help other team members and colleagues with issues of professionalism; is able to identify potential triggers, and uses this information in conduct as part of her duty to help others	Others look to this person as a model of professional conduct; has smooth interactions with patients, families, and peers; maintains high ethical standards across settings and circumstances; has excellent emotional intelligence about human behavior and insight into self, and uses this information to promote and engage in professional behavior as well as to prevent lapses in others and self
	Level 1	Level 2	Level 3	Level 4	Level 5
Self-Awareness	PROF4. Self-awareness of one's own knowledge, skill, and emotional limitations that leads to appropriate help-seeking behaviors				
	Has a lack of insight into limitations that results in the need for help going unrecognized, sometimes resulting in unintended consequences	Shows concern that limitations may be seen as weaknesses that will negatively impact evaluations results in help-seeking behaviors, typically only in response to external prompts rather than internal drive	Recognizes limitations, but has the perception that autonomy is a key element of one's identity as a physician, and the need to emulate this behavior to belong to the profession may interfere with internal drive to engage in appropriate help-seeking behavior	Recognizes limitations and has matured to the stage where a personal value system of help-seeking for the sake of the patient supersedes any perceived value of physician autonomy, resulting in appropriate requests for help when needed	Beyond recognizing limitations, has the personal drive to learn and improve results in the habit of engaging in help-seeking behaviors and explicitly role modeling and encouraging these behaviors in residents

	Level 1	Level 2	Level 3	Level 4	Level 5
Trust-worthiness	PROF5. Trustworthiness that makes colleagues feel secure when one is responsible for the care of patients				
	Has significant knowledge gaps or is unaware of knowledge gaps and demonstrates lapses in data-gathering or in follow-through of assigned tasks; may misrepresent data (for a number of reasons) or omit important data, leaving others uncertain as to the nature of the learner's truthfulness or awareness of the importance of attention to detail and accuracy; overt lack of truth-telling is assessed in a professionalism competency	Has a solid foundation in knowledge and skill, but is not always aware of or seeks help when confronted with limitations; demonstrates lapses in follow-up or follow-through with tasks, despite awareness of the importance of these tasks; follow-through can be partial, but limited due to inconsistency or yielding to barriers; when such barriers are experienced, no escalation occurs (such as notifying others or pursuing alternative solutions)	Has a solid foundation in knowledge and skill with realistic insight into limits with responsive help seeking; data-gathering is complete with consideration of anticipated patient care needs, and careful consideration of high-risk conditions first and foremost; requires little prompting for follow-up	Has a broad scope of knowledge and skill and assumes full responsibility for all aspects of patient care, anticipating problems and demonstrating vigilance in all aspects of management; pursues answers to questions, and communications include open, transparent expression of uncertainty and limits of knowledge	Same as Level 4, but any uncertainty brings about rigorous search for answers and conscientious and ongoing review of information to address the evolution of change; may seek the help of a master in addition to primary source literature

Appendix 5: Medical Professionalism: Using Literary Narrative to Explore and Evaluate... 277

	Level 1	Level 2	Level 3	Level 4	Level 5
Ambiguity	PROF6. The capacity to accept that ambiguity is part of clinical medicine and to recognize the need for and to utilize appropriate resources in dealing with uncertainty					
	Feels overwhelmed and inadequate when faced with uncertainty or ambiguity; communications with patients/families and development of therapeutic plan are rigid and authoritarian, with assumption that the patient can manage information and participate in decision-making; patient/family numeracy presumed; seeks only self or self-available resources to manage response to this uncertainty, resulting in a response characterized by their (individual) pre-existing state of risk aversion or risk taking; does not regard patient need for hope; feels compelled to make sure that patients understand full potential for negative outcome (defensive/protective of physician)	Recognizes uncertainty and feels tension/pressure from not knowing or knowing with limited control of outcomes; explains situation to the patient in framework most familiar to the physician, rather than framing it with terms, graphics, or analogies familiar to the patient; seeks rules and statistics and feels compelled to transfer all information to the patient immediately, regardless of patient readiness, patient goals, and patient ability to manage information	Anticipates and focuses on uncertainty, looking for resolution by seeking additional information; aims to inform the patient of the more optimal outcome(s), framed by physician goals; does not manage overall balance of patient/family uncertainty with quality of life, need for hope, and ability to adhere to therapeutic plan; focuses on own risk management position for a given problem and does not suggest that more or less risk taking (different from physician's position) could be chosen; still seeks patient/parent recitation of uncertainty/morbidity as proof that patient/family understands the uncertainty; has an unresolved balance of expectations with physician expectations taking precedence	Anticipates that uncertainty at the time of diagnostic deliberation will be likely; uses such uncertainty or larger ambiguity as a prompt/motivation to seek information or understanding of unknown (to self or world); balances delivery of diagnosis with hope, information, and exploration of individual patient goals; works through concepts of risk versus hope using conceptual framework that includes cost (e.g., suffering, financial) versus benefit, framed by patient health care goals; expresses openness to patient position and patient uncertainty about his or her position and response	Is aware of and keeps own risk aversion or risk-taking position in check; seeks to understand patient/family goals for health and their capacity to achieve those goals, given the uncertain treatment options; engages in discussion with high sensitivity towards numeracy, emphasizing patient/family control of choices with initial plan development and ongoing information sharing through changes as knowledge and patient health status evolve; remains flexible and committed to engagement with the patient/family throughout the patient's illness, serving as a resource to gather information so that degree of uncertainty is minimized; openly and comfortably discusses strategies and outcomes anticipated with the patient/family, emphasizing that all plans are subject to the imperfect knowledge and state of uncertainty; balances constant revisiting of knowledge, uncertainty, and developed plans acceptance of what is unknown; transparent communication of limits of treatment plan outcomes

Bibliography

Accreditation Council for Graduate Medical Education. Available at: https://www.acgme.org
———. *Advancing Education in Medical Professionalism.* Available at: http://www.acgme.org/outcome/implement/profm_resource.pdf. Published 2004.
———. *Milestones.* Available at: https://www.acgme.org/acgmeweb/tabid/430/ProgramandInstitutionalAccreditation/NextAccreditationSystem/Milestones.aspx
———. *Pediatrics Milestones.* Available at: http://www.acgme.org/acgmeweb/tabid/143/ProgramandInstitutionalAccreditation/MedicalSpecialties/Pediatrics.aspx
Charon, Rita. 2006. *Narrative Medicine: Honoring the Stories of Illness.* New York: Oxford UP.
Aristotle. *Nicomachian Ethics.* Trans. W.D. Ross. http://classics.mit.edu/Aristotle/nicomachean.html
Schleifer, Ronald, and Jerry Vannatta. 2013. *The Chief Concern of Medicine: The Integration of the Medical Humanities and Narrative Knowledge into Medical Practices.* Ann Arbor: University of Michigan Press.
Savitt, T.L. 2002. *Medical Readers' Theater: A Guide and Scripts.* Iowa City: University of Iowa Press.
Selzer, Richard. 1996. *Letters to a Young Doctor.* New York: Harvest Books.
Stern, D.T., ed. 2006. *Measuring Medical Professionalism.* New York: Oxford Press.

Appendix 6: Teaching Literature to Medical Students: Ernest Hemingway, Nick Adams, and the "Unsaid" in Narrative

For the past sixteen years, we have been team-teaching a course entitle "Literature and Medicine" for pre-med and medical students as well as conducting workshops for physicians and healthcare workers. The purpose of these teachings is to get healthcare workers more closely in touch with the human resources of narrative understanding, empathy, and shared vicarious experiences that to a large degree formal training in medicine and healthcare mitigate against. Doctors are trained to develop a broad data-base of human ailments and conditions, to systematically understand the biochemistry of life processes and illness processes with such intensity that, in many cases, what we have called "humanistic understanding" is lost. Humanistic understanding, we argue (see Vannatta et al. 2005), can grasp the experience of a person in distress, and the vital information for health that is contained in that narrated experience. The aim of our teaching engagements, then, is to train people committed to careers in healthcare to recover an array of human resources—empathy and the ability to grasp and respond to stories that patients almost always bring to their encounter with medicine. These abilities will help medical students more fully understand and engage with the patients they encounter. Among these resources is the ability to grasp what we call "narrative knowledge."

We begin this discussion with a small presentation by Dr. Rita Charon on the function of narrative knowledge in healthcare. Dr. Charon, who has a Ph.D. in literary studies as well as an MD in internal medicine, developed a program in Narrative Medicine at Columbia University College of Physicians and Surgeons and wrote a book some years ago entitled *Narrative Medicine*. The goal of both is to encourage the inclusion of training in engagements with narrative in medical education. Here, then, is her explanation of her programs.

> This interest we [medical educators pursuing "narrative medicine"] have in narrative knowledge and narrative methods is not an abstract, scholarly interest alone. It's a very practical interest. There is a very concrete, direct relationship between narrative knowledge and clinical action. Indeed, we are interested in helping our students and doctors understand things for their own purposes. We're even interested in helping them reflect on their experience and feel better for it. I'm happy when my students or the doctors who study with us feel better by virtue of their narrative training, but that's not enough. My goal in giving them narrative training is to enable them to act more effectively with their patients. So, the increase in the narrative skills of recognizing there's a story to be heard, eliciting it, being curious about what's unsaid, putting it together in some way, trying provisional hypotheses

to see "Did I get this right?", and being moved oneself by what's heard, all of these things culminate in the doctor then being able to act on the patient's behalf with more vigor, with more purpose, with more investment than they otherwise would.

> I talk sometimes about how we have to honor the narratives we hear, and this is a very active thing. People tell us very private, frightening things about themselves, and we, because we have skill and also because we have power, are privileged to hear these things. Sometimes they are things we don't want to know about, like child abuse, nonetheless, we hear about these things. We have duties toward these things we hear, and for doctors, I think there are twin duties. One duty is to honor what's been said, which is to say, not to trivialize it, not to dismiss it, not to forget it; and then we have the duty to act. By virtue of knowing what I now know, what must I do? I think this is where narrative training increases the professionalism of doctors. (Vannatta et al. 2005: Chap. 4, screen 8 [video])

Note her careful listing of the *skills* in engagement with patient narratives that she enumerates for a medical education that will allow physicians "to act," as she says, "more effectively with their patients": skills in

- recognizing there's a story to be heard,
- eliciting it,
- being curious about what's unsaid,
- putting it together in some way,
- trying provisional hypotheses to see "Did I get this right?", and
- being moved oneself by what's heard, with
- all of these things culminating in the doctor then being able to act on the patient's behalf.

These skills are important because the stories patients bring to the clinic, like the stories that Ernest Hemingway developed early in his career, depend upon our ability to grasp, as Dr. Charon says, what is "unsaid." This is because, as Dr. Charon says elsewhere, "narratives that emerge from suffering differ from those born elsewhere.... Not restricted to the linear, the orderly, the emplotted, or the clean, these narratives that come from the ill contain unruly fragments, silences, bodily processes rendered in code. The language is deputized to point to things not ordinarily admitted into prose or poetry or texts of other kinds—shameful, painful, prelingual limitations, absences, breath-taking fears" (2005: vi).

Before we turn to Hemingway in earnest—and his "theory of omission" that he developed in Paris about the time he wrote "Indian Camp," the first of his published Nick Adams stories—let us share with you another physician explaining the necessity of strong training in narrative understanding in the work of healthcare. Dr. John Stone, like Hemingway's father, is a physician, and like Hemingway himself, he is a writer. Here is his comparison between reading poetry and listening to patients:

> No one comes easily to any poem because poems are full of slippery words, but that's exactly what our patients are full of. Patients are full of slippery words. They don't know what the diagnosis is, they don't know what's important out of all this morass of information, but they tell it to us, and it comes flowing out across the desk or at the bedside; and just as we look at a poem, we inspect it, as we live with it a little longer, as we memorize it, we learn what the essential elements are and what to pay attention to in the next poem we hear. (Vannatta et al. 2005: Chap. 3, screen 70 [video])

Appendix 6: Teaching Literature to Medical Students: Ernest Hemingway...

Dr. Stone was the co-editor of an anthology of literature for medical students, *On Doctoring*, that up until recently was given to all American first-year medical students by the Robert Wood Johnson Foundation as they began their medical education. It includes Hemingway's stories "Indian Camp" and "Hills Like White Elephants."

Reading Hemingway with Medical Students

It is "Indian Camp," published in *In Our Time* in 1924 and 1925, that we focus upon here. The story describes a make-shift cesarean childbirth, performed without anesthesia and proper equipment by Dr. Adams, with the help of his brother George,[1] while his young son watches, and the pregnant woman's husband lies wounded in a bunk in the room of the operation. In the middle of the operation, the son, Nick, asks his father

> "Oh, Daddy, can't you give her something to make her stop screaming" asked Nick.
> "No. I haven't any anaesthetic," his father said. "But her screams are not important. I don't hear them because they are not important." (1972: 19)

Directly after this the patient's husband "rolls over in bed." Dr. Adams successfully delivers the baby, and afterwards he and George discover the patient's husband dead in the bunk with his neck slashed. Dr. Adams and Nick take the boat away from the Indian camp while George remains behind to help clean up.

What is most striking about teaching this story to pre-med students is how much they simply do not notice. They are so fascinated by the medicine of it—the make-shift caesarian section in the middle of the night at a poverty-stricken Indian camp—that they pay little attention to the details of the story: why Dr. Adams performs the operation "with a jack-knife and sewing it up with nine-foot, tapered gut leaders" (1972: 19), why there is no anesthetic (1972: 18), why the woman's husband is in the same room, why Nick is there at all, functioning, as his father says, as "an intern" (1972: 19). Pre-med and medical students are particularly oblivious to Nick's presence, since almost all of them have "shadowed" physicians as they work so that the presence of a young watcher hardly seems strange at all. Perhaps for similar reasons, they do not notice in any important way how young Nick is in this story. Moreover, like Dr. Adams himself, they are impressed, in a matter of course way, by the achievement of the hero-physician under these circumstances.

> [Dr. Adams] bent over the Indian woman. She was quiet now and her eyes were closed. She looked very pale. She did not know what had become of the baby or anything.
> "I'll be back in the morning," the doctor said, standing up.
> "The nurse should be here from St. Ignace by noon and she'll bring everything we need."

---

[1] In "Indian Camp" we are not sure whether George is Nick's father's brother or brother-in-law, though in a passage from Hemingway's original draft, deleted ("omitted") from the published story and posthumously reproduced as the first of the *Nick Adams Stories*, "Three Shots," we learn they are brothers (1972: 14).

> He was feeling exalted and talkative as football players are in the dressing room after a game.
> "That's one for the medical journal, George," he said. "Doing a Caesarian with a jack-knife and sewing it up with nine-foot, tapered gut leaders."
> Uncle George was standing against the wall, looking at his arm.
> "Oh, you're a great man, all right," he said. (1972: 19–20)

Just after this conversation, however, in the next moment, Dr. Adams and his brother George check on the woman's husband and find that he has committed suicide in the bunk bed in the "operating" room.

What are students to make of this? What does Nick make of this? Many critics, like Meyly Hagemann, note that "Nick emerges from the shanty no longer a boy, but fully awake, facing his father in a rowboat—man to man" (1979: 108–09), and while this is perhaps arguable, it takes the narrative too literally—as some physicians take their patients too literally—in its final sentence: "In the early morning on the lake sitting in the stern of the boat with his father rowing, he felt quite sure that he would never die" (1972: 21). That is, the repeated readings of Nick's so-called initiation by critics of "Indian Camp" fail to interrogate the insistence of the modifier "quite sure" and the oddness of the term "stern" from a young boy who needs every technical term explained to him. In this, as throughout "Indian Camp" and Hemingway more generally, we are presented with the art of "omission" that Hemingway described in *A Moveable Feast* where he describes his "theory that you could omit anything if you knew that you omitted and the omitted part would strengthen the story and make people feel something more than they understood" (1964: 8; see also Smith 1983 and Wyatt 2014). But Hemingway's omission is very much like the omissions physicians face with patients every day. Here's how Dr. Stone puts it. (He refers to Dr. William Carlos Williams' story, "The Use of Force.")

> Well, I think the artist is always struggling with ways to apprehend, to grasp nature, to grasp human relationships, and that's really the biggest element of our problem in terms of dealing with a difficult patient, a silent patient, a hostile patient. We have to find the redeeming qualities that are in every human being, and we have to realize that their storytelling at the moment is a byproduct of being sick. So often, we have the possibility of neglecting a patient, and that's the real diagnosis, that they have not come to grips with the disease they have, with the symptoms they have. They don't want you to know. It's like "The Use of Force" [William Carlos Williams' story]. A little girl didn't want to tell her story, either in terms of words or in terms of a physical diagnosis. And that's what these patients are doing, they are withholding themselves. They want to see how smart the physician is. (Vannatta et al. 2015: Chap. 3, screen 68 [video])

In this passage, Dr. Stone sounds much like Hemingway when he describes the artist grasping nature and character, and in fact there is a significant body of work discussing Hemingway's aim at "grasping nature" as much as character in his fiction (e.g., Hagemann 1979). Nick as a young boy—seven? eight? ten?, it's hard to tell, though the initial conversation with his father about birthing makes him sound quite young—Nick as a young boy has not, in Dr. Stone's words, "come to grips" with the disease he has, with the symptoms he has.

So in teaching medical students how to notice things, we have found that encountering Hemingway is particularly useful. This is notable in relation to the title of the book we wrote together, *The Chief Concern of Medicine: The Integration of the Medical Humanities and Narrative Knowledge into Medical Practices*. The first thing American physicians record on the patient's "chart" is what is called her "chief complaint"—the condition that motivated her to seek out a physician's care—which, in our country, is recorded in the patient's own words: "I have had a continuous headache for ten days," for instance. And the aim of our book is to encourage healthcare workers to add an additional item to the medical protocols, namely to ask the patient's "chief concern." Such a concern is various, such as "I am afraid I have brain cancer"; "I fear I'll lose my job"; "I fear my partner won't understand"; etc. Adding the chief concern to the interview creates a moment early in the patient-physician encounter where the doctor is not fully in charge: it creates a moment where patient and physician together can discover—most importantly, can *negotiate*—what counts as "health" for this patient and her situation. Throughout the book, we offer what we call "schemas" of narrative to busy healthcare workers so they might develop habits of attention in their interactions with patients. In the book, we develop a number of schemas, including one we call the "interview encounter schema" (2013: 377–78) that offers a checklist for physicians to help develop habits of attention to the patient's story, to which we give the odd acronym **WET C²**. Here it is:

Interview Encounter Schema
    WET C²
        W   *Who* is this person?—repeat their name
        E   Recognize and acknowledge the ***Emotion*** the patient exhibits
        T   *Tell* me a story—about the chief complaint
        C¹   Articulate the ***Chief Complaint***
        C²   Encourage the Patient to articulate the ***Chief Concern***

WET C² is an acronym that can remind younger doctors, but also experienced physicians, how to begin an interview that will consistently solicit the patient's agenda—his overall goals for consulting a physician, or, to put this differently, what "health" might mean under these specific circumstances. This works because the doctor, using the checklist, reminds herself that the patient's name is important and that his primary emotion needs to be addressed so as to make the story telling easier. Physicians almost always use a statement like "What brought you in today?", which solicits the chief complaint from the patient. However, the explicit request for the patient's chief concern allows the patient to define—or to work with her doctor in defining—both the *meaning* of her ailment and the wished-for end, its resolution into what may count as "health" in these circumstances. This is how stories work: a story *requires* a teller and a listener, and the explicit request for a story almost always makes it difficult for a physician to interrupt his patient.

We can ask of Ernest Hemingway's story the same questions we can bring to the patient interview, namely:

Who is this person?: a young boy, Nick, probably around eight years old.
What is his Emotion: fear and bewilderment at confronting suffering and death, probably for the first time.
Tell me a story—about the chief complaint: This is the story we read.
What IS his Chief Complaint: "I feel bad because I saw a man kill himself tonight and a baby cut out of a screaming lady's stomach."
What is his Chief Concern: "I don't want to die." [Note: in his own words.]

Like so many of Hemingway's narratives—and like so many patient narratives—this story offers little information beyond the experience of the young protagonist. That is, the story offers events without reflective commentary that makes explicit—that abstractly describes—what is going on in a vocabulary outside the events themselves. The job of understanding the story entails understanding the unspoken context for its events, what Hemingway omitted. Thus, we have to infer the child's age from the way he talks and the way his father talks to him. In a similar fashion none of the background of the story is explicitly stated (the boy does not have to think about it), and it is the job of the listener to figure out and piece together a narrative context from the small details. Why does the doctor bring his son and brother to this caesarian operation? Why doesn't he have any anaesthetic, suture, a scalpel? Why does the woman's husband kill himself? Why does the story end the way it does? We can understand the story by supplying what is unsaid: that the doctor, Nick's father, was called to the Indian camp from a fishing-camping trip with his son and brother and for that reason does not have any medical supplies. The woman's husband probably kills himself because he sees a stranger with a knife cutting his screaming wife and is helpless to stop it. After the operation and its aftermath, Nick's uncle George has to stay back at the Indian camp to clean up the "awful mess" of the events (1972: 20), and Nick and his father take their boat back to their camp, talking in the boat. The chief concern of the story is simply its remembered significance: events seemingly etched in the boy's mind as he struggles with his first encounters with suffering and death.

This is a schematic reading of the story, hardly detailed or focused in important ways on the language and significance of Hemingway's story. Rather, it is simply an attempt to gather together its elements in outline, to figure out what is going on and why it was important to be said. The Interview Encounter Schema (WET C^2) offers a framework to engage this story on this basic level so that its elements can be grasped as a meaningful whole that includes both the events and the motivation—the concern—that inhabits all its parts. That is, in class we can ask of Ernest Hemingway's story the same *kind* of questions that healthcare workers can bring to their patients. We can even ask, to use the language of the medical interview, what is Nick's chief complaint? And we can ask what is his chief concern and what is the "concern" of the story as a whole. As with so many of Hemingway's narratives—and as with so many patient narratives—the job of understanding the story entails understanding the unspoken context for its events; it is the job of the listener to figure out and piece together a narrative context from the small details.

Earlier, we quoted Hemingway's description of how he discovered—he suggested in *A Moveable Feast* that it was a "secret" he learned from Cezanne (1964: 3; see Hagemann 1979 for a thorough discussion of Hemingway's engagement with

Cezanne)—"that you could omit anything if you knew that you omitted and the omitted part would strengthen the story and make people feel something more than they understood" (1964: 75). The goal of teaching "narrative knowledge" to people who have committed themselves to the care, healing, and comfort of healthcare is precisely what Hemingway describes here, the possibility of making healthcare workers, in their encounters with people in distress, feel something more than they understand. That "more," as Paul Smith discusses it in his fine essay on Hemingway's "Theory and Practice of Omission," is "the commonplace that the structures of literature, like the sentences of the language, imply more than they state and make us feel more than we know" (1983: 271). The increase in knowledge and feeling that Hemingway provokes in his readers offers wonderful training for professionals who engage with ailing people. As we note in our conclusion to *The Chief Concern of Medicine*, "Medicine and doctoring are built around this human relationship between patient and physician; they are grounded in storytelling, good listening, and the sense—which can always be improved and shared—of how stories work; and because they touch on the great crises of our shared lives, they are always, in their smallest gestures as well as largest decisions, a profoundly ethical enterprise" (2013: 356). These things—human relationships, storytelling, attention to the great crises of our shared lives—are found throughout Hemingway's work, and finding them there can make our medical students do better by the patients they serve.

Bibliography

Charon, Rita. 2006. *Narrative Medicine: Honoring the Stories of Illness*. New York: Oxford University Press.

Charon, Rita, and Maura Spiegel. 2005. On Conveying Pain/On Conferring Form. *Literature and Medicine* 24: vi–ix.

Hagemann, Meyly Chin. 1979. Hemingway's Secret: Visual to Verbal Art. *Journal of Modern Literature* 7: 87–112.

Hemingway, Ernest. 1964. *A Moveable Feast*. New York: Scribner's.

———. 1972. *The Nick Adam Stories*. Ed. Philip Young. New York: Scribner's.

Schleifer, Ronald, and Jerry Vannatta. 2013. *The Chief Concern of Medicine: The Integration of the Medical Humanities and Narrative Knowledge into Medical Practices*. Ann Arbor: University of Michigan Press.

Smith, Paul. 1983. Hemingway's Early Manuscripts: The Theory and Practice of Omission. *Journal of Modern Literature* 10: 268–288.

Vannatta, Jerry, Ronald Schleifer, and Sheila Crow. 2005. *Medicine and Humanistic Understanding: The Significance of Narrative in Medical Practices*. Philadelphia: University of Pennsylvania Press. (A DVD-ROM publication.)

Wyatt, David. 2014. Awkwardness and Appreciation in *Death in the Afternoon*. *The Hemingway Review* 33: 81–98.

Bibliography

Bal, P. Matthijs, and Martijn Vektkamp. 2013. How Does Fiction Reading Influence Empathy? An Experimental Investigation of the Role of Emotional Transportation. *PLoS One* 8 (1): E55341.
Becker, Ernest. 1973. *The Denial of Death*. New York: Free Press.
Benjamin, Walter. 1969. *Illuminations*. Trans. Harry Zohn. New York: Schoeken.
Biro, David. 2010. *Listening to Pain: Finding Words, Compassion, and Relief*. New York: Norton.
Blythe, Ronald. 1979. *The View in Winter: Reflections on Old Age*. New York: Harcourt Brace Jovanovich.
Boyd, Brian. 2009. *On the Origin of Stories: Evolution, Cognition, and Fiction*. Cambridge: Harvard University Press.
Broyard, Anatole. 1992. Doctor Talk to Me. *New York Times*. https://www.nytimes.com/1990/08/26/magazine/doctor-talk-to-me.html
Cassel, Eric. 1982. The Nature of Suffering and the Goals of Medicine. www.ericcassell.com/download/NatureOfSuffering.pdf
Charon, Rita. 2006. *Narrative Medicine: Honoring the Stories of Illness*. New York: Oxford University Press.
Charon, Rita, and Maura Spiegel. 2005. On Conveying Pain/On Conferring Form. *Literature and Medicine* 24: vi–ix.
Con Davis-Undiano, Robert. 2017. *Mestizos Come Home!: Making and Claiming Mexican American Identity*. Norman: University of Oklahoma Press.
Djkic, Maja, Keith Oatley, Sara Zoeterman, and Jordan Peterson. 2009. On Being Moved by Art: How Reading Fiction Transforms the Self. *Creativity Research Journal* 21 (1): 24–29.
Donald, Merlin. 1991. *The Origin of the Modern Mind*. Cambridge: Harvard University Press.
Doyle, Roddy. 1996. *The Woman Who Walked into Doors*. New York: Viking.
Dunbar, Robin. 1996. *Grooming, Gossip, and the Evolution of Language*. Cambridge: Harvard University Press.
Edson, Margaret. 1999. *Wit*. New York: Faber and Faber. The film version by Mike Nichols, staring Emma Thompson, is also available.
Fadiman, Anne. 1998. *The Spirit Catches You and You Fall Down: A Hmong Child, Her American Doctors, and the Collision of Two Cultures*. New York: Noonday.
Fishman, Scott (with Lisa Berger). 2000. *The War on Pain*. New York: Quill.
Gawande, Atul. 2010. *The Checklist Manifesto: How to Get Things Right*. Amazon Kindle ed. New York: Metropolitan Books.
Gerrig, R.J. 1993. *Experiencing Narrative Worlds: On the Psychological Activities of Reading*. New Haven: Yale University Press.
Green, Melanie. 2004. Transportation into Narrative Worlds: The Role of Prior Knowledge and Perceived Realism. *Discourse Processes* 38 (2004): 247–266.
Green, Melanie, and Timothy Brock. 2000. The Role of Transportation in the Persuasiveness of Public Narratives. *Journal of Personality and Social Psychology* 79: 701–721.
Hemingway, Ernest. 1972. *The Nick Adam Stories*. Ed. Philip Young. New York: Scribner's.
Heshusius, Lous. 2009. *Inside Chronic Pain: An Intimate and Critical Account*. Ithaca: Cornell University Press.

Hickok, Gregory. 2014. *The Myth of Mirror Neurons*. New York: Norton.
Hilfiker, David. 1984. Facing Our Mistakes. *New England Journal of Medicine*. http://www.davidhilfiker.com/index.php?option=com_content&view=article&id=51:facing-our-mistakes&Itemid=41
Iacoboni, Marco. 2009. *Mirroring People: The Science of Empathy and How We Connect with Others*. New York: Picador.
Jakobson, Roman. 1987. Linguistics and Poetics. In *Language in Literature*, ed. Krystyna Promorska and Stephen Rudy, 62–94. Cambridge: Harvard University Press.
Kidd, David, and Emanuele Castano. 2013. Reading Literary Fiction Improves Theory of Mind. *Sciencexpress*. http://www.sciencemag.org/content/early/recent/3 october 2013/Page 1/10.1126/science.1239918. Accessed 8 Mar 2015.
LaCombe, Michael. 2010. *Bedside: The Art of Medicine*. Orono: University of Maine Press.
Leys, Ruth. 2012. 'Both of Us Disgusted in My Insula': Mirror Neuron Theory and Emotional Empathy. *Nonsite* 5, March 18, 2012. At https://nonsite.org/article/"both-of-us-disgusted-in-my-insula"-mirror-neuron-theory-and-emotional-empathy. Accessed 16 May 2015.
Mar, Raymond, Keith Oatley, Jacob Hirsh, Jennifer dela Paz, and Jordan Peterson. 2011. Bookworms Versus Nerds. *Journal of Research in Personality* 40 (2011): 694–712.
Miall, D., and D. Kuiken. 1994. Foregrounding Defamiliarization, and Affect: Response to Literary Stories. *Poetics* 22: 389–407.
———. 2002. A Feeling for Fiction: Becoming What We Behold. *Poetics* 30 (2002): 221–241.
Milner, B. 1966. Amnesia Following Operations on the Temporal Lobes. In *Amnesia*, ed. C. Whitty and O. Zangwill. New York: Butterworth.
———. 1975. Psychological Aspects of Focal Epilepsy and Its Neurosurgical Treatment. In *Advances in Eurology*, ed. D.O. Purpura, J.K. Penry, and R.D. Walter, vol. 8. New York: Raven Press.
Morrison, Toni. 1988. *Beloved*. New York: Plum.
Neff, D.S. 1983. 'Extraordinary means': Healers and Healing in 'A Conversation with My Father'. *Literature and Medicine* 2: 118–124.
Peirce, Charles Sanders. 1931–1935, 1958. *Collected Papers*, vols. 1–6. Ed. C. Hartshorne and P. Weiss; vols. 7–8. Ed. A. Burks. Cambridge: Harvard University Press.
———. 1992. Deduction, Induction, and Hypothesis. In *The Essential Peirce, Volume I (1867–1893)*, ed. Nathan Houser and Christian Kloesel. Bloomington: Indianapolis University Press.
Phelan, James. 1996. *Narrative as Rhetoric: Techniques, Audiences, Ethics, Ideology*. Columbus: Ohio State University Press.
Polkinghorne, Donald. 1988. *Narrative Knowing and the Human Sciences*. Albany: SUNY Press.
Safehorizon. 2018. Domestic Violence Statistics and Fact. https://www.safehorizon.org/get-informed/domestic-violence-statistics-facts/#description/
Schleifer, Ronald. 2011. *Modernism and Popular Music*. Cambridge: Cambridge University Press.
———. 2014. *Pain and Suffering*. New York: Routledge.
———. 2018a. *A Political Economy of Modernism: Literature, Post-Classical Economics, and the Lower Middle-Class*. Cambridge: Cambridge University Press.
———. 2018b. The Aesthetics of Pain: Semiotics and Affective Comprehension in Music, Literature, and Sensate Experience. *Configurations* 26 (2018): 471–491.
Schleifer, Ronald, and Jerry Vannatta. 2013. *The Chief Concern of Medicine: The Integration of the Medical Humanities and Narrative Knowledge into Medical Practices*. Ann Arbor: University of Michigan Press.
Shakir, Mubeen, Jerry Vannatta, and Ronald Schleifer. 2017. Effect of College *Literature and Medicine* on the Practice of Medicine. *Journal of the Oklahoma State Medical Association* 110 (November 2017): 593–600.
Sherry, D.F., and D.L. Schacter. 1987. The Evolution of Multiple Memory Systems. *Psychology Review* 94: 439–454.
Shklovsky, Viktor. 1989. Art as Technique, Trans. Lee T. Lemon and Marion J. Reis. In *Contemporary Literary Criticism*, 2nd ed., ed. Robert Con Davis and Ronald Schleifer. New York: Longman.

Steen, Francis. 2005. The Paradox of Narrative Thinking. *Journal of Cultural and Evolutionary Psychology* 3: 87–105.
Stern, D.T., ed. 2006. *Measuring Medical Professionalism*. New York: Oxford Press.
Stroud, Scott R. 2008. Simulation, Subjective Knowledge, and the Cognitive Value of Literary Narrative. *Journal of Aesthetic Education* 42: 19–41.
Trilling, Lionel. 1950. *The Liberal Imagination*. New York: Anchor.
Tweedy, Damon. 2015. *Black Man in a White Coat: A Doctor's Reflections on Race and Medicine*. New York: Picador.
Van Laer, Tom, Ko de Ruyter, Luca Visconti, and Martin Wetzels. 2014. The Extended Transportation-Imagery Model: A Meta-Analysis of the Antecedents and Consequences of Consumers' Narrative Transportation. *Journal of Consumer Research* 40: 797–817.
Vannatta, Jerry, Ronald Schleifer, and Sheila Crow. 2005. *Medicine and Humanistic Understanding: The Significance of Narrative in Medical Practices*. Philadelphia: University of Pennsylvania Press. A DVD-Rom publication.
Verghese, Abraham. 1994. *My Own Country*. New York: Viking.
Williams, William Carlos. 1938. The Use of Force. https://bookpdf.services/downloads/the_use_of_force_by_william_carlos_williams.pdf
———. 1967. *The Autobiography of William Carlos Williams*. New York: New Directions.

Index

A
Abduction, xxvii, 33, 35, 36, 47, 249, 259, 260
Ability, vii, xi, xxx, xxxii, xxxvii, 3–5, 14, 26, 55, 57, 70, 123, 125, 139, 156, 159, 244, 269, 279, 280
Abse, Dannie, xxxii, 194–195, 253
Accreditation Council of Graduate Medical Education (ACGME), xxviii, 263
Acute ailments, 159
Affective aspect, xxxv
Affective engagement, xxxiii, 107
African American slaves, 69
Ageing, xxx, xxxi, 139, 171–179
Agents, xxx, 13, 19, 23–24, 27
AIDS, 8, 9, 149
Alliteration, 18
Allusion, 19, 168, 216
Ambiguity, vii, 55, 155, 172, 265, 271
Anaesthetic, 159, 194, 281, 284
Anesthesia, 59, 162
Anesthetic, 281
The Annunciation: Lupe (Detretria Martinez), xxx, 139, 142–144, 251
Anomaly, 15, 26, 33, 35, 153, 176
Araby, xxix, 84–87, 247–250
Aristotle, xxx, 123, 199, 264, 265
Assonance, 18
Autobiography (Williams, William Carlos), 203
Awareness
  the elements, 84
  of personal/professional boundaries, 123, 269

B
Bal, P. Matthijs, 245
Becker, Ernest, 199, 200
Bell, Dr. Joseph, 8, 35, 37, 164, 165

Beloved (Toni Morrison), xi, xxviii, 69–71, 119, 244
Benjamin, Walter, 17
Bentham, Jeremy, 123
Berenice (Edgar Allan Poe), xxxi, 150–156, 158, 252
Biomedical information, xxxvii, xxxviii
Biomedical understanding, vii, viii, xxxiii
Biopsy, 200
Biro, John, xxxiii
Birthing, xxxi, 139, 142, 146
Black Man in a White Coat (Tweedy), xxix, 109–112, 114, 141, 183, 203, 251, 259
Blake, William, xxx, 134–135, 239
Blythe, Ronald, 172
The Body in Pain, 161
Boundaries, 61, 123, 193, 271
Boyd, Brian, 16
Breast cancer, 91–93, 96
Broyard, Anatole, xi, xxix, 91–92, 96, 133, 199
Brute, 110, 157
Buck, Pearl S., 273

C
Campo, Rafael, xxix, 91, 107–108, 251
The Cancer Journals, xxix, 92–97
Carelessness, 185, 187
Cassel, Eric, 160
Castano, Emanuele, 245
Charon, Rita, vi, xxxiii, xxxvi, xxxvii, 3, 4, 6, 13, 14, 28, 29, 81, 83, 89, 149, 265, 279, 280
Chattel slavery, 69, 78, 119
Chauvinism, 112, 113
The Checklist Manifesto, 53

© The Author(s) 2019
R. Schleifer, J. B. Vannatta, *Literature and Medicine*,
https://doi.org/10.1007/978-3-030-19128-3

Chekhov, Anton, xxix, xxx, 11, 17, 20, 27, 48, 71–78, 91, 126–134, 239, 250, 251, 273
Chemotherapy, 200, 201
Chief complaint, xxxiii–xxxv, xxxvii, 7, 87, 89, 148, 283
Chief concern, xxxiv, xxxv, xxxvii, 7, 28, 87, 89, 133, 140–142, 203, 208, 241, 283
The Chief Concern of Medicine, 285
   The Integration of Medical Humanities and Narrative Medicine into Medical Practices, 266
   The Integration of the Medical Humanities and Narrative Knowledge into Medical Practices, vi, xxxiii, xxxiv, 37, 70–71, 82–84, 110, 124, 133, 140–142, 200, 239, 283
Chronic illness, 124, 125, 159
Chronic pain, 97, 139, 159–161
Classification, 48
Cognitive/affective responses and literature, 3–6
Cognitive psychology, vii, x, xxvii, xxxv, xxxvi, 3–7, 10, 13–15, 20, 24, 33, 70–71, 243
The Cognitive Science of Literary Reading, xxxiii, 243–245
Comedy, 19, 22, 23, 248, 267
Common errors in diagnostic medicine, 261–262
Compassion, 133
Compassionate physician, 125
Competence, 133
The complexity of clinical medicine, xxv, xxvii, xxxiii–xxxviii, 47, 69
The complex nature of clinical medicine, xxxiii
Condescension, 113
The conscientious action of verbal empathy, 125
Conscientiousness, 133
Conscientious physician, 125
Constant fear of recurrence, 94
Core competencies, 263
Cosmetic sham, 96
Coumadin, 184
The Couple, xxix, 107–108, 251
Crises in life, 133
Critical reading, xxvi, xxvii
Critical thinking, v, viii, ix, xxxiv, 6, 29
Cultural filters, 110
Cultural values, 22, 139
Culture, v, x, xxxi, 11, 12, 73, 97, 109, 119, 120, 139–147, 157, 186, 199

D

Daily writings, vii, xxxii, 255–257
de Maupassant, 11, 20
Death and dying, xi, xxv, xxxii, 199–236
Death Be Not Proud, xxxii, 235–236
The Death of Ivan Ilych, xxxii, 17, 26, 202, 204, 234, 248, 253, 256
Deaths in American hospitals, 185
Decency, 133
Deduction, 35, 36, 38, 39, 47, 152
Defamiliarization, xxxi, xxxii, 6, 25–27, 33, 155, 204, 234, 244
Definition of health, 80
The Denial of Death, 199
Detective story, xxvii, 21, 26, 33, 35, 47, 48, 150
Diabetic Ketoacidosis, xxx, 124, 126, 183, 184, 193
Diagnosis, x, xi, xxvii, xxix, xxxiii, xxxiv, xxxvii, xxxviii, 15, 26, 33–49, 53, 82, 89, 110, 112, 113, 148, 193, 201
Dickinson, Emily, xxxi, 168–169, 252
Differential diagnosis, xxxvii, 48
Discern human virtues, 125
Discerning physician, 124
Discernment, 133
Disparaging/dismissing, 141
Disparities of Ageing, 172–173
Doctor babble, 203
A Doctor's Reflections on Race and Medicine, 110–112
"A Doctor's Visit," 17, 71–78, 250
"Doctor Talk to Me," xi, xxix, 96, 199
Donne, John, xxxii, 235–236
Doyle, Arthur Conan, xxvii, 17, 19, 21, 35, 37, 47, 48, 55, 91, 249, 259, 273
Doyle, Roddy, xi, xxxi, 19, 147–149, 156, 162, 252
Dunbar, Paul Laurence, xxix, 27, 113–119, 251
Dunbar, Robin, 5, 16
Duty, 57, 58, 73, 87, 99, 115, 116, 123, 162, 190, 206, 209–214, 223, 231, 232, 271
Dynamic of form and content, xxxi, xxxv, xxxvi, 14–17, 155, 157

E

Edson, Margaret, 235
Elements of narrative, 7, 16, 89
Ellison, Ralph, 114
Emotion, xxv, xxviii, xxx, 5–7, 9, 27, 38, 43, 47, 70, 87, 89, 95, 107, 117, 128, 148, 163, 166, 174, 189, 193, 200, 207, 222, 234, 239, 283

Emotional filters, 110
Empathy, v, ix, xxvii–xxix, xxxiv–xxxvi, 3–7, 9, 10, 13, 15, 29, 63, 69–80, 91, 107, 119, 123, 125, 161, 169, 203, 241, 243, 269, 271, 279
Encounter Schema, 283
Enemies, xxx, 27, 126–133, 251
Enumerated virtues, 133
Ethics, xi, xxvii, xxx, 27, 54, 65, 123–135, 150, 183, 187
Everthing is Going to be All Right (Derek Mahon), xxxii
Everyday ethics, ix, xxv, xxx, xxxi, 109, 123–135, 183
Everyday life, felt-experiences of, 142
Everyday nature of prejudice, 114
Everything Is Going To Be All Right, 240–241, 254
Explanation, 20, 33, 36, 47, 48, 88, 111, 217, 232
Explicit racism, 114

F

Facing Our Mistakes, xi, xxxii, 183, 185, 253
Fadiman, Anne, xi, 139
Failure of judgment, 187
Feature analysis, 13–29
Features of narrative, 5–7, 9, 10, 14–17, 25, 54, 84, 89, 155
Figurative explanation, 20
Fishman, Scott, 160
Flaubert, Gustav, xxxii, 17, 27, 91, 187–194, 253
Foregrounding, 6, 17, 18, 33, 83, 203, 244
Freeman, Mary E. Wilkins, 273
Frenzied death (Ivan), 203
The Fulfillments of Healthcare, xxv

G

Gaspar de Alba, Alicia, xxxi, 140, 142, 144–146, 251
Gawande, Atul, 53, 185, 186
The genres of narrative, 22, 267
Gerasim, 203, 206, 208, 218, 224–226, 228–230, 232
Gilman, Charlotte Perkins, xxix, 27, 91, 96–106
Goals of clinical medicine, xxv, xxxvii–xxxviii
The Goals of the Book, v–xi
Goodside, I., 24

Green, Melanie, xxxv, 5, 13, 14, 59
Guide for discussing diagnosis and diagnosis errors, x, 259–262
Gunn, Dr. C. G., 34

H

Habits
 of attention, xxxvi, 6, 7, 9, 28
 of thought, 114, 139
Hagemann, Meyly, 282
Hardy, Thomas ("I Look into My Glass"), xxxi, 172, 173, 178, 179, 253
Helper, 23, 24, 269
"He Makes a House Call," 78–80
Hemingway, Ernest, xxix, xxxiii, 88, 159, 279–285
Heroic Melodrama (epic), 267
Heshusius, Lous, xxxi, 97, 160–161, 168, 252
Hester, Casey, xxviii, xxxiii, 263
Hilfiker, David, xi, xxxii, 183–187, 193, 195, 253, 273
History and physical exam, xxxiii, 34, 171
History of Present Illness (HPI), xxxiii, xxxvi, 259
Hopkins, Gerard Manley, 18
Hospice, 201
The House of God, 172
The House of Seven Gables (Hawthorne, Nathaniel), xxxi, 17, 18, 172–178, 253
Humanistic understanding, viii, 279
Hyponatremia, 34–36, 48

I

Iacoboni, Marco, 4
I Look into My Glass, xxxi, 178, 179, 253
Ilych, Ivan, 91, 202–235, 239
Imaginative grasping of experience, 149
Imelda, xxviii, 27, 53, 54, 56–64, 91, 250, 267, 270
Impersonal procedures in the administration of medications, 185
Implicit
 bias, xxix, 112, 125
 discrimination, 112
 racism, 114
Imposed silence, 92
In Dahomey, 114
Indian Camp, xxix, xxxiii, 88, 159, 160, 280, 281
Induction, 36, 47
Inference to the best explanation, xxvii, 35

In medicine, the stakes are much greater, 112
Inside Chronic Pain: An Intimate and Critical Account, xxxi, 160–161
The Interview Encounter Schema (WET C^2), 284
In the Theatre, 194–195, 253
"Invisibility" of pain, 160, 161
Irish Civil War, 156
Ironic narrative, 248
Irony, 22, 178, 267
Ivan Ilych, 256

J
Jakobson, Roman, 18, 19, 25
Joyce, James, xxix, xxxiii, 20, 24, 25, 84–88, 247, 250

K
Kaiser Family Foundation, 110, 114
Kidd, David, 13, 203, 245
Kuhl, David, 161
Kuiken, Don, 6, 14, 17, 26, 243, 244, 247

L
l5, 79
"Lack of knowledge," categories of, 187
Lack of skill, 187, 195
Lack of will to do the right thing, 187
LaCombe, Michael, xxviii, xxix, 53–55, 112–114
Lamb, Margaret, 273
Leda and the Swan (Yeats), xxxi, 29, 156–158, 252
The legal system, 54
Likert scales, 269, 270
Linguistic strategies, 71
Listening, xxviii, 91
Listening to Pain: Finding Words, Compassion, and Relief, xxxiii
Literary fiction, 6, 243
Literary semiotics, 243
Literature and Medicine, viii, xi, xxxvii
Literature and virtues, 125
Logic of Making a Diagnosis, ix, xxv, xxx, 33, 84
Loneliness and grief–of ageing, 178
Lorde, Audre, xxix, 91–97, 107, 251
"The Lynching of Jube Benson" (Paul Laurence Dunbar), xxix, 113–119, 251

M
Madame Bovary, xxxii, 17, 27, 186–194, 253
Mahon, Derek, xxxii, 240–241, 254
Makeshift caesarian section, 281
Making Tortillas, xxxi, 145–146, 251
Manners, xxx, xxxi, xxxiii, xxxv, 14, 15, 17, 20, 21, 26, 39, 47, 48, 55, 65, 81, 82, 89, 91, 99, 103, 107–109, 114, 130, 134, 140, 141, 144, 151, 152, 155, 158, 168, 183, 193, 210, 215
Mar, Raymond, xxxv, 13, 244, 245
Martinez, Demetria, xxx, 27, 139, 142–145, 251
Mastectomy, 92, 93, 95
Mayo Clinic, 53
Measuring Medical Professionalism, 263
Medical drama, 266
Medical Paternalism, 193
Medical pedagogy, 243
Medical professionalism, ix, 263–273
Medical Readers' Theater, 269–271
Medicine, 112
Melanie Green, 245
Melodrama, 248, 272
"Melodramatic" narrative, 268
Melville, Herman, xxxi, 91, 115, 162, 168, 186, 203, 252
Memento mori, 165, 168
Mestizo culture, 141
Miall, David, xxxv, 6, 13, 14, 17, 26, 244, 247
Milestones, 133, 263
Mirror neurons, 4, 161
Misdiagnosis, 113
Mistakes, ix, xxxii, 8, 81, 183–187, 192, 193, 195, 223
  in medicine, ix, xi, xxv, xxxii, 27, 183–195
  systematic institutions of professional healthcare, 184
Moral character, 123, 125, 135
Moral education, 6, 26–28, 156, 158
Morris, David, 161
Morrison, Toni, xi, xxviii, 69–71, 119, 244
A Moveable Feast, 282, 284
Mukařovský, Jan, 244
The Murders in the Rue Morgue, 21, 23, 36, 47, 48
My Own Country, 8–10, 14, 15, 110, 149

N
Nabokov, Vladimir, 256
Narrative, 81, 266
  agents and concern, xxviii, xxx, 23–24
  knowledge, vii, ix, xxx, xxxiv, xxxv, 13–14, 26, 89, 265, 285

and medicine, xxv–xxvii, xxxvii
    as moral education, xxviii, 26–28
    organization of experience is an evolutionary adaptation, xxvii
    roles in, 83, 266
    structure, 5, 6, 14, 33–49, 266
    understanding and knowledge, xxxv
Narrative genres, xxx, xxxiii, 19, 22
Narrative Medicine, vi, xxxiii, xxxvi, 265, 279
    Honoring the Stories of Illness, 265
Narrative Transportation Theory, 5–6
Narratology, 243, 266
National survey of physicians, 110, 114
Neff, D. S., 20–22
Negative stereotypes, 112
Neuro-imaging, 243
The New England Journal of Medicine, 160, 183, 185
Nick Adams stories, 280
"Normative" ("deontological" or "principle-based") ethics, 123, 125

O

On Doctoring, 78, 281
The Operation (Herman Melville), xxxi, 162, 252
Opponent, narrative role, 83, 84
Ordinary decency, 149
The overall meaning, 9, 21, 28–29, 89, 125

P

Pain, xxx, 4, 34, 58, 63, 81–84, 92–95, 97, 98, 111, 112, 116, 126, 129, 130, 150, 152, 154, 159–169, 172, 184, 188, 189, 200, 201, 203, 214, 218–221, 223–227, 229–231, 233, 234, 239
    and suffering, 27, 81, 97, 159–161, 168, 199
"Pain has an Element of Blank," xxxi, 168–169
Paley, Grace, xxvii, xxxvi, 7, 9–28, 71, 91, 239, 244
Palliative care, 161, 202
Parallel acts of telling, 16
Patient
    agenda, xxxiv, xxxv, 283
    concern, xxv, xxxiv, 9
    role of, 91
    understanding, xxxiii, xxxiv
Patterned repetition, xxviii, xxx, xxxi, 14, 17–20, 25, 29, 155
    of narrative events (syntax), 18–19
    in narrative themes (semantics), 19–20
    of sound, 155
    in the sounds of language (phonics), 18
Pediatric Professionalism Milestones, 123, 274–278
Pediatric Residency Program Director, 264
Pediatrics, 107, 270
Peirce, Charles Sanders, xxvii, 33, 34, 36, 249
Percolate, 149
Personal information, xxxvii
Phronesis, 265
Physician "infomercial," 142
Pierce, Charles Sanders, xxvii, 33, 259
Poe, Edgar Allan, xxxi, 18, 19, 21, 25, 29, 36, 37, 47, 55, 149–156, 252
"A Poison Tree," xxx, 134–135
Porter, Katherine Anne, 273
Post-mastectomy woman, 92
Postpartum depression, 97
Postpartum psychosis, xxix, 97
Power of narrative, 14
Practical mistakes, 183
Prague School of Linguistics, 244
Pre-anesthetic surgery, 168
Pressure and stress physicians, xxv
Principle-based (or "normative") ethics, xxx
Professional behaviors, 53, 264
Professional duty, 232, 269
Professionalism, v, ix–xi, xxv, xxvii, xxx, 27, 53–65, 123, 183
Professionalization, xxviii, xxxiii, 54, 63
Professionalization workshop, xxviii, xxxiii, 63
Proust, Marcel, 256
Psychosis, 19, 106

R

Racism, 27, 109, 113, 114, 183
Racist, 109, 110, 113, 114
Racist stereotyping, xxix
Rage, sadness, and loneliness, 92, 96
Rapport, xxviii, xxix, 7, 9, 38, 69–80, 91
Recognize and recover narrative knowledge, 89
The Red Wheelbarrow, xxix, 9, 84, 88–89, 250
Relational facts, 21–22, 33, 35–36, 157
Reluctant to bring up questions about domestic abuse, 149
The Resident Patient, xxvii, 8, 19, 21, 23, 35–37, 43, 47, 48, 249, 261
Rhetorical power, 256
Rhyme, 18, 49, 80, 157
Robert Wood Johnson Foundation, 78, 281
Roles in Narrative, 266
The Russian Formalists, 25

S

Savitt, T. L., 269
The Scarlet Letter, 173
Scarry, Elaine, 161
Schemas of narrative, 266
Schematic reading, 7, 284
Schleifer, Ronald, v, 3, 21, 97, 147, 160, 162, 200, 266
Scientific explanation, 20
Self-awareness, 123, 269, 271
Self-conscious life, 95
"Self-evident" truths, 139
Selzer, Richard, xxviii, 27, 53–64, 91, 110, 133, 186, 250, 267, 273
Senseless violence, 156, 157
Sense of manners, 144
Sensitivity to ambiguity, 55, 270
Sexism, 27, 97, 109, 112, 113
Sexist stereotyping, xxix
Sexual and domestic, 147–158
Sexuality, 58, 139, 140, 145, 146
Shafer, Audrey, xxviii, 64–65, 250
Shakespeare, William, xxviii, 48, 49, 235, 249
Shem, Samuel, 172
Sherlock Holmes, xxvii, 8, 9, 17, 19, 24, 35, 37–47
Shklovsky, Viktor, 25, 26
Situation of a narrative, 156
Six "elements" of narrative, 266
Skills in engagement with patient narratives, 280
Smith, Paul, 285
Social psychology, 243
Social role of caretaker, 109
Sometimes I Feel like a Motherless Child, 69, 119–120, 251
Sonnet, xxviii, 48, 49, 157, 235
Spiegel, Maura, 81, 83, 89
The Spirit Catches You and You Fall Down, xi, 139
Steen, Francis, 26, 27
Stereotyping, xxix
Stern, D. T., 53, 263
Stone, John, xxix, xxxii, 78–80, 240, 280, 282
Story filters, 110
Strategies of listening, 69
Strikingness, 247
Stroud, Scott, 27
Stylistics and linguistics, 243
Systematic repetition, 157
Systematic study of literary narrative, xxxiv, 15

T

Theory of Mind (ToM), xxvii, xxxvi, 3–6, 10, 13, 15, 156, 243, 245
Tolstoy, Leo, ix, xxxii, 17, 23, 26, 91, 199, 202–204, 234–235, 248, 253, 256
Tragedy, xxx, 13, 19, 22, 24, 128, 176, 199, 248, 267, 272
Transportation, xxxvi, 3, 5, 6, 10, 15, 19, 20, 24, 26, 27, 186, 203, 247
Transportation states, 243
Transportation Theory, 5, 243, 244
Trilling, Lionel, 140
Trust and honesty, xxxvii
Trust, honesty, and goodwill, xxxv, xxxvii
Trustworthiness, 53, 123, 133, 183, 269, 271
Tsar Alexander II, 119
Tweedy, Damon, xxix, 109–114, 141, 183, 203, 251, 259
Twice-told stories, xxx, xxxvi, 17–18, 47–48, 55, 77, 78, 84, 155, 157, 173, 234
Two time-frames in narrative, xxxvi

U

Unconscious (implicit) bias, xxix, 113
The unsaid, xxviii, xxx, xxxiv, 4, 20–21, 83, 87–88, 157
The unspoken context, 88, 284
"The Use of Force" (Williams, William Carlos, Dr.), xi, 110, 193, 282
Utilitarian (or "cost-benefit") ethics, xxx, 123, 125

V

Van Laer, Tom, 5
Vannatta, Jerry B., v, xxviii, xxix, xxxi, xxxii, 3, 4, 18, 26, 69, 70, 80, 108, 148–150, 156, 171–172, 184–187, 200–202, 240, 266
Veltkamp, Martijn, 245
Verghese, Abraham, 7–10, 14–16, 20, 28, 35, 89, 110, 149
Vicarious experience, ix, xxv, xxx, xxxii, xxxv, xxxvii, 4–6, 10, 19, 27, 28, 63, 80, 109, 139, 147, 149, 155–156, 158, 178, 183
Virtue ethics, xxx, 123, 125, 168, 264
Virtue
   in action, 124–125
   and character, 123, 124
   of compassion, 125
   of decency, 133
   in healthcare practices, 133

W

Weekly writing assignments, 255
Weir Mitchell, Silas, 97, 100
What Dying People Want, 161
White Jacket (Melville), xxxi, 162, 203
"Whole" narrative, xxxiv, 81
Williams, William Carlos, xi, xxix, 9, 88–89, 110, 193, 203, 239, 250, 273, 282
Wit (Margaret Edson), 235
Witnesses of violence, 158
The witness who learns, 24–25, 157
The Woman Who Walked into Doors, xi, xxxi, 19, 147, 148, 150, 156, 162, 252

Women, 72, 85, 91–97, 142, 146–148, 150, 157, 177, 188, 201, 215, 251
  with hyponatremia, xxviii, 34–36, 84, 183, 260
The Workshop, 126, 269–272

Y

Yeats, W. B., xxxi, 29, 48, 156–158, 235, 252
  "Leda and the Swan," 29
The Yellow Wallpaper, 19, 96–106, 108, 125, 251

The manufacturer's authorised representative in the EU is Springer Nature Customer Service Centre GmbH, Europaplatz 3, 69115 Heidelberg, Germany. If you have any concerns regarding our products, please contact ProductSafety@springernature.com

Printed and bound by CPI Group (UK) Ltd, Croydon, CR0 4YY
23/03/2026
02076735-0020